Microelectronic Systems N3

Checkbook

by the same author

Microelectronic Systems 1 Checkbook
Microelectronic Systems N2 Checkbook
Microprocessor Based Systems for the Higher Technician
Microprocessor Interfacing

The Checkbook Series

Microelectronic Systems N3 Checkbook

R. E. Vears

Heinemann Newnes

Heinemann Newnes
An imprint of Heinemann Professional Publishing Ltd
Halley Court, Jordan Hill, Oxford OX2 8EJ

OXFORD LONDON MELBOURNE AUCKLAND SINGAPORE
IBADAN NAIROBI GABORONE KINGSTON

First published 1990

British Library Cataloguing in Publication Data
Vears, R. E.
Microelectronic Systems N3
1. Microelectronic devices
I. Title II. Series
621.38152

ISBN 0 434 92334 6

Typeset by Key Graphics, Aldermaston, Berkshire.
Printed and bound in Great Britain by
Courier International Ltd, Tiptree, Essex

Contents

Notes to readers

Checkbooks are designed for students seeking technician or equivalent qualifications through the courses of the Business and Technician Council (BTEC), the Scottish Technical Education Council, Australian Technical and Further Education Departments, East and West African Examinations Council and other comparable examining authorities in technical subjects.

Checkbooks use problems and worked examples to establish and exemplify the theory contained in technical syllabuses. *Checkbook* readers gain real understanding through seeing problems solved and through solving problems themselves. *Checkbooks* do not supplant fuller textbooks, but rather supplement them with an alternative emphasis and an ample provision of worked and unworked problems, essential data, short answer and multi-choice questions (with answers where possible).

Preface

This textbook of worked problems provides further coverage of the Business and Technician Education Council (BTEC) unit in Microelectronic Systems N (syllabus U86/333), dealing principally with topics remaining after the extraction of a single unit at NII level (see *Microelectronic Systems N2 Checkbook*). However, it can be regarded as a textbook in microelectronic systems for a much wider range of studies.

The aim of this book is to extend the range of hardware, software and interfacing techniques developed at level NII. Each topic considered in the text is presented in a way that assumes in the reader only the knowledge attained at BTEC level NII in Microelectronic Systems.

The subject matter of principal objective D (*Classification and packaging of VLSI elements*) has, for convenience, been divided into three sections (Chapters 3, 4 and 5).

This book concentrates on the popular 6502, Z80 and 6800 microprocessors and contains sample programs which may be used with little or no modification on most systems based on these microprocessors. The text includes over 140 worked problems followed by some 170 further problems.

The author would like to express his thanks to the general editors, J. O. Bird and A. J. C. May for their helpful advice and careful checking of the manuscript. Finally, the author would like to add a special word of thanks to his wife Rosemary, for her patience and encouragement during the preparation of this book.

The publisher and author would also like to thank the following firms for permission to reproduce diagrams and data in this book: Intel; Zilog; MOS Technology Inc.; Mostek UK Ltd; Motorola Semiconductor Products Inc.

R. E. Vears
Highbury College of Technology
Portsmouth

1 Microcomputer bus systems

A MAIN POINTS CONCERNED WITH MICROCOMPUTER BUS SYSTEMS

1 Microcomputer memories consist of a very large number of individual storage locations. Each storage location contains a **memory cell** which is capable of storing a single binary digit (0 or 1). The construction of a typical memory cell is similar to that used for the **bistable** or **flip-flop** circuit.

2 When a microprocessor transfers data to and from memory, it is important that only one storage location is selected at any given time. Therefore, each location is assigned a unique identifying label called an **address**. Selection of one particular location within memory is achieved by applying the correct address to the memory address input lines.

3 Common types of microprocessor operate on an 8-bit (1 byte) word. Therefore a microcomputer memory system must be capable of storing **groups of 8 bits** at each addressable location. Memory ICs are commonly available containing multiples of 1K (1024) bits, arranged with 1, 4 or 8 bits at each addressable location, and typical arrangements are shown in *Fig. 1(a) to (c)*.

4 Microcomputer memories make use of **binary addresses** since this reduces the number of address lines necessary for a given size of memory. If 'n' address lines are available on a memory circuit, 2^n different binary inputs are possible, therefore 'n' address lines permits 2^n different locations to be addressed. For example, if a microcomputer memory has 12 address lines, then:

the number of different addressable locations $= 2^{12} = 4096_{10}$

To enable selection of a single location to take place, the binary address applied to a memory must first be **decoded**. Most proprietary memory ICs include some **address decoding circuits** in addition to the memory cells, as illustrated in *Fig. 2*.

5 Decoding circuits of this type are known as **n line to 2^n line** decoders, and their operation may be studied by referring to *Fig. 3*.

6 In order to minimize the amount of interconnecting wiring in a microcomputer, a bus structure is used in which groups of parallel conductors, called **busses**, are used to convey information from one part of the system to another. All major components in a microcomputer are connected between three busses:
(i) a **data bus** (8 bits); (ii) an **address bus** (16); (iii) a **control bus**.
The organization of a bus-structured microcomputer is illustrated in *Fig. 4*. Note the bidirectional nature of the data bus, which is necessary since data must pass freely between microprocessor and memory in either direction.

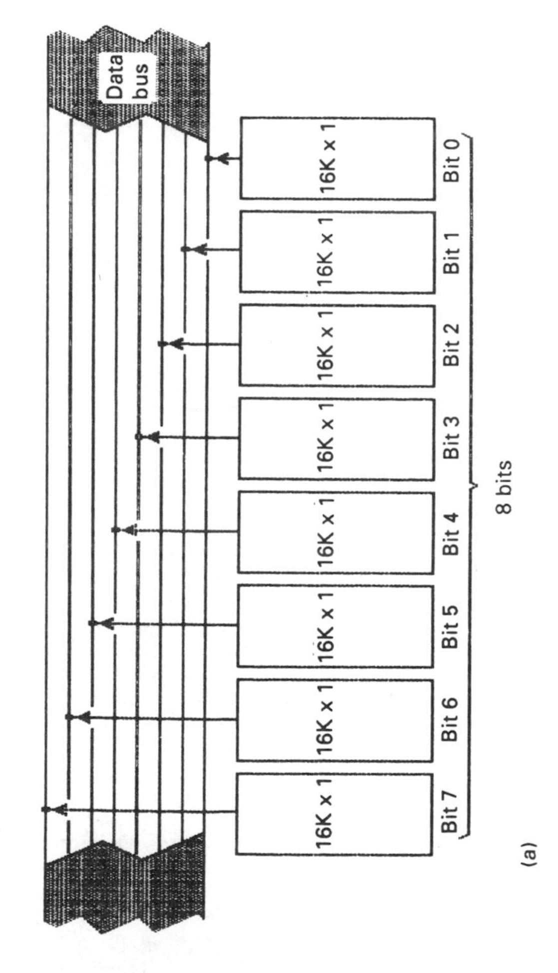
Data bus
16K x 1
16K x 1
16K x 1
16K x 1
16K x 1
16K x 1
16K x 1
16K x 1
Bit 7
Bit 6
Bit 5
Bit 4
Bit 3
Bit 2
Bit 1
Bit 0
8 bits
(a)

Fig. 1

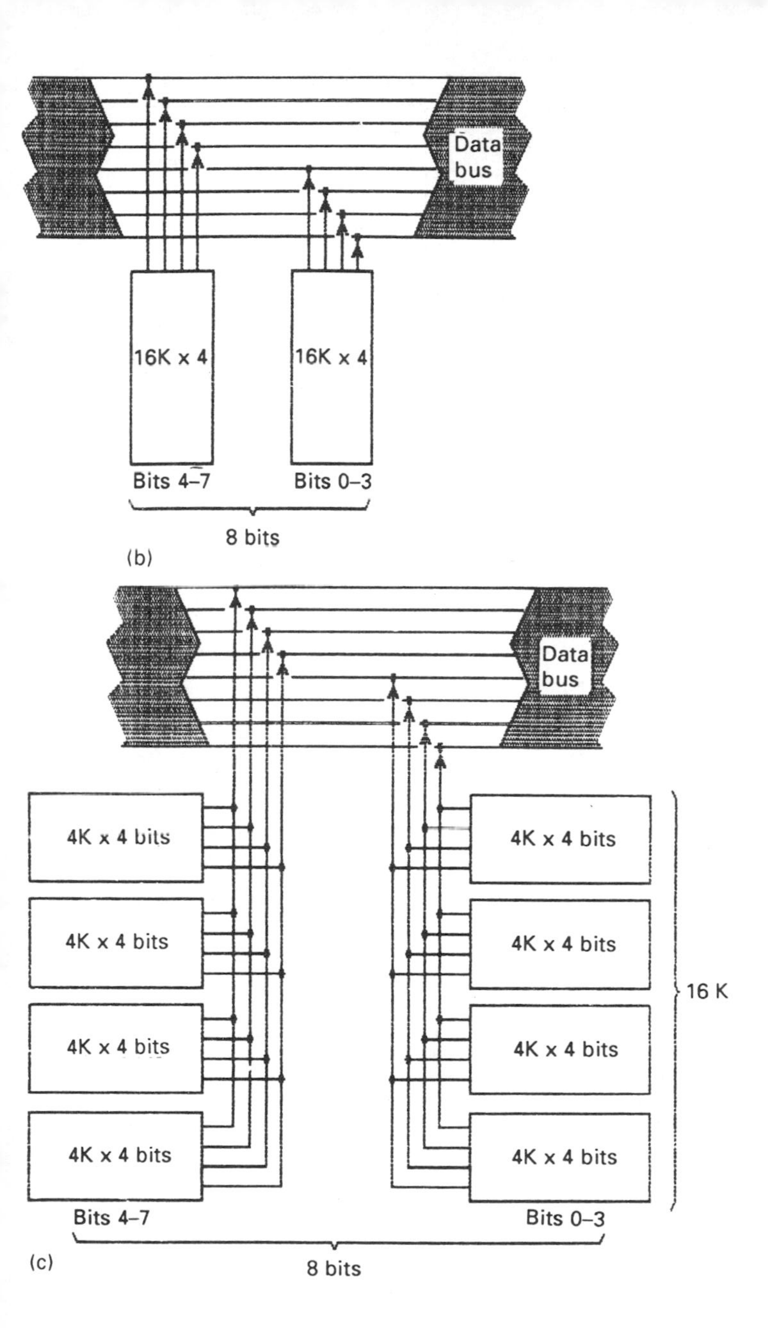
Data bus
16K x 4
16K x 4
Bits 4–7
Bits 0–3
8 bits
(b)
Data bus
4K x 4 bits
4K x 4 bits
4K x 4 bits
4K x 4 bits
4K x 4 bits
4K x 4 bits
4K x 4 bits
4K x 4 bits
16 K
Bits 4–7
Bits 0–3
(c)
8 bits

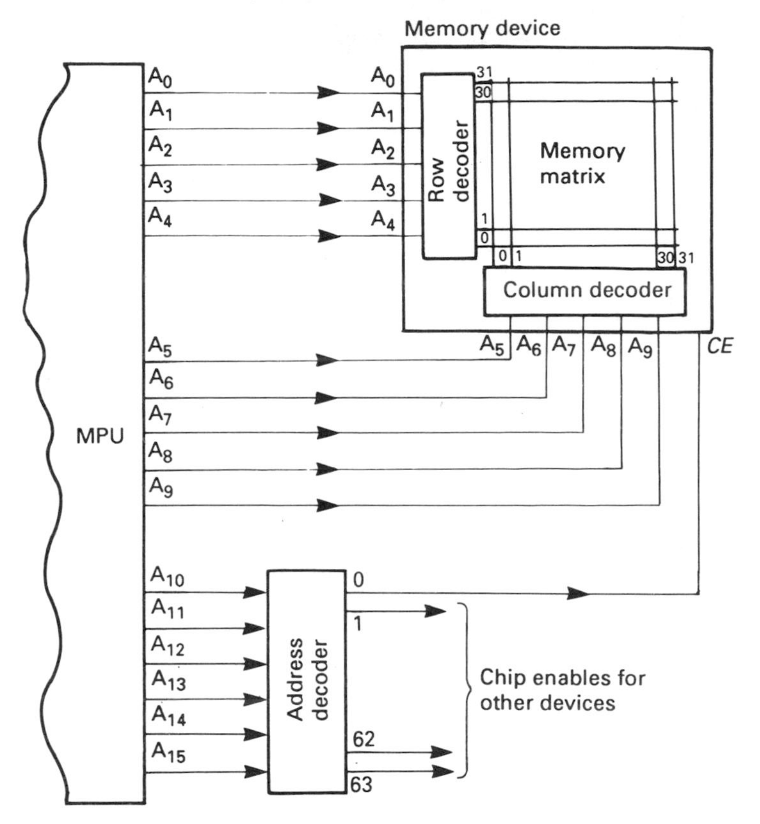

Fig. 2

7 When using a bus structure, only **one device** is permitted to transfer information to the data bus at any instant in time, otherwise a situation may arise where two devices simultaneously try to drive a data line to opposite logic levels. Such a situation is called a **bus conflict** or **bus contention** which leads to incorrect operation and possible damage to components.

A less serious, but equally undesirable situation occurs when two inputs are actively connected to a data line at the same instant in time. Therefore, only one device is permitted to receive information from the data bus at any instant.

8 The normal method of avoiding bus conflicts in a bus structured system is to make use of devices which have **tri-state buffers** (in this context, a buffer may be considered as a device which prevents unwanted interaction between two parts of a system). The operation of a tri-state buffer may be studied by reference to *Fig. 5(a) and (b)*.

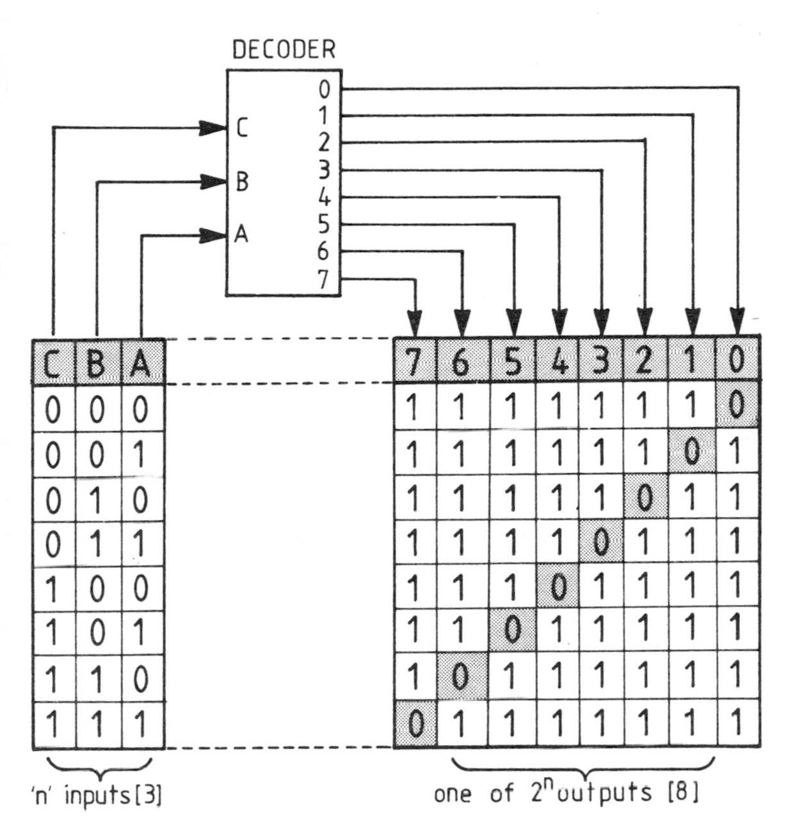

C	B	A
0	0	0
0	0	1
0	1	0
0	1	1
1	0	0
1	0	1
1	1	0
1	1	1

7	6	5	4	3	2	1	0
1	1	1	1	1	1	1	0
1	1	1	1	1	1	0	1
1	1	1	1	1	0	1	1
1	1	1	1	0	1	1	1
1	1	1	0	1	1	1	1
1	1	0	1	1	1	1	1
1	0	1	1	1	1	1	1
0	1	1	1	1	1	1	1

Fig. 3

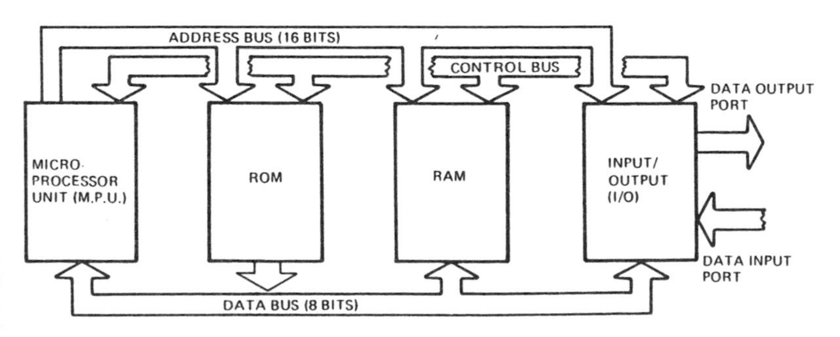

Fig. 4

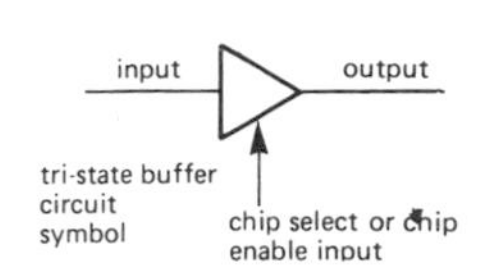

INPUTS	CHIP SELECT	OUTPUT
Ø	Ø	HIGH Z
1	Ø	HIGH Z
Ø	1	Ø
1	1	1

HIGH Z = high impedance or open circuit

tri-state buffer truth table

(a) Tri-state buffer circuit symbol

(b) Tri-state buffer truth table

Fig. 5

Devices with tri-state buffers are physically connected to the busses of a system, but are not actively connected until **enabled** or **selected** by means of appropriate signals applied to their **chip enable** (CE) or **chip select** (CS) inputs. When not selected, the data inputs and outputs of tri-state devices effectively become **open circuit** or **high impedance**, and electrically may be considered to be disconnected from the data bus. The use of tri-state buffers in a bus structured system is illustrated in *Fig. 6*.

Enabling of the tri-state buffers is under the control of the microprocessor which ensures that:

(a) only **one output buffer** is actively connected to the data bus at any given instant;
(b) only **one input buffer** is actively connected to the data bus at any given instant;
(c) input and output buffers of the **same device** are not simultaneously enabled, and one of the devices is the MPU.

Since the system illustrated in *Fig. 6* uses **read/write memory** (RAM), two select signals are required. These are:

(i) a **chip select** (CS) signal to select one particular memory IC; and
(ii) a **read/write** ($R/\overline{W}$) signal to determine whether the input or output buffer of the selected IC is enabled.

Chip select signals are obtained by decoding the **higher order address bits**.

The $R/\overline{W}$ signal is available as a microprocessor output which is set to a logical 1 when the microprocessor wishes to **read data** from memory (e.g. during 'Load Accumulator' instructions), and is reset to logical 0 when the microprocessor wishes to **store** data in memory (e.g. during store accumulator instructions). The use of these signals is shown in *Fig. 7*.

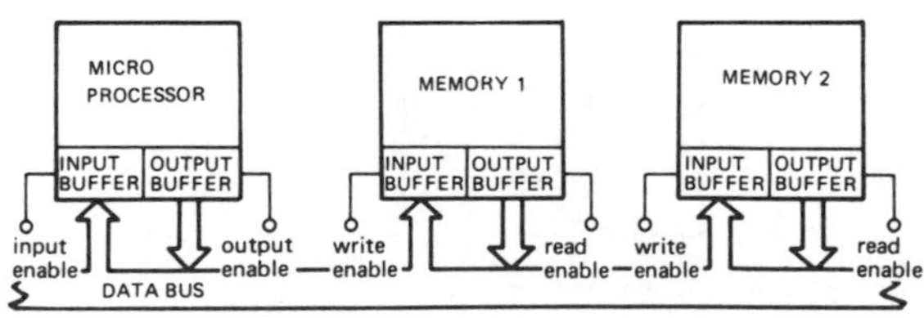

note: all enabling signals are generated by the microprocessor

Fig. 6

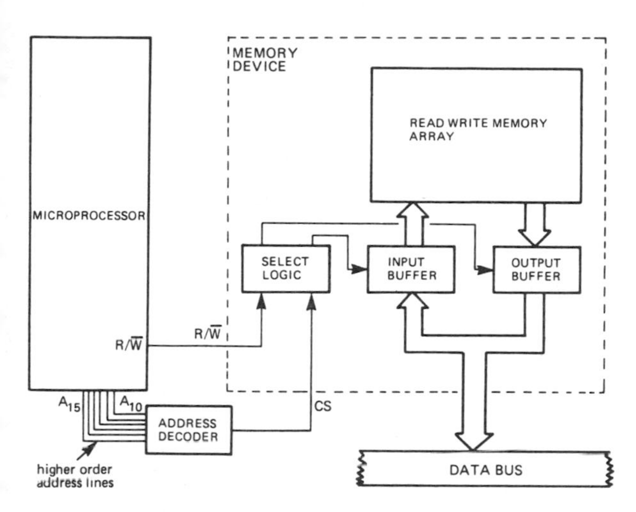

Fig 7

When using **read-only memory (ROM)**, a R/$\overline{W}$ signal is not required. This is because ROMs have output buffers only (since they are read-only) and therefore may be enabled by chip select signals alone.

9 The storage capacity of proprietary memory chips is organized as **'m' words**, each **'n' bits wide**, e.g. 1 K × 1 bit or 256 × 4 bits. The total capacity of such storage devices is **m × n bits**.

Most microprocessors operate on an 8-bit or 16-bit word, therefore it is frequently necessary to combine several memory devices to obtain the required word length, e.g. two 1 K × 4-bit memory chips to obtain a 1 K × 8-bit memory, as shown in *Fig. 8*.

10 In many applications more data words are required than are available in a single memory chip, or in a combination of memory chips arranged as in *Fig. 8* for example, a microcomputer may require 2 K × 8 bits of memory which may be constructed by combining two groups of 1 K × 4 bits, as shown in *Fig. 9*.

It can be seen from *Fig. 9* that all memory chips are connected to address lines A_0–A_9 and A_{10} and A_{11} are decoded and used to select one pair of chips at any instant in time. By this means, n lines may be used to select any one of 2^n locations, and each 1 K block of memory may be located within any specified address range.

11 In addition to memory devices, **input/output** (I/O) devices are also connected across the busses of a microcomputer so that communication with the outside world may take place. In many systems, I/O devices are allocated addresses and are treated in an identical manner to memory devices, i.e. they are said to be **memory mapped** I/O devices. Data transfers to and from the outside world are achieved by using load accumulator and store accumulator instructions. Some microprocesses, e.g. the Z80, have special instructions for I/O operations, and in this case, I/O devices are not memory mapped but are known as **isolated I/O**.

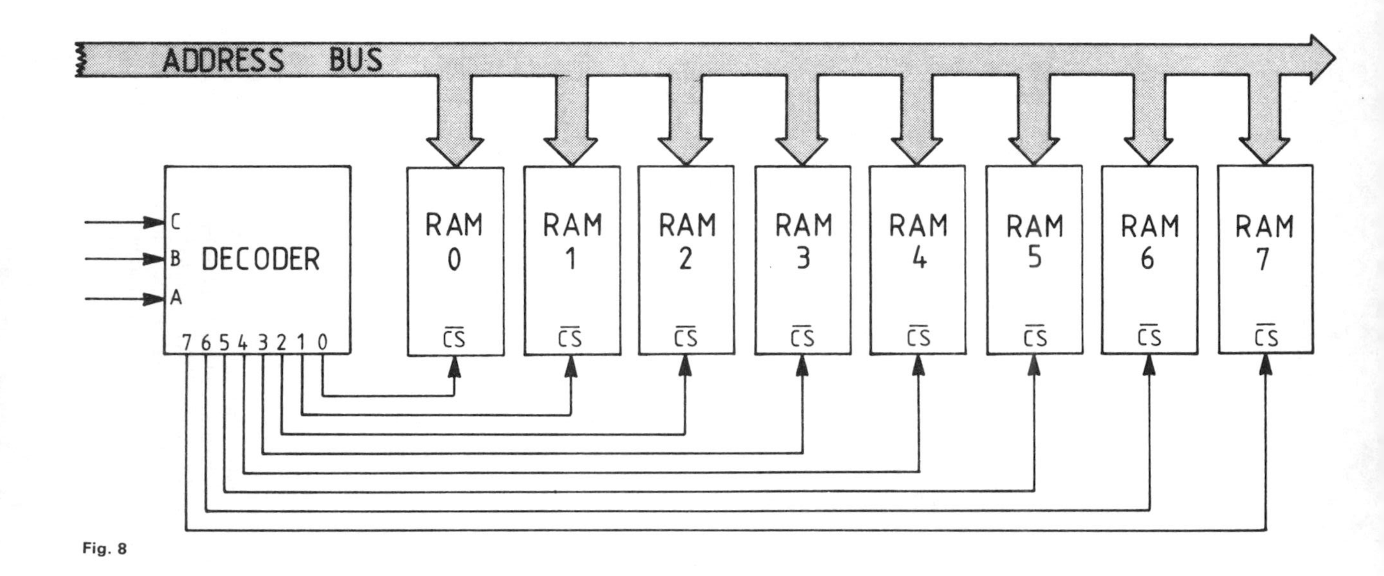
ADDRESS BUS
C
B
A
DECODER
7 6 5 4 3 2 1 0
RAM 0
RAM 1
RAM 2
RAM 3
RAM 4
RAM 5
RAM 6
RAM 7
$\overline{CS}$

Fig. 8

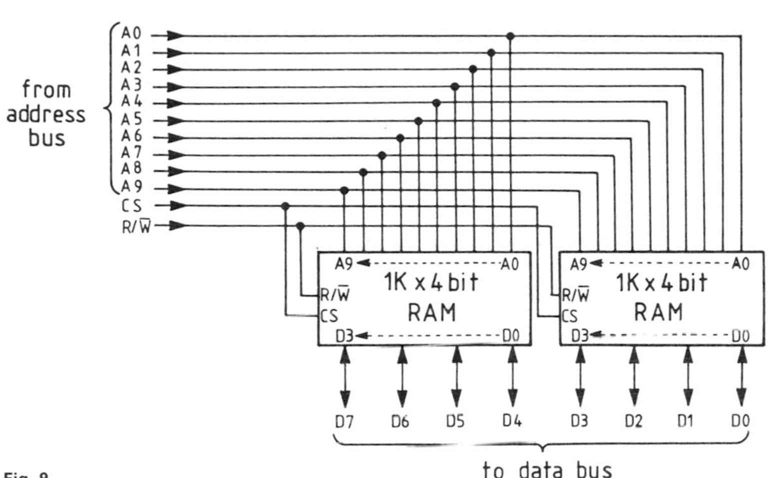

Fig. 9

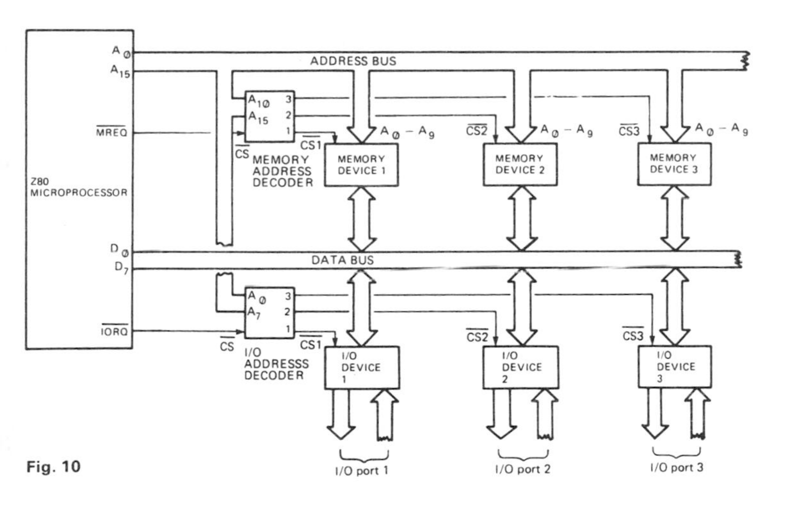

Fig. 10

The organization of memory and I/O in a Z80 system is shown in *Fig. 10*.

A Z80 microprocessor generates two relevant signals, $\overline{\mathbf{MREQ}}$ and $\overline{\mathbf{IORQ}}$ which operate as follows:

(a) $\overline{\text{MREQ}}$ is a tri-state, active low **memory request** signal which indicates that the address bus holds a valid address for memory read/write operations;

(b) $\overline{\text{IORQ}}$ is a tri-state, active low **input/output request** signal which indicates that the lower 8 bits of the address bus holds a valid address for I/O read/write operations.

In simple systems with only one I/O device, $\overline{IORQ}$ alone may be used as an I/O chip select signal which is activated during the execution of **IN** and **OUT** instructions only. In systems with multiple I/O devices, $\overline{IORQ}$ is used to enable a device which is responsible for decoding the lower 8 bits of the address but. The outputs from this decoder then select the particular I/O device specified by the **IN** or **OUT** instruction.

B WORKED PROBLEMS ON MICROCOMPUTER BUS SYSTEMS

Problem 1 Draw the circuit of a simple tri-state logic buffer and explain how tri-state operation is obtained.

The circuit diagram of a simple tri-state logic buffer is shown in *Fig. 11*. Operation of this circuit may be described by considering the circuit under three conditions:

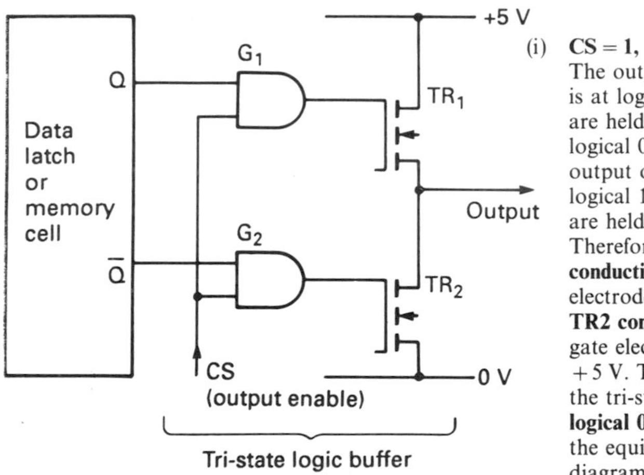

(i) **CS = 1, Q = 0 and $\bar{Q}$ = 1**
The output of AND gate G1 is at logical 0 since its inputs are held at logical 1 and logical 0 respectively. The output of AND gate G_2 is at logical 1 since both its inputs are held at logical 1. Therefore **TR1 is non-conducting** since its gate electrode is held at 0 V, and **TR2 conducts heavily** since its gate electrode is held at +5 V. Thus, the output of the tri-state buffer is held at **logical 0 (0 V)** as shown by the equivalent switching diagram *Fig. 12(a)*.

(ii) **CS = 1, Q = 1 and $\bar{Q}$ = 0**
The output of AND gate G_1 is at logical 1 since both its inputs are held at logical 1. The output of AND gate G_2 is at logical 0 since its inputs are held at logical 1 and logical 0 respectively. Therefore, **TR2 is non-conducting** since its gate electrode is held at 0 V, and **TR1 conducts heavily** since its gate electrode is held at +5 V. Thus, the output of the tri-state buffer is held at **logical 1** (+5 V) as shown by the equivalent switching diagram *Fig. 12(b)*.

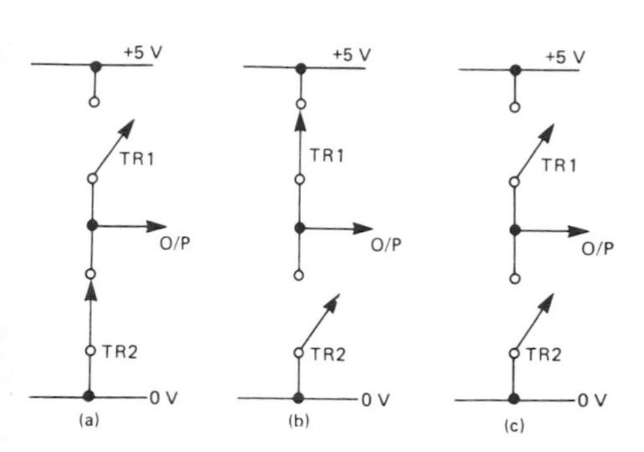

Fig. 12

(iii) **CS = 0, Q = 0 and $\bar{Q}$ = 1, or Q = 1 and $\bar{Q}$ = 0**
The outputs of both AND gates, G_1 and G_2, are at logical 0 regardless of the states of the Q and $\bar{Q}$ inputs since CS is at logical 0. Therefore **TR1 and TR2** are **both non-conducting** since their gate electrodes are both held at 0 V. TR1 and TR2 have a very high resistance when in the non-conducting state, thus the output of the tri-state buffer circuit is in an **open circuit** or **high impedance** condition, as shown by the equivalent switching diagram *Fig. 12(c)*.

Problem 2 Explain:
(a) the necessity to use an address decoder in a microcomputer;
(b) the operation of a typical address decoder.

(a) The number of address lines on a memory device is **fixed**, and is determined by the **number of addressable locations** contained within the device. For example, a 1 K memory chip has 10 address lines (A_0 to A_9) which are connected to the 10 lower order lines of a microcomputer address bus.
When several 1 K memory chips are so connected, addressing one memory location results in this location in every memory chip putting information out onto the data bus. This situation is clearly undesirable and causes **bus conflicts**. A solution to this problem is to **decode** some of the **higher order address lines** (A_{10} upwards) and use the decoded outputs to activate chip select inputs on the individual memory chips (see *Fig. 13*).

(b) An address decoder is an arrangement of logic gates which form an **n line to 1 of 2^n line** decoder (see *Fig. 14*). For '*n*' input lines there are 2^n individual output lines, arranged such that each of its possible binary inputs causes one of the output lines to change its logic state.
This system is used in *Fig. 13* so that only **one** of the memory devices is selected depending upon which combination of A_{10} and A_{11} is present, thus 4 K (4096) different locations to be individually addressed, as indicated in *Fig. 14*. Further decoding circuits are required if more memory or I/O devices are added.

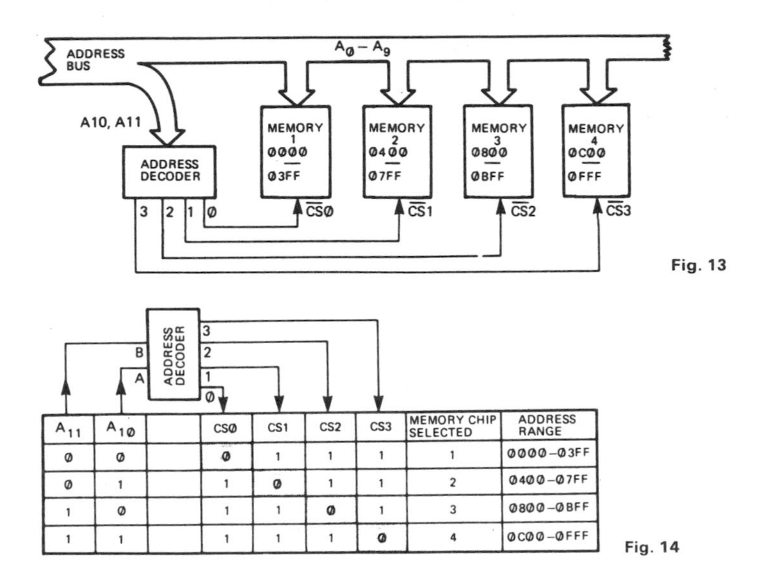

Fig. 13

A_{11}	A_{10}		CS0	CS1	CS2	CS3	MEMORY CHIP SELECTED	ADDRESS RANGE
0	0		0	1	1	1	1	0000–03FF
0	1		1	0	1	1	2	0400–07FF
1	0		1	1	0	1	3	0800–0BFF
1	1		1	1	1	0	4	0C00–0FFF

Fig. 14

Problem 3 With the aid of diagrams, explain how memory devices are arranged internally to reduce the number of external address lines.

The internal organization of a typical read/write memory is illustrated in *Fig. 15*. The address input lines are divided into **two groups** which are decoded separately within the memory device to form two sets of memory cell select signals. These are known as **row select (X)** and **column select (Y)** signals, and they are arranged in the form of a matrix. A small memory matrix is shown in *Fig. 16*. A memory cell (or group of memory cells) is connected to the X and Y lines at each intersection of the matrix, and a memory cell is selected when both its X and Y lines are energized (see *Fig. 17*).

Using this system, it can be seen that 64 different locations may be addressed by the use of only 6 address input lines. This principle may be used for larger memories, for example, a 32 × 32 matrix enables 1 K (1024) memory locations to be addressed by the use of only 10 address input lines.

Problem 4 (a) Show how two 4 K × 8 ROMs of the type shown in *Fig. 18* may be connected, without using external decoding components, such that they are assigned addresses in the range 0000_{16} to $1FFF_{16}$.
(b) Show how your answer to (a) must be modified if the CS input of each ROM is active low.
(c) Explain the main disadvantage of using this method decoding.

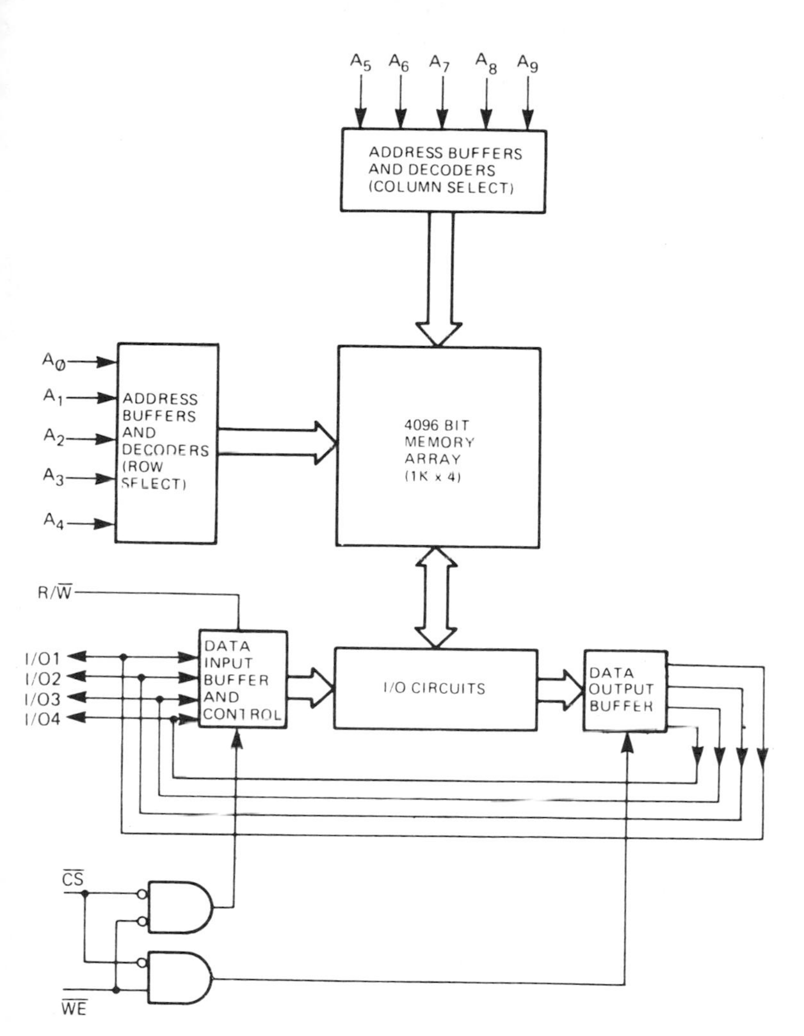

Fig. 15

(a) Two ROM chips of the type shown in *Fig. 18* may be assigned to addresses 0000_{16} to $1FFF_{16}$ by connecting them into circuit as shown in *Fig. 19*.

For addresses in the range 0000_{16} to $0FFF_{16}$, A_{12} is at logical 0 and IC1 is selected, but IC2 is not selected since it requires a logical 1 on its CS input to enable it. For addresses in the range 1000_{16} to $1FFF_{16}$, A_{12} is at logical 1 and IC2 is selected, but IC1 is not selected since it requires a

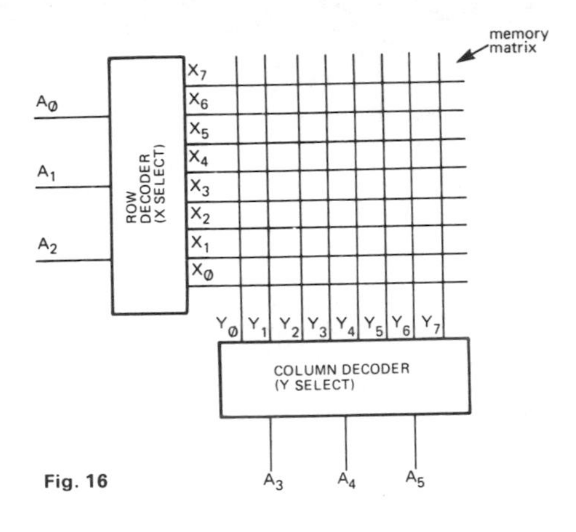

Fig. 16

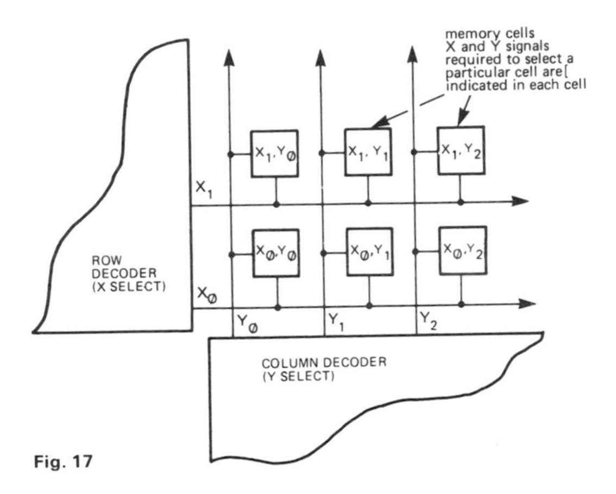

Fig. 17

logical 0 on its $\overline{CS}$ input to enable it. Therefore, IC1 responds to addresses in the range 0000_{16} to $0FFF_{16}$ and IC2 responds to addresses in the range 1000_{16} to $1FFF_{16}$.

(b) If the CS inputs to IC1 and IC2 are both active low (or both active high), the circuit shown in *Fig. 19* must be modified by including an inverter circuit, as shown in *Fig. 20*.

(c) The main disadvantage of using this method of decoding is that each ROM responds to eight different address ranges, as shown in *Fig. 21*. This is due to the fact that address lines A_{13} to A_{15} are not decoded. Unless further memory expansion is contemplated at a later stage, this effect is unlikely to cause major difficulties.

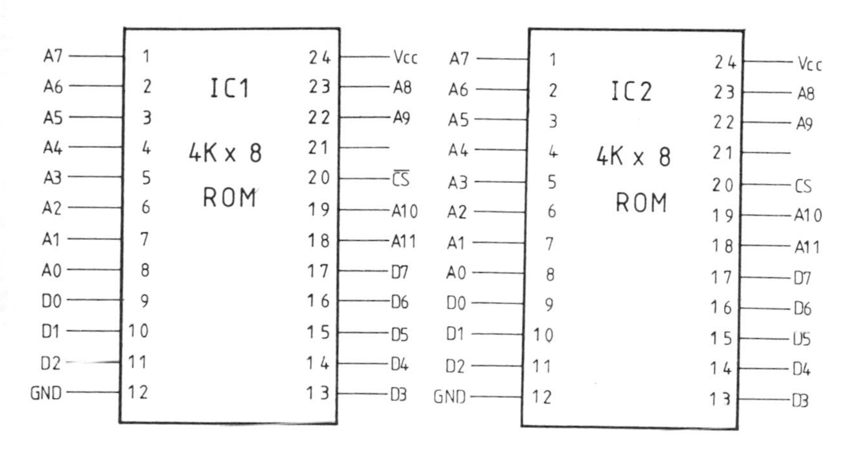
A7
A6
A5
A4
A3
A2
A1
A0
D0
D1
D2
GND
IC1
4K x 8
ROM
Vcc
A8
A9
CS
A10
A11
D7
D6
D5
D4
D3
A7
A6
A5
A4
A3
A2
A1
A0
D0
D1
D2
GND
IC2
4K x 8
ROM
Vcc
A8
A9
CS
A10
A11
D7
D6
D5
D4
D3

Fig. 18

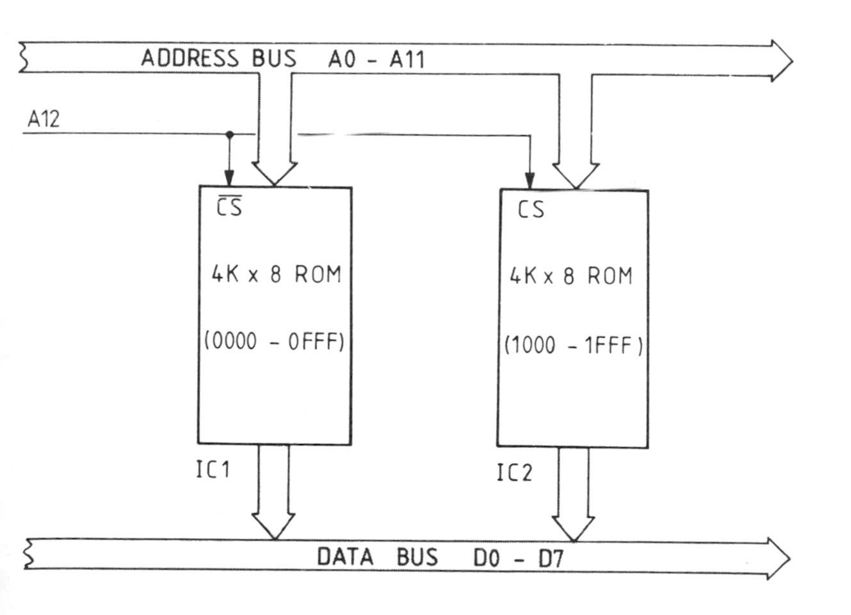
ADDRESS BUS A0 - A11
A12
CS
4K x 8 ROM
(0000 - 0FFF)
IC1
CS
4K x 8 ROM
(1000 - 1FFF)
IC2
DATA BUS D0 - D7

Fig. 19

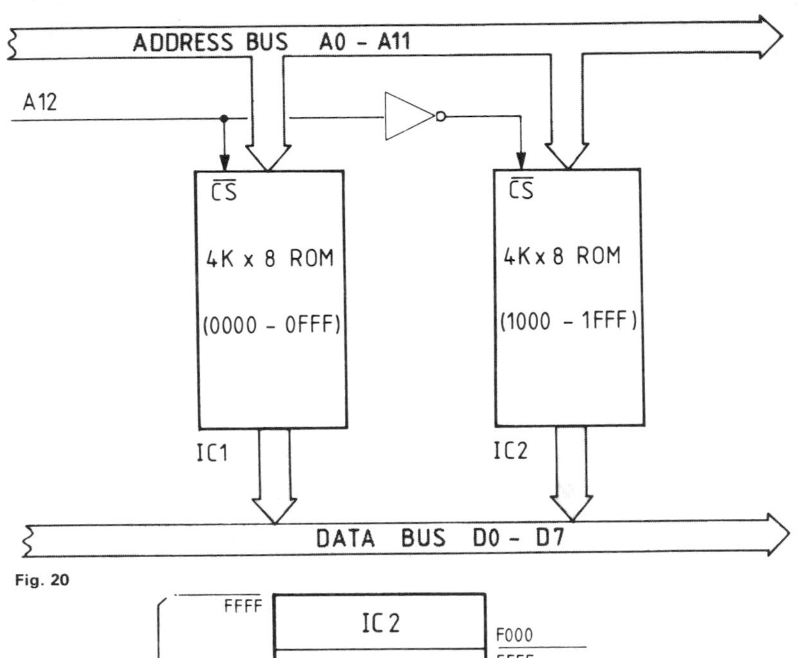
ADDRESS BUS A0 - A11
A12
CS
4K x 8 ROM
(0000 - 0FFF)
IC1
CS
4K x 8 ROM
(1000 - 1FFF)
IC2
DATA BUS D0 - D7
Fig. 20

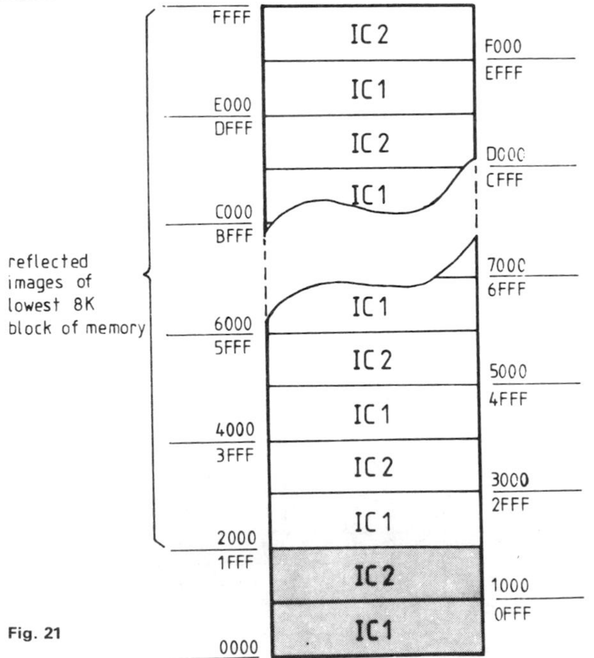
FFFF
IC2
F000
EFFF
IC1
E000
DFFF
IC2
D000
CFFF
IC1
C000
BFFF
7000
6FFF
IC1
6000
5FFF
IC2
5000
4FFF
IC1
4000
3FFF
IC2
3000
2FFF
IC1
2000
1FFF
IC2
1000
0FFF
IC1
0000
reflected images of lowest 8K block of memory
Fig. 21

Problem 5 (a) Show how four 1 K × 8 RAMs of the type shown in *Fig. 22* may be connected into circuit without the use of any external decoding components.
(b) Describe two disadvantages of using this method of connection.

(a) Address lines only may be used as select signals, as shown in *Fig. 23*. IC1, 2, 3 and 4 are selected, in turn, when A_{10}, A_{11}, A_{12} and A_{13} are each at logical 1. Therefore IC1 to IC4 are assigned addresses as shown in *Table 1*.

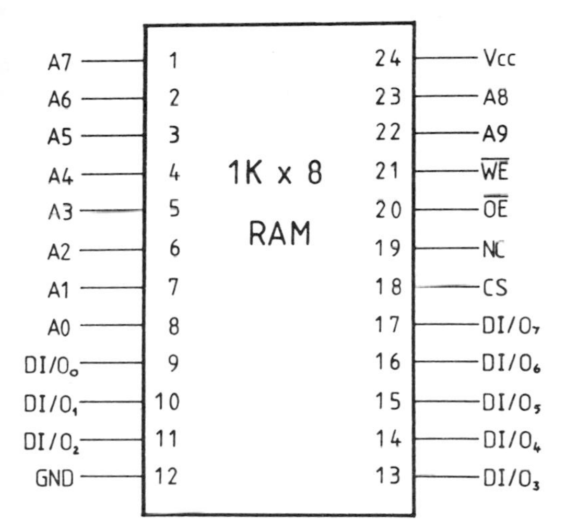

Fig. 22

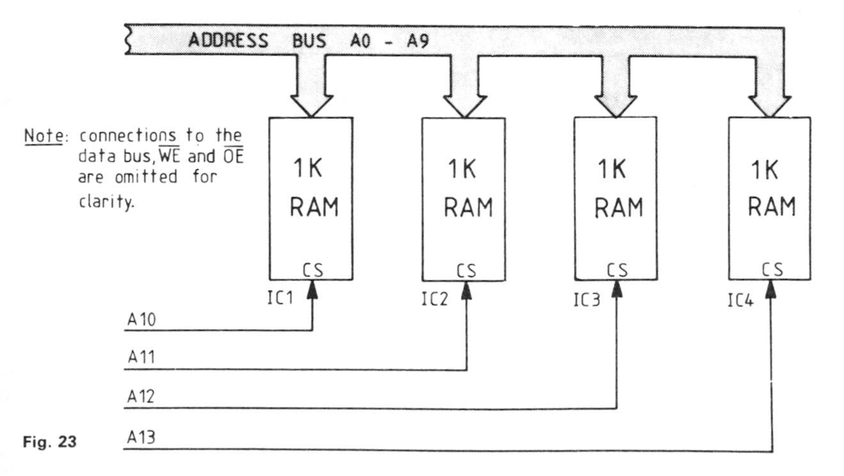

Fig. 23

Table 1

IC	*Address range*
1	0400–07FF
2	0800–0BFF
3	1000–13FF
4	2000–23FF

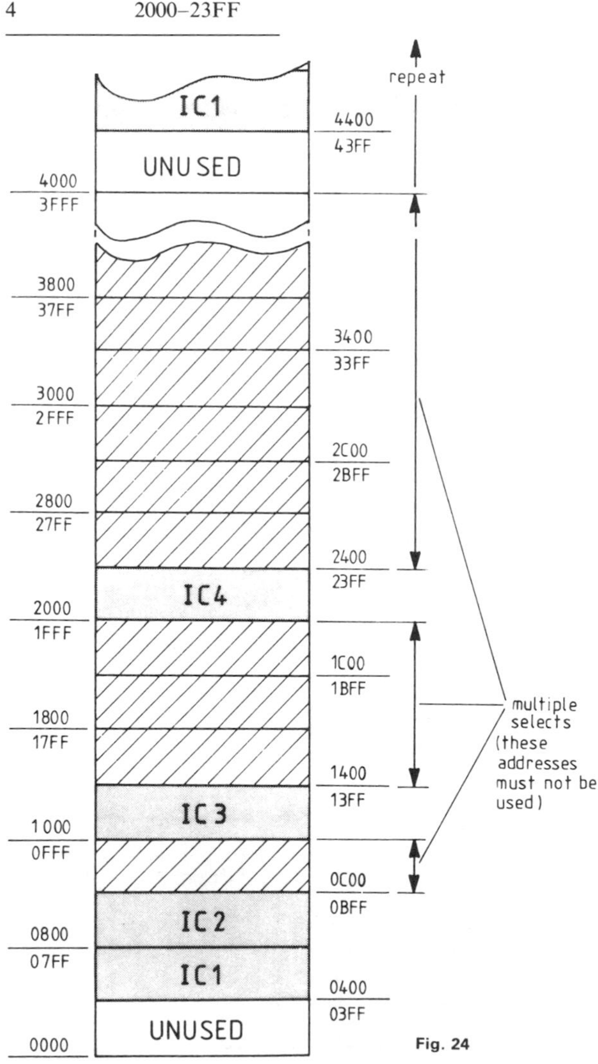

Fig. 24

(b) One disadvantage of using this method of connection is that memory is not contiguous, i.e., there are gaps in the memory map, as can be seen from *Table 1*. A second, more serious disadvantage is that certain addresses may result in two or more memory devices being selected simultaneously. For example, if an address in the range 0800_{16} to $0BFF_{16}$ is used, both A_{10} and A_{11} are at logical 1 which results in both IC1 and IC2 being selected simultaneously. This results in a **bus conflict**, and care must be taken in the software to avoid using such addresses. A memory map for the arrangement in *Fig. 23* is shown in *Fig. 24*.

Problem 6 With the aid of a diagram, explain how **combinational logic** may be used to assign two ROMs of the type shown in *Fig. 25* uniquely to the address range $F800_{16}$ to $FFFF_{16}$.

Two ROMS of the type shown in *Fig. 25* may be uniquely assigned to the address range $F800_{16}$ to $FFFF_{16}$ by means of the circuit shown in *Fig. 26*. AND gates IC1 (a) and (b) are used to detect when address lines A_{11} to A_{15} are all at logical 1, and this condition causes IC1 (b) output to become a logical 1.

The output from IC1 (b) is applied to IC2 (a) and (b) as shown, and this enables A_{10} to select IC3 (ROM 1) when it is at logical 0, and IC4 (ROM 2) when it is at logical 1 (IC2 (a) acts as an inverter for A_{10}). Therefore, IC3 and IC4 respond to addresses $F800_{16}$ to $FFFF_{16}$, as shown in *Fig. 27(a) and (b)*.

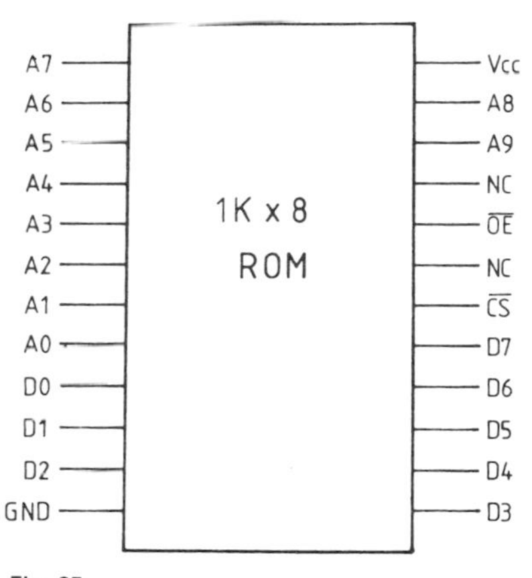

Fig. 25

Problem 7 Describe the operation of a **demultiplexer** circuit and explain how it may be used as an address decoder.

A demultiplexer circuit behaves as a single pole multiway switch, whose position is determined by binary inputs A and B, as shown in *Fig. 28*.

A demultiplexer circuit may be implemented by using logic gates arranged as shown in *Fig. 29(a)*. The operation of this circuit may be studied by considering the circuit under the following conditions:

(i) Input G at logical 0

The inverting action of G_1 causes one input of each of the NAND gates G_6 to

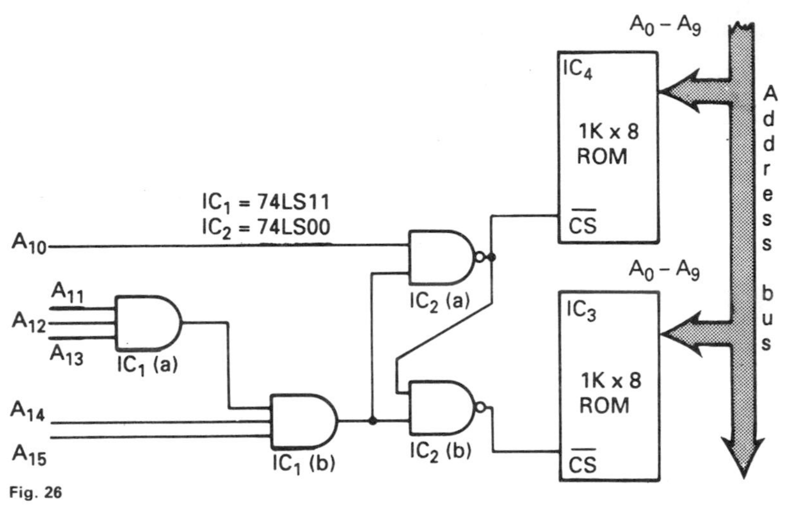
A0 – A9
IC4
1K x 8
ROM
CS
Address bus
IC1 = 74LS11
IC2 = 74LS00
A10
A11
A12
A13
IC1 (a)
IC2 (a)
A0 – A9
IC3
1K x 8
ROM
A14
A15
IC1 (b)
IC2 (b)
CS

Fig. 26

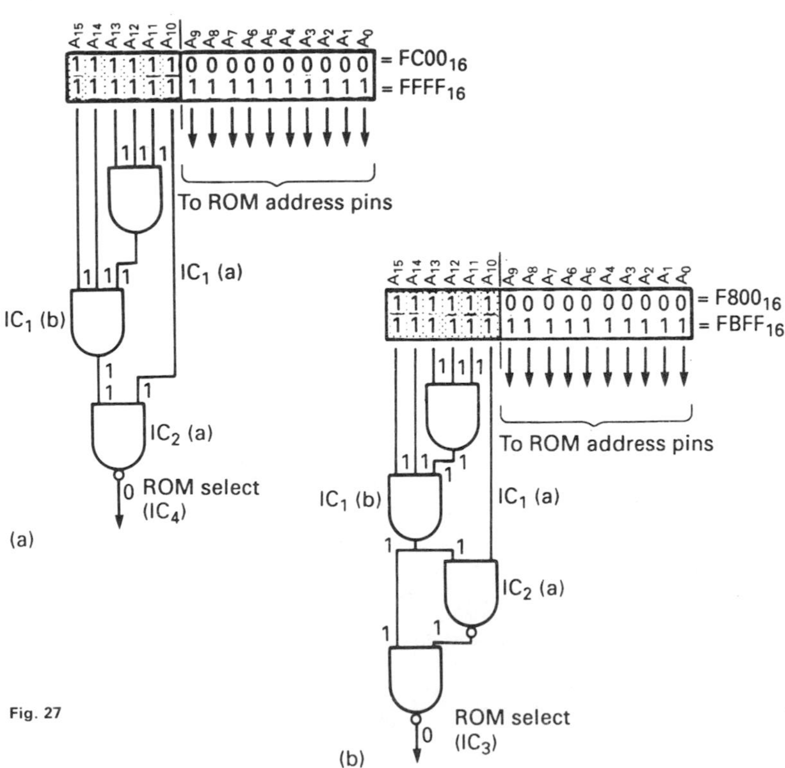
A15 A14 A13 A12 A11 A10 A9 A8 A7 A6 A5 A4 A3 A2 A1 A0
1 1 1 1 1 1 0 0 0 0 0 0 0 0 0 0 = FC00 16
1 1 1 1 1 1 1 1 1 1 1 1 1 1 1 1 = FFFF 16
To ROM address pins
IC1 (a)
IC1 (b)
IC2 (a)
0 ROM select (IC4)
(a)
A15 A14 A13 A12 A11 A10 A9 A8 A7 A6 A5 A4 A3 A2 A1 A0
1 1 1 1 1 1 0 0 0 0 0 0 0 0 0 0 = F800 16
1 1 1 1 1 1 1 1 1 1 1 1 1 1 1 1 = FBFF 16
To ROM address pins
IC1 (b)
IC1 (a)
IC2 (a)
ROM select (IC3)
(b)

Fig. 27

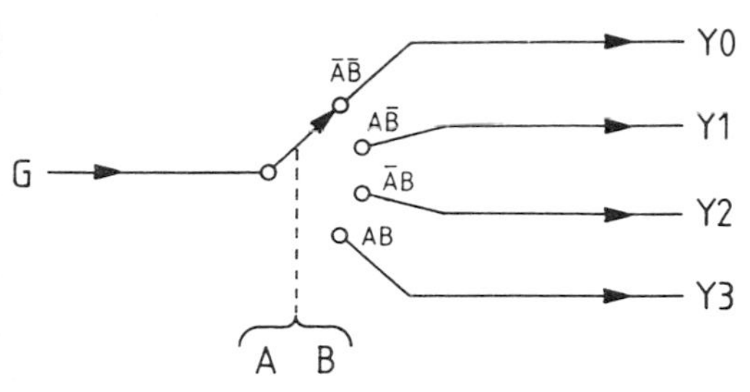

Fig. 28

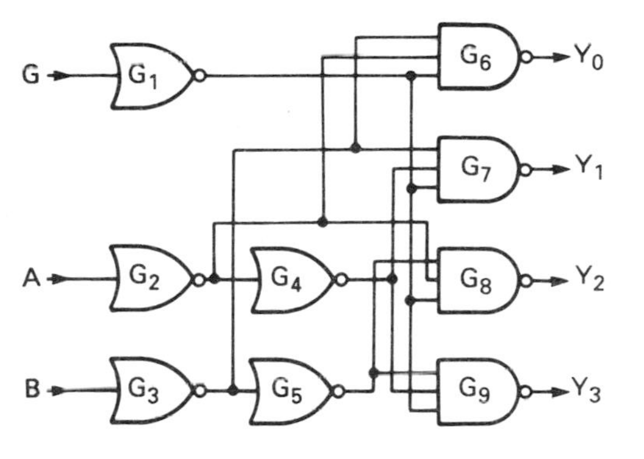

INPUTS			OUTPUTS			
ENABLE	SELECT					
G	B	A	Y3	Y2	Y1	Y0
1	X	X	1	1	1	1
0	0	0	1	1	1	0
0	0	1	1	1	0	1
0	1	0	1	0	1	1
0	1	1	0	1	1	1

(b)

Fig. 29

G_9 to be maintained at logical 1. Inverter circuits G_2 to G_5 are used to apply all possible combinations of A and B, in true or complemented form, to the remaining inputs of G_6 to G_9, such that only one of these gates has all of its inputs at logical 1 for a particular combination of A and B. Therefore the Y output which corresponds to the output of this particular gate becomes a logical 0, and the remaining outputs are at a logical 1.

(ii) Input G at logical 1

The inverting action of G_1 causes one input of each of the NAND gates G_6 to G_9 to be at logical 0. Therefore, irrespective of the states of inputs A and B, the outputs of all NAND gates G_6 to G_9 are at a logical 1.

Therefore, when this circuit is used as a demultiplexer, one of the outputs Y_0 to Y_3, selected by inputs A and B, changes logic state in step with the input at G. If, however, input G is held permanently at logical 0, a demultiplexer circuit may be used as an address decoder, operating as shown in *Fig. 29(b)*.

Problem 8 (a) Describe the operation of a **logic comparator** circuit.
(b) Show how a logic comparator may be used to decode memory into 4 K blocks.
(c) State where it would be advantageous to use a logic comparator as an address decoder.

(a) A logic comparator circuit may be used to determine whether two binary numbers are equal or not equal to one another. The circuit of a simple logic comparator is shown in *Fig. 30*, and is based upon exclusive OR gates.

Bits A_0 to A_3 of one binary number are exclusively ORed with their equivalent bits B_0 to B_3 of a second binary number. If corresponding bits in each number are both logical 0 or both logical 1, i.e., equivalent, then the exclusive OR gate to which they are connected has an output of logical 0. Therefore, if all bits in both numbers are equivalent, all exclusive OR gates

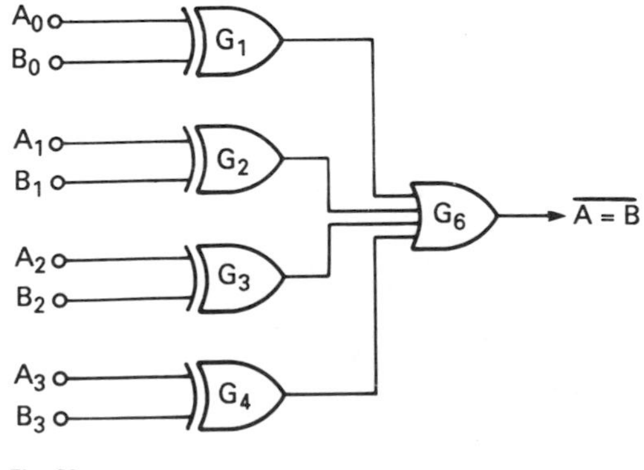

Fig. 30

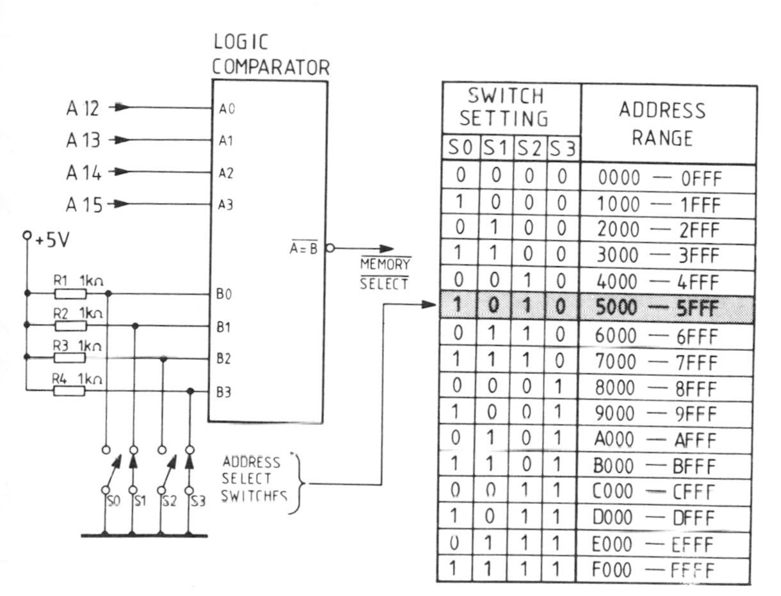

SWITCH SETTING				ADDRESS RANGE
S0	S1	S2	S3	
0	0	0	0	0000 — 0FFF
1	0	0	0	1000 — 1FFF
0	1	0	0	2000 — 2FFF
1	1	0	0	3000 — 3FFF
0	0	1	0	4000 — 4FFF
1	**0**	**1**	**0**	**5000 — 5FFF**
0	1	1	0	6000 — 6FFF
1	1	1	0	7000 — 7FFF
0	0	0	1	8000 — 8FFF
1	0	0	1	9000 — 9FFF
0	1	0	1	A000 — AFFF
1	1	0	1	B000 — BFFF
0	0	1	1	C000 — CFFF
1	0	1	1	D000 — DFFF
0	1	1	1	E000 — EFFF
1	1	1	1	F000 — FFFF

Fig. 31

G_1 to G_4 have outputs at logical 0 and the output from OR gate G_5 is at logical 0. If any of the bits in the two numbers are not equivalent, then the output from the corresponding exclusive OR gate becomes a logical 1 and the output from OR gate G_5 also becomes a logical 1.

(b) An arrangement which allows a logic comparator to decode memory into 4 K blocks is shown in *Fig. 31*.

Address lines A_{12} to A_{15} from a microprocessor are applied to inputs A_0 to A_3 of a logic comparator circuit. Address select switches S_0 to S_3 are connected to inputs B_0 to B_3 of the logic comparator and are used to determine which 4 K block of memory is selected (S_0 to S_3 are shown in the diagram in a position to select memory locations 5000_{16} to $5FFF_{16}$). When the logic levels on A_{12} to A_{15} correspond with the logic levels set by S_0 to S_3, the output from the logic comparator becomes a logical 0 level, and is used to select a particular 4 K memory block.

(c) This method of decoding is particularly useful for large memory boards or peripheral devices where the user may readily determine the address of the board in the system by setting up the switches (or using wire links on some boards).

Problem 9 (a) Explain how a **PROM** may be used as an address decoder. (b) Give **two** advantages of using a PROM for address decoding.

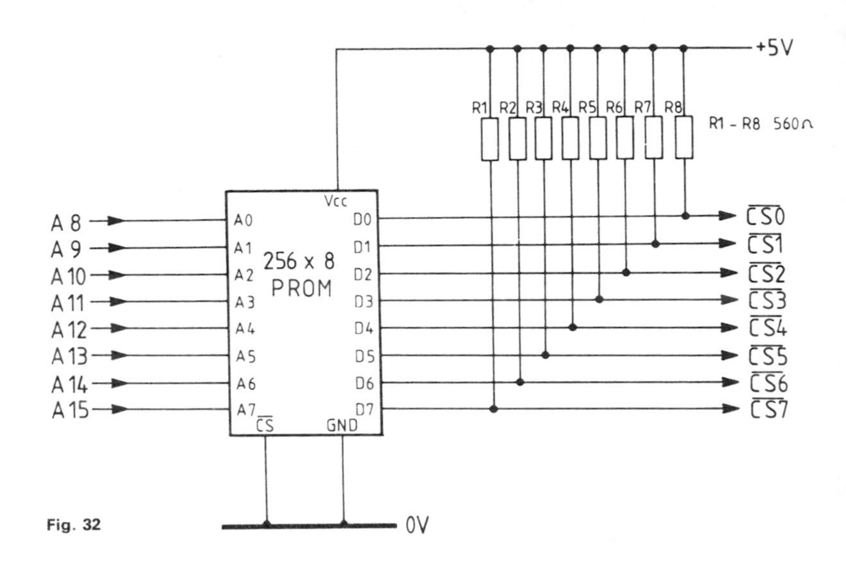

Fig. 32

(a) A PROM may be used as an address decoder by programming it to respond with the required chip select signals on its data output lines for specific address inputs. This arrangement is shown in *Fig. 32*.

The 256 × 8 bit PROM in *Fig. 32* is used to select one of eight memory devices, and may be used to decode the memory into 256 byte blocks. The memory may be decoded into larger blocks by programming the same chip select data into two or more consecutive locations in the PROM. For example, if the same chip select information is programmed into four consecutive locations in the PROM, the memory may be divided into blocks of 1 K. An arrangement of this type is illustrated by *Table 2*, where memory addresses 0000_{16} to $1FFF_{16}$ are decoded into eight 1 K blocks.

It may be seen from *Table 2* that only 32 of the 256 possible locations in the PROM are used when decoding in this manner, therefore, if 32 different locations in the PROM are programmed with the chip select information, it is possible to assign the eight 1 K blocks of memory to different addresses to those indicated. When the memory is decoded into larger blocks, more PROM locations are used since it becomes necessary to program more consecutive locations with the same chip select information, and all 256 locations are used if memory is decoded into 8 K blocks.

(b) One advantage of using this method of address decoding is that memory devices of differing capacity may be more easily mixed, e.g. 2 K × 8 ROM chips may be mixed with 1 K × 8 RAM chips by appropriate programming of the PROM. Also, the chip select outputs of the PROM do not necessarily have to respond to contiguous groups of memory addresses, i.e., ROM, RAM and I/O may be spaced out within the memory map as required.

A second advantage of using this method of address decoding is that, should modifications to the memory map of a microcomputer be required

ADDRESS INPUTS								DATA OUTPUTS								CS ACTIVE	RESPONDS TO ADDRESSES
A15	A14	A13	A12	A11	A10	A9	A8	D7	D6	D5	D4	D3	D2	D1	D0		
0	0	0	0	0	0	0	0	1	1	1	1	1	1	1	0		
0	0	0	0	0	0	0	1	1	1	1	1	1	1	1	0	$\overline{CS0}$	0000 to 03FF
0	0	0	0	0	0	1	0	1	1	1	1	1	1	1	0		
0	0	0	0	0	0	1	1	1	1	1	1	1	1	1	0		
0	0	0	0	0	1	0	0	1	1	1	1	1	1	0	1		
0	0	0	0	0	1	0	1	1	1	1	1	1	1	0	1	$\overline{CS1}$	0400 to 07FF
0	0	0	0	0	1	1	0	1	1	1	1	1	1	0	1		
0	0	0	0	0	1	1	1	1	1	1	1	1	1	0	1		
0	0	0	0	1	0	0	0	1	1	1	1	1	0	1	1		
0	0	0	0	1	0	0	1	1									
									0	1	1	1			1	$\overline{CS6}$	1800 to 1BFF
								1	0	1	1	1	1	1	1		
0	0	0	1	1	0	1	1	1	0	1	1	1	1	1	1		
0	0	0	1	1	1	0	0	0	1	1	1	1	1	1	1		
0	0	0	1	1	1	0	1	0	1	1	1	1	1	1	1	$\overline{CS7}$	1C00 to 1FFF
0	0	0	1	1	1	1	0	0	1	1	1	1	1	1	1		
0	0	0	1	1	1	1	1	0	1	1	1	1	1	1	1		

Table 2

at some later date, changes may be made by installing a differently programmed PROM rather than having to make changes to the internal wiring.

Problem 10 Show how **two demultiplexer** chips of the type shown in *Fig. 33* may be used to decode the addresses of a microcomputer into 12, 1 K blocks in the range 0000_{16} to $2FFF_{16}$.

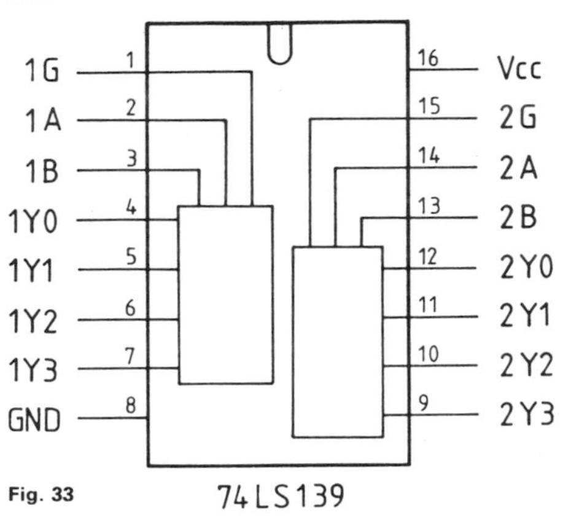

Fig. 33

Two demultiplexer chips of the type shown in *Fig. 33* may be used to decode addresses 0000_{16} to $2FFF_{16}$ as shown in *Fig. 34*.

Note that since A_{14} and A_{15} are not included in this decoding circuit, images of these addresses will appear throughout the memory map. If it is important for a particular application that this does not occur, then A_{14} and A_{15} must be decoded and used to enable 1G of IC1 (a) rather than leaving IC1 (a) permanently enabled.

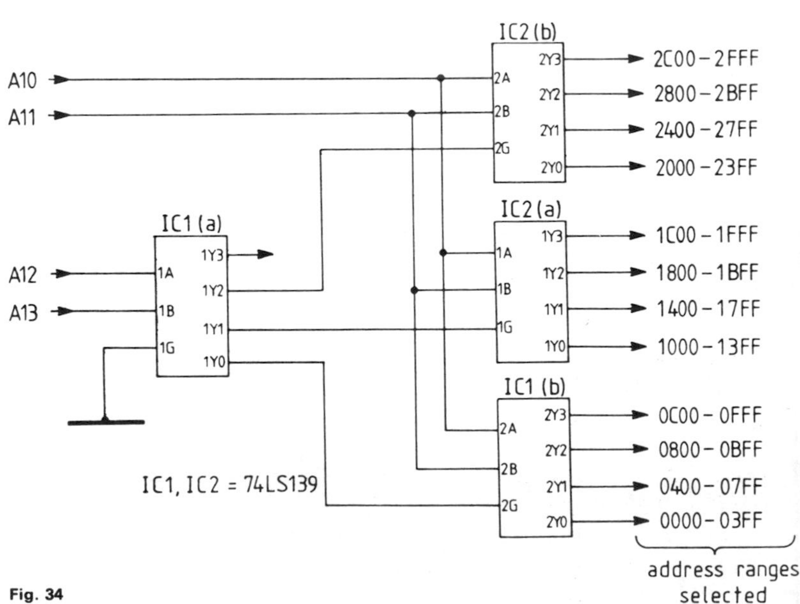

Fig. 34

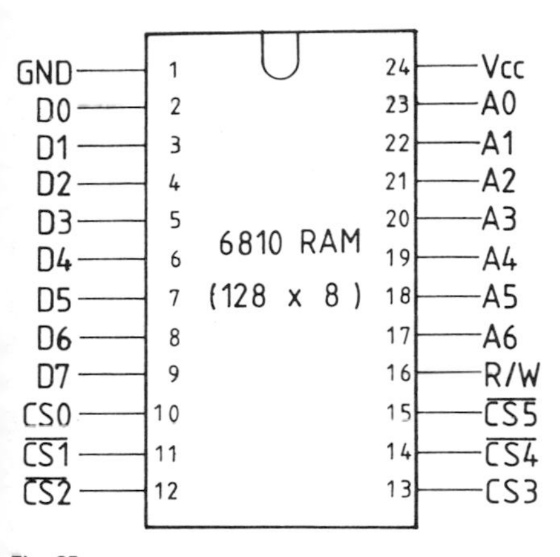

Fig. 35

Problem 11 Show how **four 128 × 8 bit RAMs** of the type shown in *Fig. 35* may be connected to form a 512 × 8 bit RAM without the use of any external decoding components.

RAMS of the type shown in *Fig. 35* have six chip select inputs, two of which are active high and the remainder are active low. Four such RAM devices may be assigned to, for example, addresses 0000_{16} to $01FF_{16}$ by connecting various parallel combinations of the chip select inputs to A_7 and A_8 of the address bus, as shown in *Fig. 36*.

Problem 12 Show how the **full 64 K address space** of a microcomputer may be decoded into 1 K blocks using demultiplexer and n line to 2^n line decoder circuits.

Full address decoding for the 64 K address space of a microcomputer may be accomplished using the circuit shown in *Fig. 37*.

Problem 13 Show how **eight** memory devices of the type shown in *Fig. 38* may be connected to form a 1 K × 8 bit RAM.

Eight 1 K × 1 bit RAMs of the type shown in *Fig. 38* may be connected to form a 1 K × 8 bit RAM as shown in *Fig. 39*.

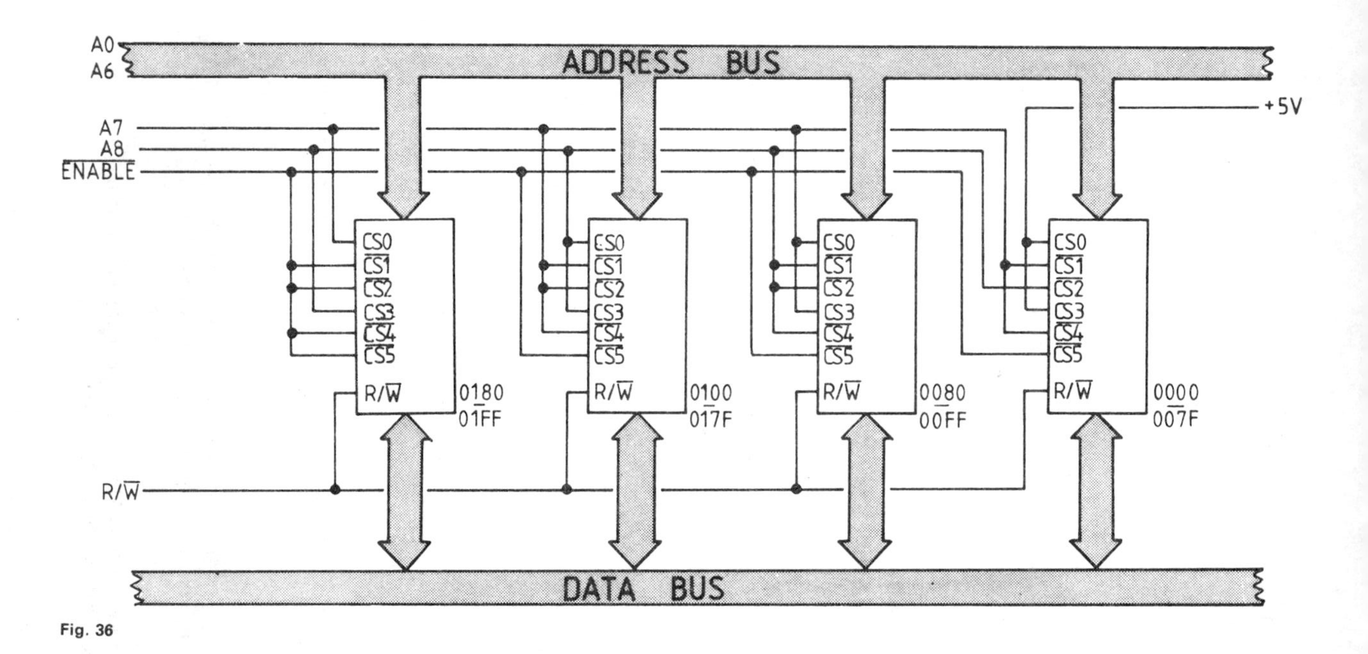
ADDRESS BUS
A0
A6
A7
A8
ENABLE
+5V
CS0
CS1
CS2
CS3
CS4
CS5
R/W
0180
01FF
0100
017F
0080
00FF
0000
007F
R/W
DATA BUS

Fig. 36

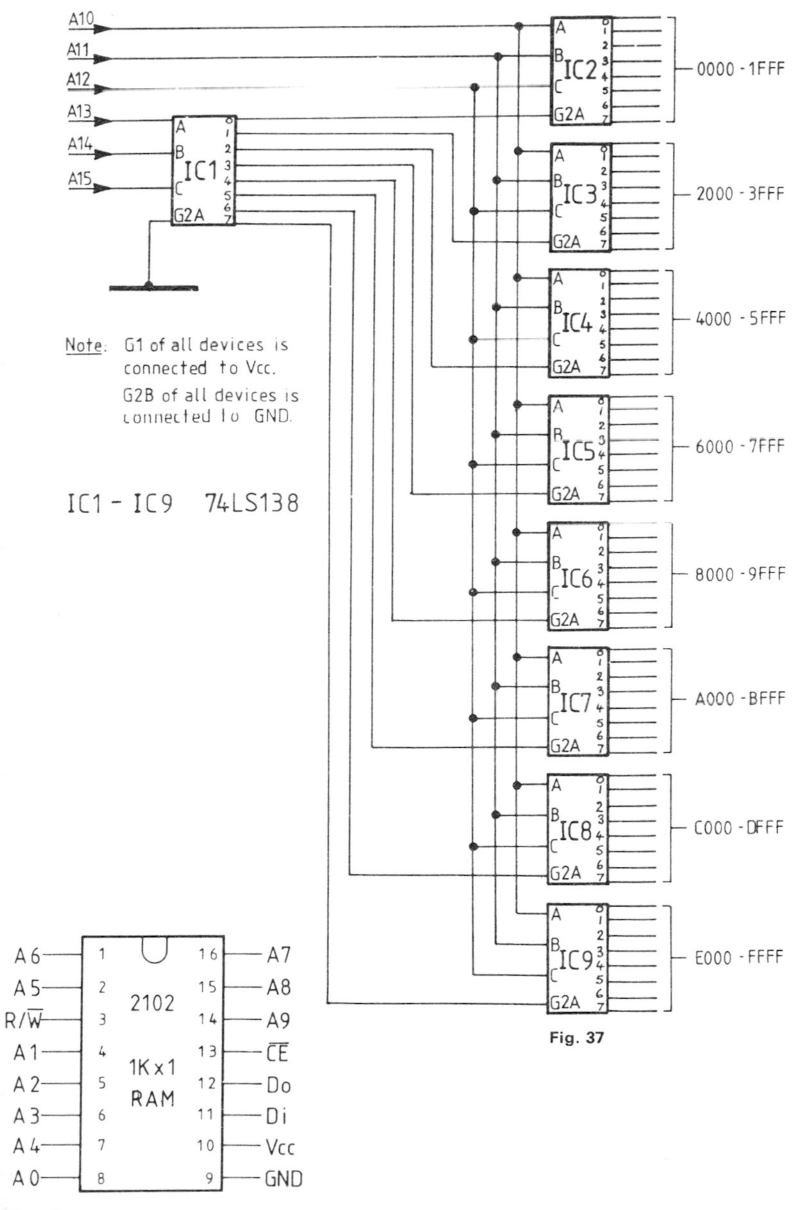
A10
A11
A12
A13
A14
A15
IC1
IC2
IC3
IC4
IC5
IC6
IC7
IC8
IC9
G2A
0000 - 1FFF
2000 - 3FFF
4000 - 5FFF
6000 - 7FFF
8000 - 9FFF
A000 - BFFF
C000 - DFFF
E000 - FFFF
Note: G1 of all devices is connected to Vcc.
G2B of all devices is connected to GND.
IC1 - IC9 74LS138
A6
A5
R/$\overline{W}$
A1
A2
A3
A4
A0
2102
1K x 1
RAM
A7
A8
A9
$\overline{CE}$
Do
Di
Vcc
GND

Fig. 37

Fig. 38

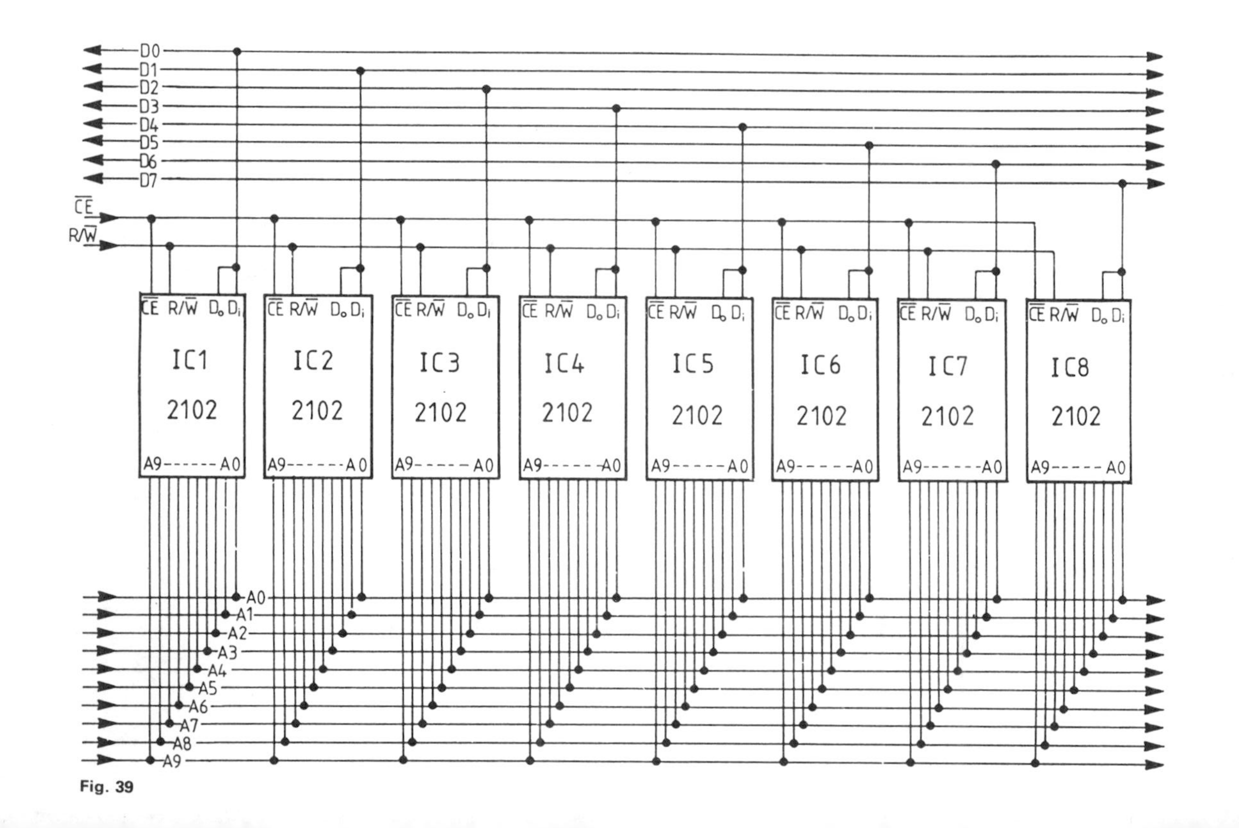
D0
D1
D2
D3
D4
D5
D6
D7
$\overline{CE}$
R/$\overline{W}$
$\overline{CE}$ R/$\overline{W}$ D_o D_i
IC1
IC2
IC3
IC4
IC5
IC6
IC7
IC8
2102
A9- - - - -A0
A0
A1
A2
A3
A4
A5
A6
A7
A8
A9

Fig. 39

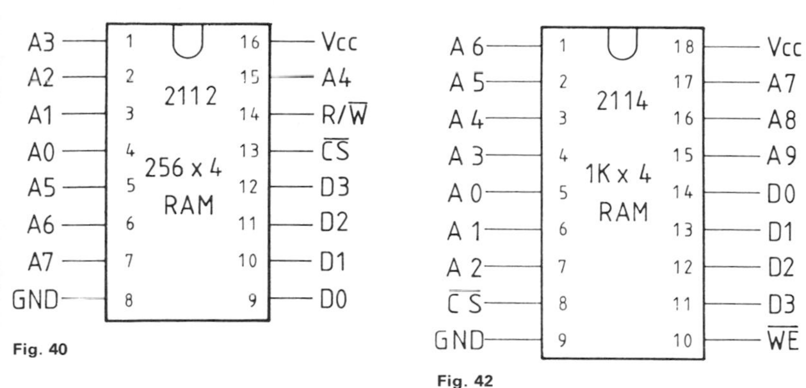

Fig. 40

Fig. 42

Problem 14 Show how **four** memory devices of the type shown in *Fig. 40* may be connected to form a 512 × 8 bit RAM.

Four 256 × 4 bit RAMs of the type shown in *Fig. 40* may be connected to form a 512 × 8 bit RAM as shown in *Fig. 41*.

Problem 15 Show how **two** memory devices of the type shown in *Fig. 42* may be connected to form a 1 K × 8 bit RAM.

Two 1 K × 4 bit RAMs of the type shown in *Fig. 42* may be connected to form a 1 K × 8 bit RAM as shown in *Fig. 43*.

Problem 16 Show how a **2 K × 8 bit EPROM** and **two 1 K × 8 bit RAMs** of the type shown in *Fig. 44* may be connected to form a contiguous 4 K block of memory, with the EPROM occupying the lower 2 K block of addresses.

It is necessary in this circuit to decode the addresses into 1 K blocks to allow the selection of each of the individual RAM chips to take place. A problem then occurs due to the fact that the EPROM must respond to a 2 K block of addresses. This problem is dealt with by decoding A_{10} and A_{11} to give four chip enable signals, each capable of selecting a 1 K block of memory, and to use two of these signals, suitably combined, to select the EPROM. An AND gate may be used for this purpose, as shown in *Fig. 45*.

If either Y_0 or Y_1 is at a logical 0, the output from the AND gate is also at a logical 0 and the EPROM is selected, thus placing the EPROM in the lower 2 K addresses of the 4 K memory block. Outputs Y_2 and Y_3 are used as 1 K block selects for each of the RAM chips.

The $\overline{\text{ENABLE}}$ input to the decoder may be connected to ground (0 V) in

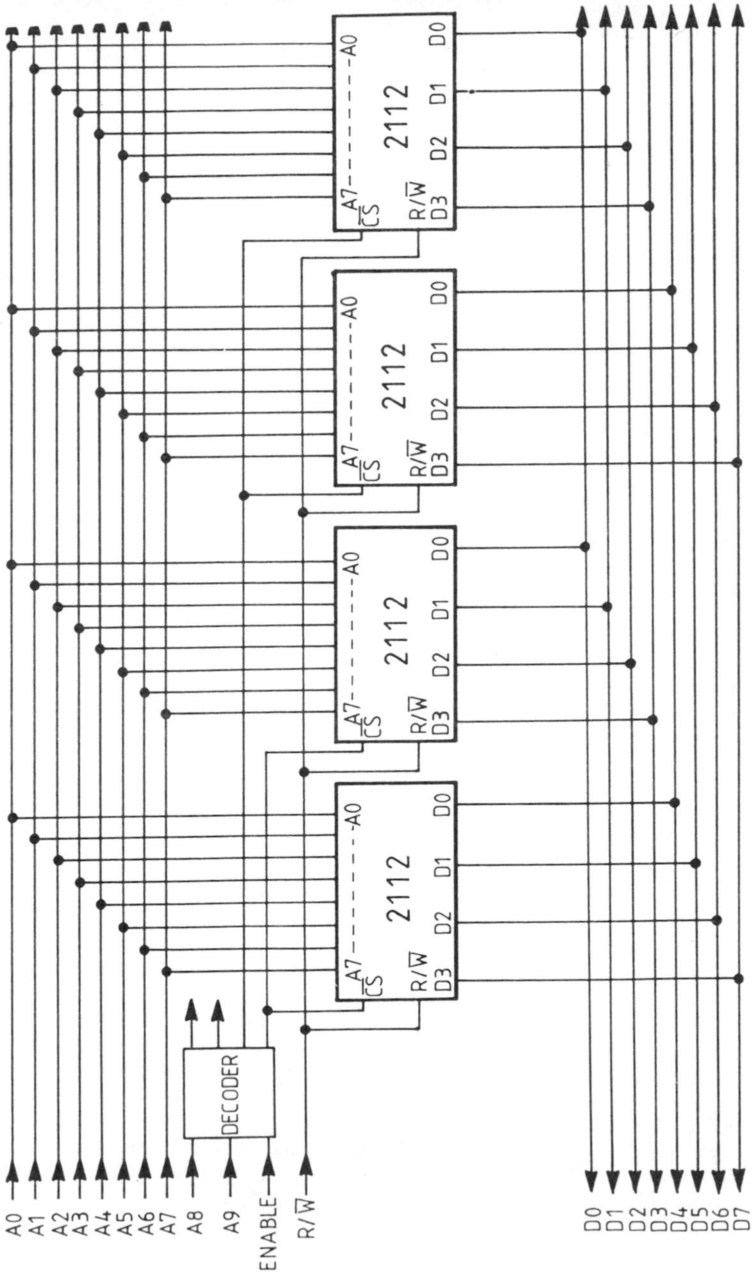

Fig. 41

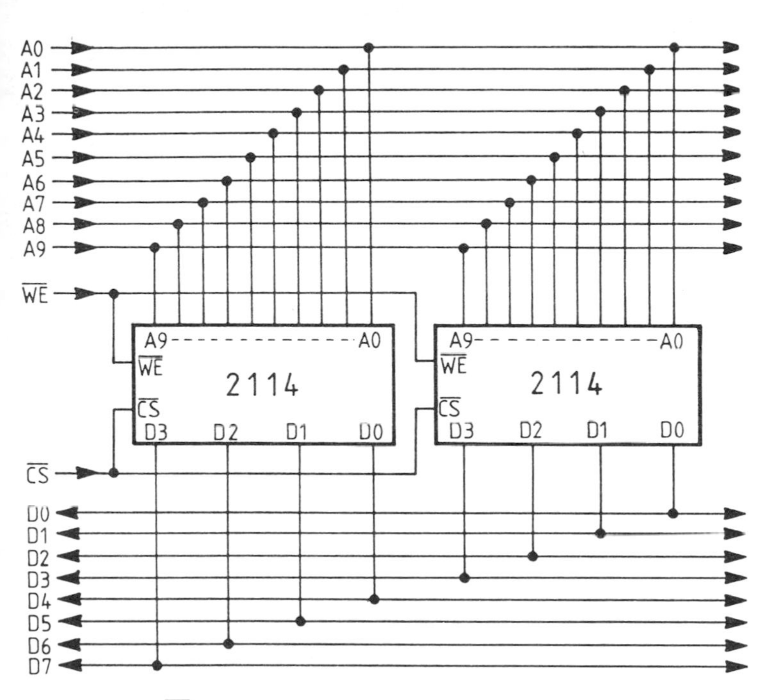
A0
A1
A2
A3
A4
A5
A6
A7
A8
A9
WE
A9 A0
WE
2114
CS
D3 D2 D1 D0
A9 A0
WE
2114
CS
D3 D2 D1 D0
CS
D0
D1
D2
D3
D4
D5
D6
D7
Note: WE = 'write enable' and is functionally the same as R/W

Fig. 43

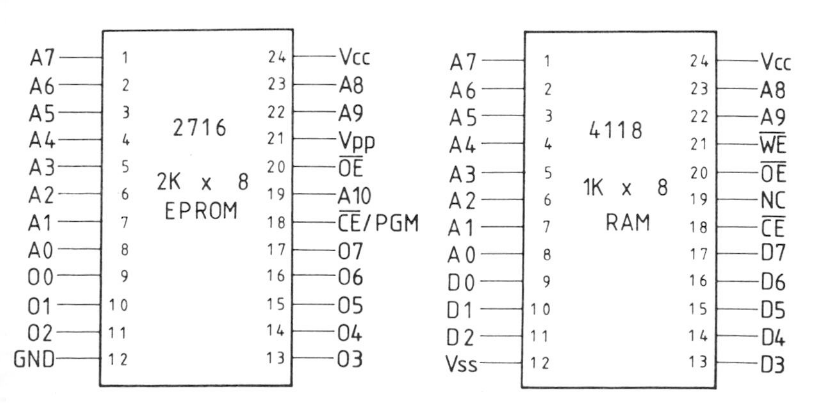
2716
2K x 8
EPROM
A7 1
A6 2
A5 3
A4 4
A3 5
A2 6
A1 7
A0 8
O0 9
O1 10
O2 11
GND 12
24 Vcc
23 A8
22 A9
21 Vpp
20 OE
19 A10
18 CE/PGM
17 O7
16 O6
15 O5
14 O4
13 O3
4118
1K x 8
RAM
A7 1
A6 2
A5 3
A4 4
A3 5
A2 6
A1 7
A0 8
D0 9
D1 10
D2 11
Vss 12
24 Vcc
23 A8
22 A9
21 WE
20 OE
19 NC
18 CE
17 D7
16 D6
15 D5
14 D4
13 D3

Fig. 44

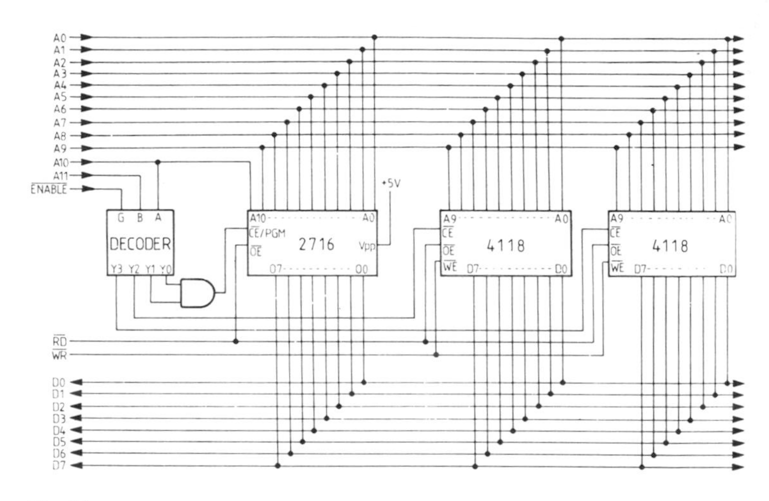

Fig. 45

small systems where full address decoding is unnecessary, or A_{12} to A_{15} may be decoded to provide this signal so that this memory circuit may be assigned to any desired 4 K block of addresses in the 64 K memory space.

C FURTHER PROBLEMS ON MICROCOMPUTER BUS SYSTEMS

(a) SHORT ANSWER PROBLEMS

The following questions refer to *Fig. 46*.

1 Block A contains an . . . .

2 Block B contains a . . . .

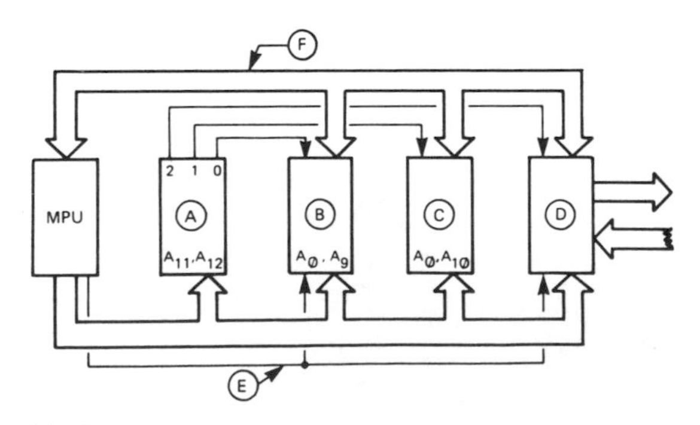

Fig. 46

3 Block C contains a . . . .

4 Block D contains an . . . .

5 Conductor E acts as a . . . .

6 F represents a . . . .

7 The system RAM has a capacity of . . . . bytes

8 The system ROM has a capacity of . . . . bytes

9 The system RAM is located between addresses . . . . and . . . .

10 The system ROM is located between addresses . . . . and . . . .

11 Decoding is a technique which may be used to . . . . in a microcomputer.

12 A decoding device with 4 address inputs may have up to . . . . .outputs.

13 In order to partition a 64 K memory system into 4 K blocks of addresses, ... to . . . . must be decoded.

14 Memory systems which do not make use of all of the available address lines have . . . . decoding.

15 One effect of not using all of the available address lines in a microcomputer memory system is that . . . .

(b) CONVENTIONAL PROBLEMS

1 Explain the need for addresses in a microcomputer memory system.

2 Explain the difference between 'memory mapped' and 'isolated' I/O.

3 Explain what is meant by the term 'bus conflict' and explain how these are avoided in bus structured systems.

4 With reference to microcomputer systems, explain the following terms:
(a) ROM
(b) RAM
(c) volatile
(d) chip enable
(e) tri-state logic

5 With the aid of a diagram, explain how tri-state buffers may be used to enable four memory devices to share common data lines.

6 Explain why it is necessary to use a bidirectional data bus in a microcomputer when a unidirectional bus is satisfactory for addresses.

7 With the aid of diagrams, explain how a 1 K × 8 memory block may be constructed using:
(a) 1 K × 4 memory chips
(b) 256 × 4 memory chips
(c) 1 K × 1 memory chips

8 With the aid of a diagram, show how a 2 K ROM and a 1 K RAM may be assigned to addresses 0000_{16} to $07FF_{16}$ and $0C00_{16}$ to $0FFF_{16}$ respectively by means of combinational logic decoding.

9 With the aid of a diagram, show how a 1 K ROM and two 1 K RAMs may be assigned to address $FC00_{16}$ to $FFFF_{16}$ and 0000_{16} to $07FF_{16}$ respectively using demultiplexer circuit as address decoders.

10 With the aid of a diagram, show how a 4 K ROM and four 1 K RAMs may be assigned to addresses 0000_{16} to $1FFF_{16}$ using *n*-line to 2^n-line decoder circuits.

11 With the aid of a diagram, show how an I/O port may be uniquely assigned to address $8C00_{16}$ by means of combinational logic decoding.

12 Explain the difference between full address decoding and partial address decoding, and with the aid of memory maps, show the main disadvantage of partial address decoding.

13 With the aid of a diagram, show how eight 256 × 4 bit memory devices may be connected to form a 2 K × 8 bit RAM.

14 With the aid of a diagram, show how four 2 K × 8 bit RAMs may be connected to form an 8 K × 8 bit RAM.

15 *Fig. 47* shows the circuit diagram of a 6502-based microcomputer. Carry out a case study of this circuit to include:
(a) analysis of address decoding circuits
(b) determination of memory assignments
(c) use of control lines
(d) description of the clock circuit
(e) operation of the reset circuit

16 *Fig. 48* shows the circuit diagram of a Z80-based microcomputer. Carry out a case study of this circuit to include:
(a) analysis of address decoding circuits
(b) determination of memory assignments
(c) use of control lines
(d) description of the clock circuit
(e) operation of the reset circuit

17 *Fig. 49* shows the circuit diagram of a 6800-based microcomputer. Carry out a case study of this circuit to include:
(a) analysis of address decoding circuits
(b) determination of memory assignments
(c) use of control lines
(d) description of the clock circuit
(e) operation of the reset circuit

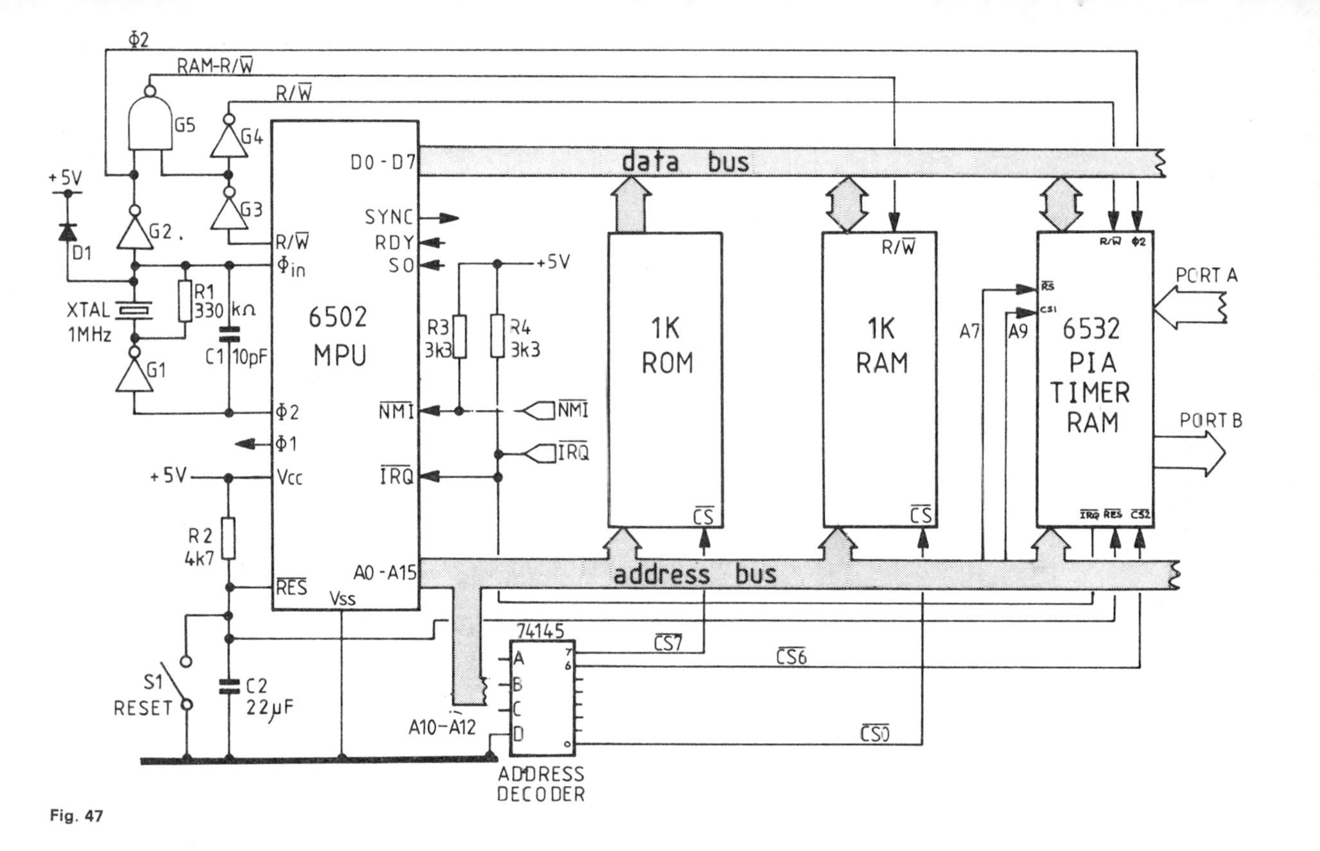
Φ2
RAM-R/W̄
R/W̄
G5
G4
G3
G2
G1
+5V
D1
XTAL
1MHz
R1
330 kΩ
C1 10pF
6502
MPU
D0 - D7
SYNC
RDY
SO
R/W̄
Φin
Φ2
Φ1
Vcc
+5V
NMĪ
IRQ̄
RES̄
Vss
A0 - A15
+5V
R3
3k3
R4
3k3
NMĪ
IRQ̄
R2
4k7
S1
RESET
C2
22µF
data bus
address bus
1K
ROM
CS̄
1K
RAM
R/W̄
CS̄
6532
PIA
TIMER
RAM
R/W̄
Φ2
RS̄
CS1
IRQ̄
RES̄
CS̄2
A7
A9
PORT A
PORT B
74145
A
B
C
D
A10–A12
ADDRESS
DECODER
CS̄7
CS̄6
CS̄0

Fig. 47

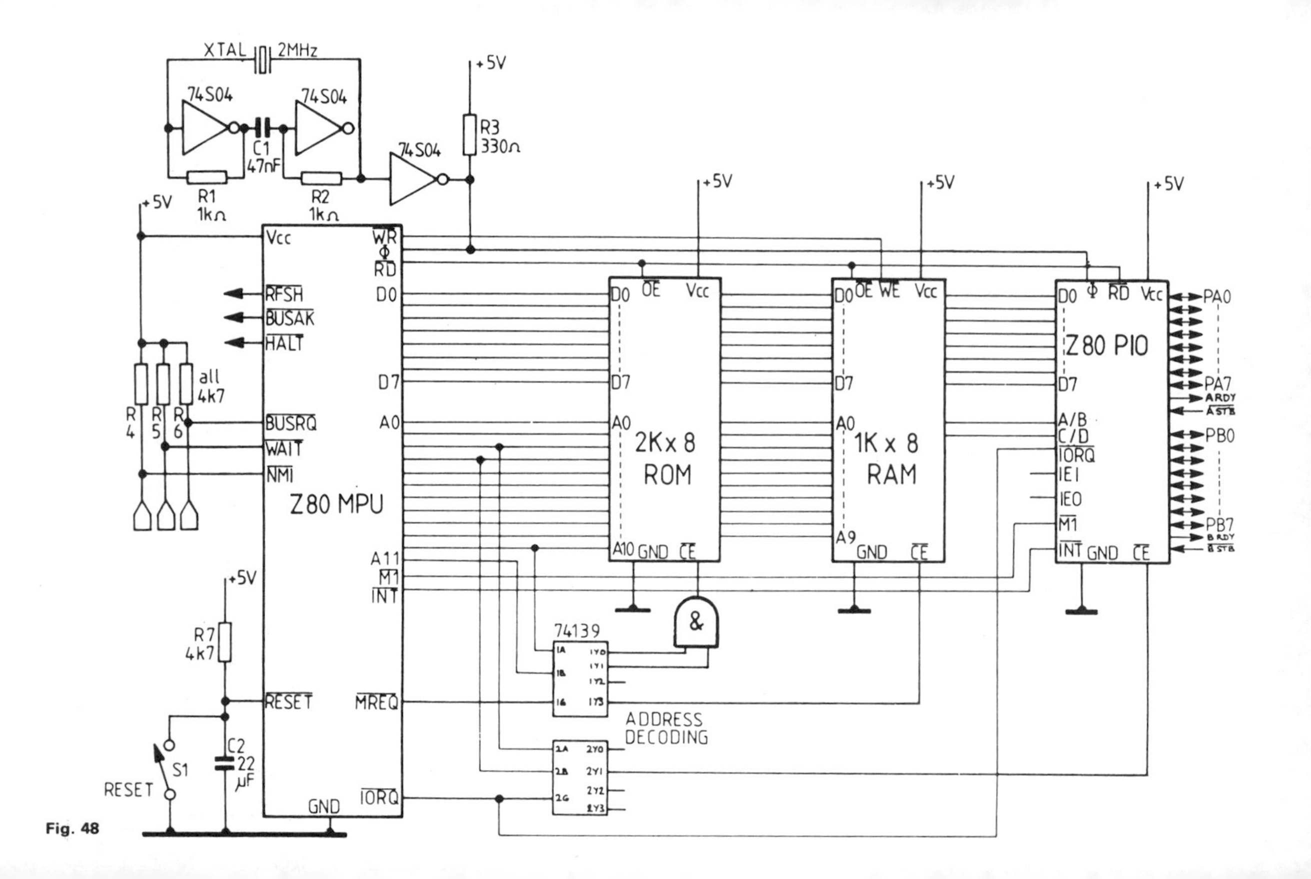

XTAL
2MHz
74S04
74S04
74S04
C1
47nF
R1
1kΩ
R2
1kΩ
R3
330Ω
+5V
Z80 MPU
Vcc
WR
Φ
RD
D0
D7
A0
A11
M1
INT
MREQ
IORQ
GND
RFSH
BUSAK
HALT
BUSRQ
WAIT
NMI
RESET
R4
R5
R6
all
4k7
R7
4k7
C2
22
µF
S1
RESET
2K x 8
ROM
OE
A10
CE
1K x 8
RAM
OE
WE
A9
Z80 PIO
PA0
PA7
ARDY
ASTB
PB0
PB7
BRDY
BSTB
A/B
C/D
IORQ
IEI
IEO
M1
INT
&
74139
ADDRESS
DECODING

Fig. 48

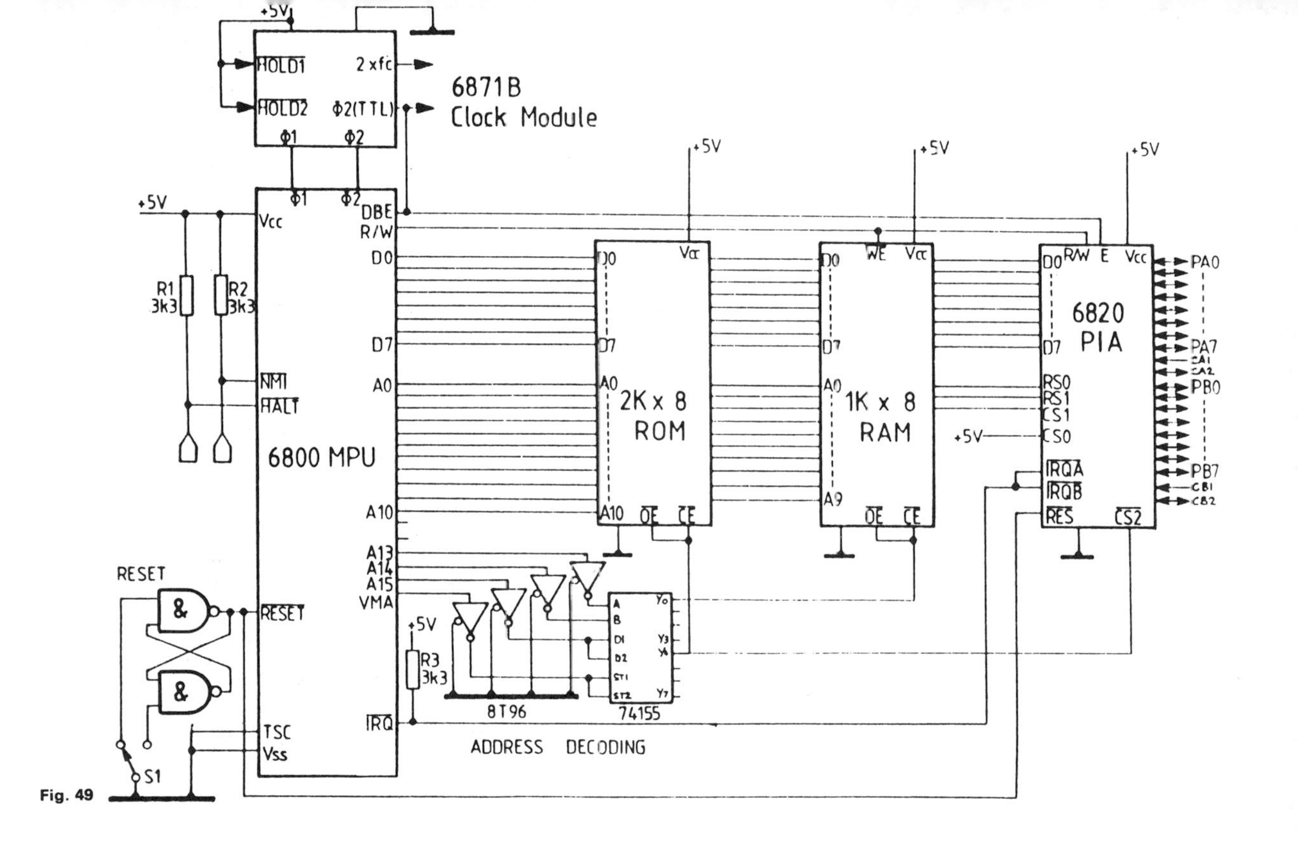
6871B
Clock Module
6800 MPU
2K x 8
ROM
1K x 8
RAM
6820
PIA
8T96
74155
ADDRESS DECODING
RESET
S1
R1
3k3
R2
3k3
R3
3k3
+5V

Fig. 49

2 Interrupts

A MAIN POINTS CONCERNED WITH INTERRUPTS

1 Parallel data transfers between a microcomputer and its peripheral devices take place via the parallel I/O port of a microcomputer, as shown in *Fig. 1(a) and (b)*.
Actual data transfers are achieved by means of the following instructions:

(a) **LOAD** and **STORE** instructions in systems which use **memory mapped I/O**, or

(b) **IN** and **OUT** instructions in systems which use **isolated I/O**.

In either case, the rate at which data may be transferred to and from a microcomputer depends upon the time that it takes to execute these instructions. Using a typical microcomputer, execution times of between 2 μs and 10 μs may be achieved, depending upon the clock frequency of the microcomputer and the addressing mode used. This means (in theory) that data transfer rates of between 100 and 500 K bytes per second could be achieved, although due to the need to include other instructions within the data transfer program, much slower data transfer rates are more likely to be realized.

2 Most peripheral devices do **not** operate at the **same speed** as a microcomputer. Some devices operate at considerably lower speeds than those obtainable with a microcomputer, for example a printer may require data to be transferred at a rate of only 30 bytes per second. Therefore some form of common timing between a microcomputer and its peripheral devices is required. Two techniques which are available for timing of data transfers are: (a) **software polling**, and (b) **external interrupts**.

3 A system of **software polling** involves a microcomputer interrogating each peripheral device, in turn, to see if it is ready to send or accept a new byte of data. This technique is illustrated in *Fig. 2(a) and (b)*.

4 An external interrupt is **hardware initiated**. Each peripheral device is connected to an interrupt input pin on the microprocessor, and indicates that it requires attention by changing the logic state on this input. This action causes a temporary break in the main program and execution of a routine to deal with the peripheral device that initiated the interrupt.
This routine is known as an **interrupt service routine** (ISR), and upon completion of the ISR, a return to the main program occurs which then continues as though it had never been interrupted. This process is shown in *Fig. 3*.
A return to the main program is accomplished by the use of **RETI** or **RTI** as the last instruction of an ISR.

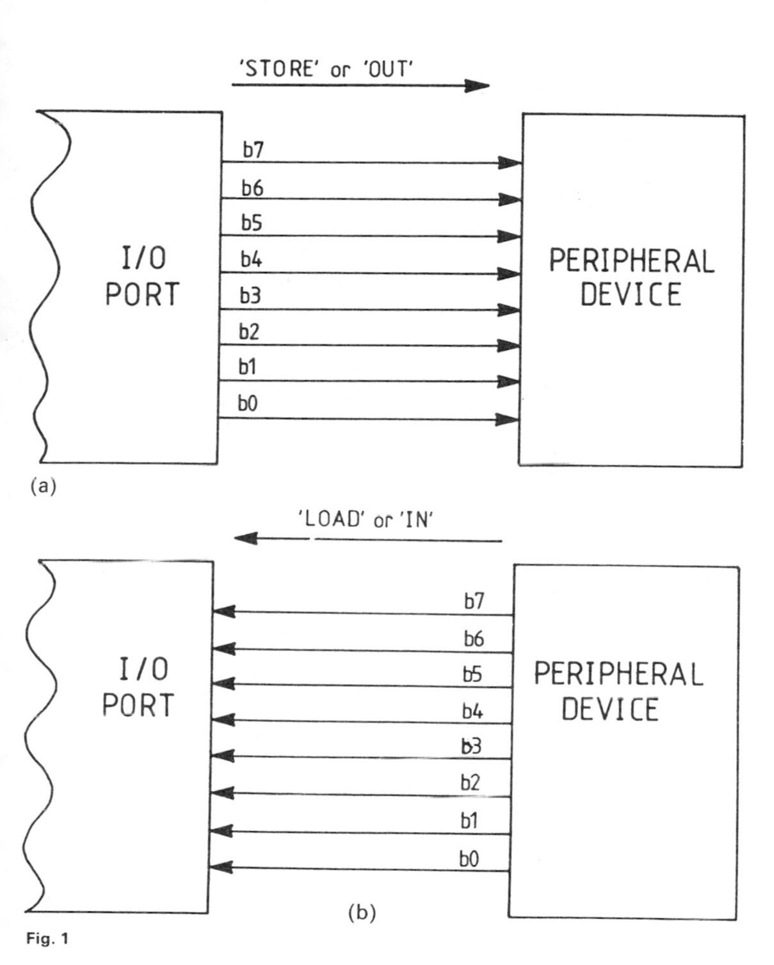

Fig. 1

5 Registers in use in the main program at the time an interrupt occurs may be required for use in an ISR. Therefore the contents of **all** registers may need to be saved, since it is not usually possible to predict which registers are in use at the time an interrupt occurs, due to the random nature of an interrupt. Register contents may be saved by using one of the following methods:

(a) transfer data to **other registers** which are not in use in either the main program or the ISR;
(b) save the data in **specified memory locations**;
(c) transfer data to an **alternate register set**, if available (e.g. Z80),
(d) transfer data to the **stack**.

Register contents must be saved immediately the ISR is entered, and restored just prior to returning to the main program. This is illustrated in *Fig. 4*.

B WORKED PROBLEMS ON INTERRUPTS

Problem 1 Explain the advantage of using **external interrupts** rather than **software polling** as a means of communicating with certain peripheral devices.

Events associated with certain peripheral devices occur at totally random and therefore unpredictable times. For example, sensors associated with detecting excessive temperature or pressure in a system may seldom be activated – perhaps once a year. When such sensors are activated, however, immediate action must be taken to prevent damage from occurring to the system.

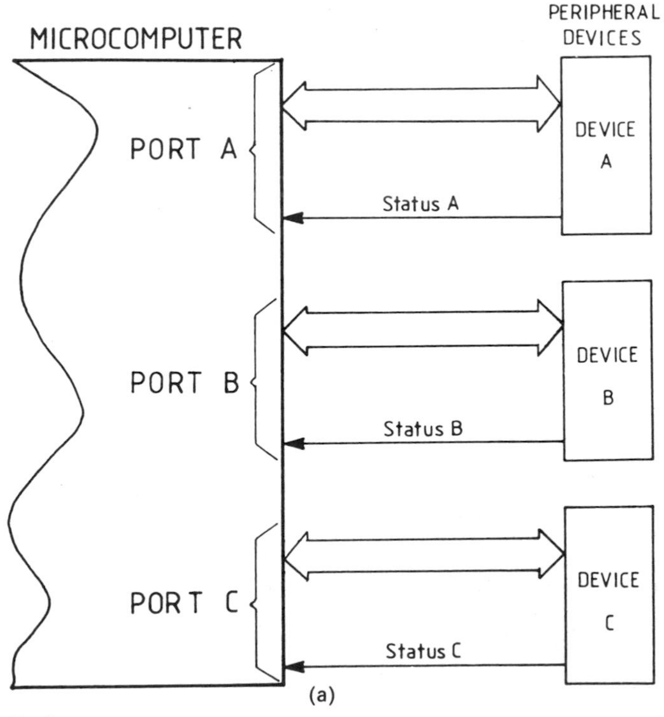

Fig. 2

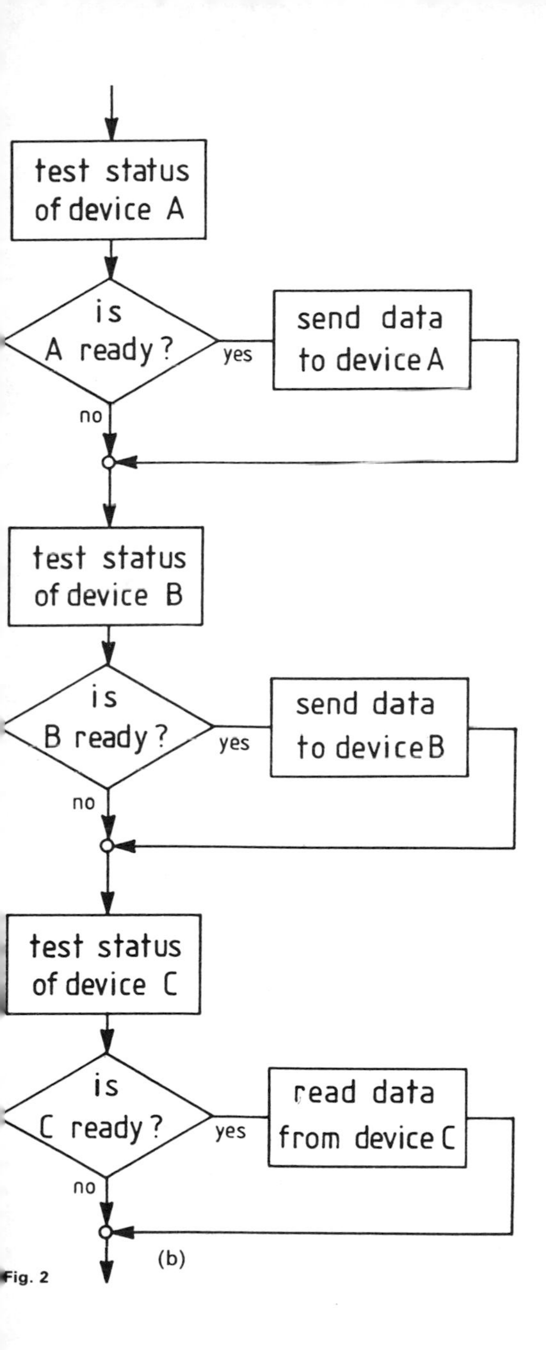
test status of device A
is A ready?
yes
send data to device A
no
test status of device B
is B ready?
yes
send data to device B
no
test status of device C
is C ready?
yes
read data from device C
no
(b)
Fig. 2

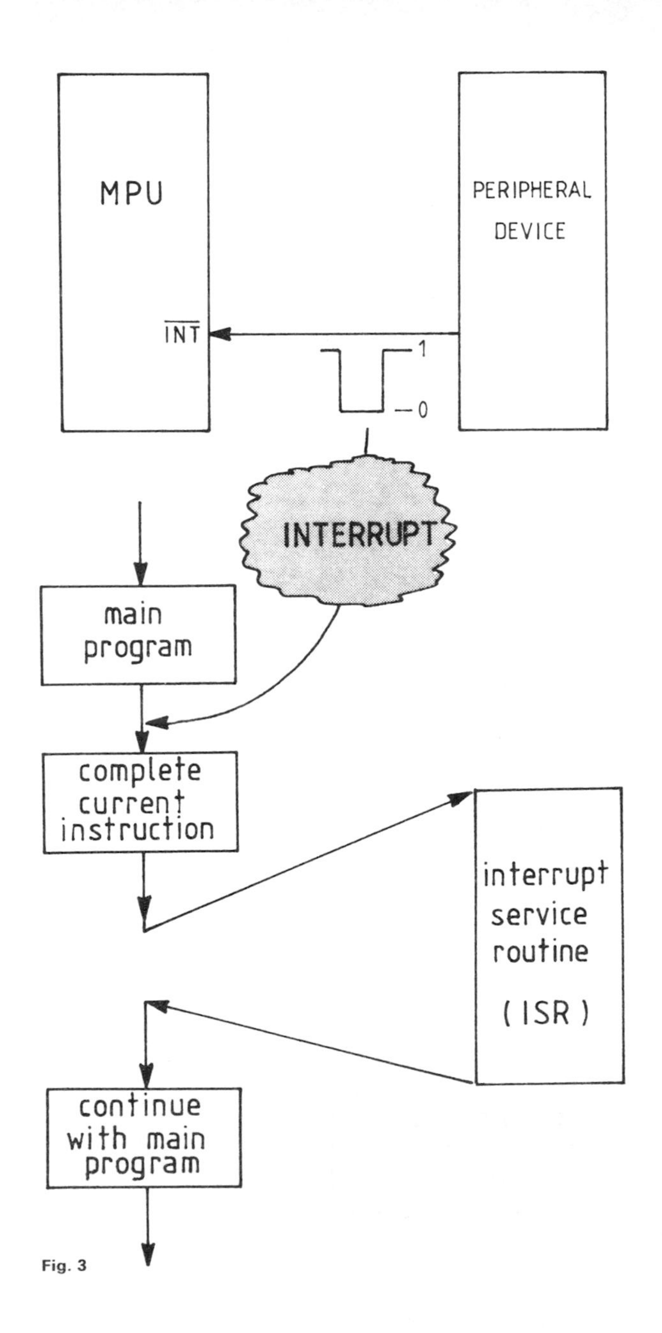
MPU
PERIPHERAL DEVICE
$\overline{INT}$
1
0
INTERRUPT
main program
complete current instruction
interrupt service routine (ISR)
continue with main program

Fig. 3

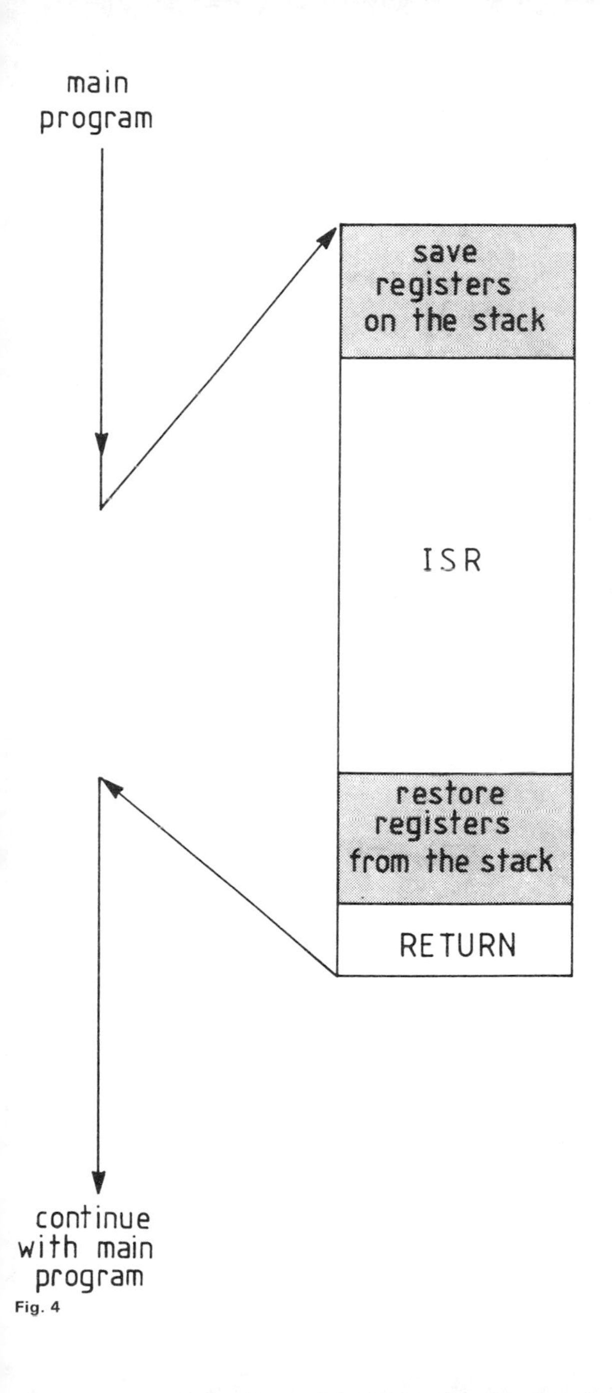
main
program
save
registers
on the stack
ISR
restore
registers
from the stack
RETURN
continue
with main
program
Fig. 4

If software polling is used, these sensors must be interrogated at frequent intervals in order to obtain an immediate response, e.g. once or twice a second. Polling for this purpose is very inefficient and reduces the rate at which a microcomputer can perform its main task, since much time is spent reading sensors which are seldom activated.

An external interrupt is initiated by hardware connected to the interrupt input pin of a microprocessor. Upon receipt of an interrupt request, a microcomputer completes its current instruction, and is then diverted from the main task to an interrupt service routine (ISR). The ISR is a program which performs the task required by the interrupting peripheral.

On completion of the ISR, the microcomputer returns to the main task at the point where it left off, and continues with the main task as though it had never been interrupted. When this method of communicating with a peripheral device is used, the microcomputer operates much more efficiently, since ths main task is suspended only when external events requires this.

Problem 2 A sensor, represented by S_1, is connected to bit 7 of port A of a **6502** based microcomputer, as shown in *Fig. 5*. A polling routine is required which waits until S_1 is closed before continuing with the next section of the program. With the aid of a flow chart, devise a **6502 machine code** polling routine to perform this task, assuming that port A is configured to accept inputs at address 1700_{16}.

If the data on port A is read into the accumulator, the N flag is set or reset according to the state of bit 7, since this action copies bit 7 into the N flag. When S_1 is open, but 7 is at logical 1 and the N flag is set, but if S_1 is closed, bit 7 is at logical 0 and the N flag is reset. Therefore a **branch if plus**

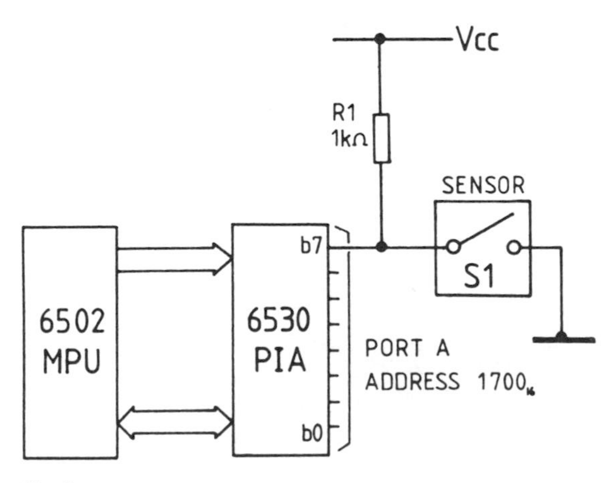

Fig. 5

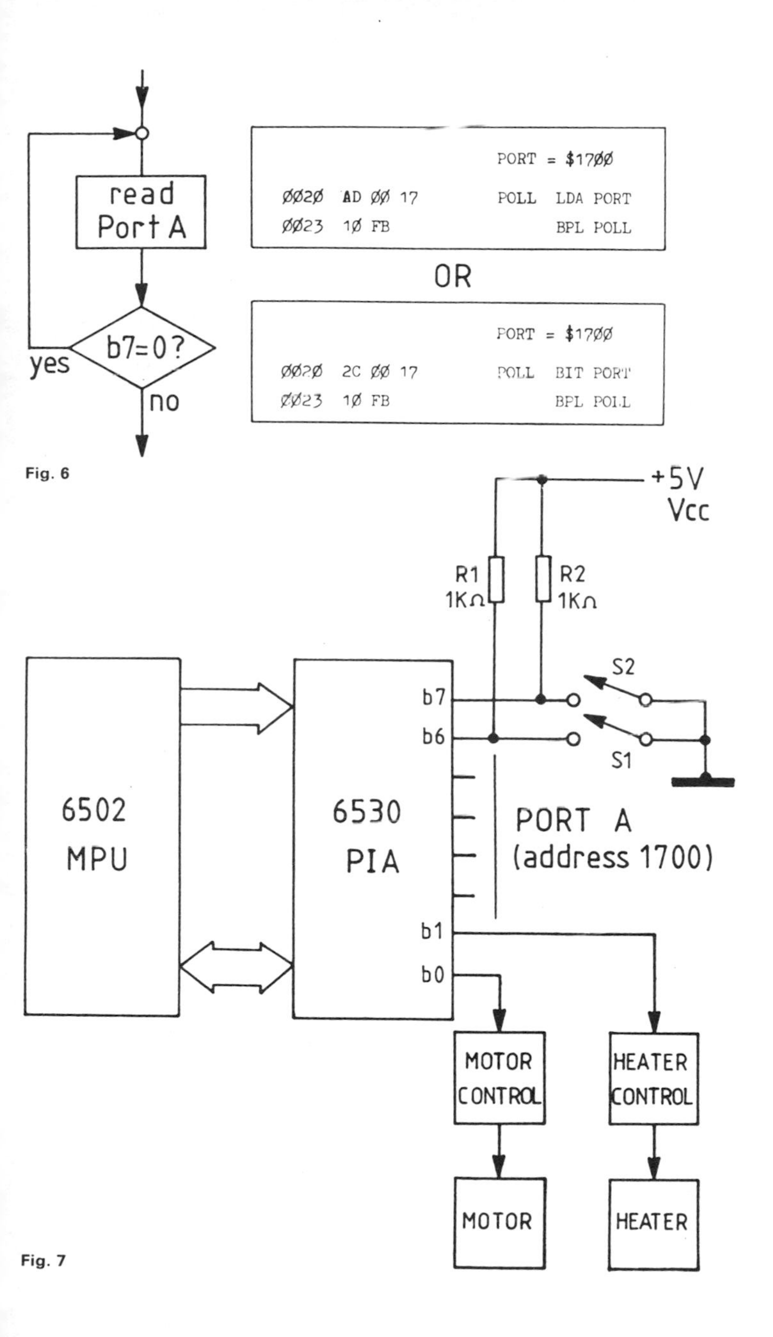
read
Port A
b7=0?
yes
no
PORT = $17ØØ
ØØ2Ø AD ØØ 17 POLL LDA PORT
ØØ23 1Ø FB BPL POLL
OR
PORT = $17ØØ
ØØ2Ø 2C ØØ 17 POLL BIT PORT
ØØ23 1Ø FB BPL POLL
+5V
Vcc
R1
1KΩ
R2
1KΩ
S2
S1
b7
b6
b1
b0
6502
MPU
6530
PIA
PORT A
(address 1700)
MOTOR
CONTROL
HEATER
CONTROL
MOTOR
HEATER

Fig. 6

Fig. 7

(BPL) instruction may be used to determine what action must be taken, i.e., keep polling if N = 1; service the peripheral device if N = 0. This action is shown in *Fig. 6*.

Alternatively, a bit instruction always copies bit 7 of the location that it specifies into the N flag, and this may be used as an alternative to the LDA instruction in *Fig. 6*.

Problem 3 An industrial process which incorporates a heating element and an electric motor is controlled by a **6502** based microcomputer with a **6530** PIA. The motor and heater circuits are connected to bit 0 and bit 1, respectively of port A of the **6530**, and are controlled by switches S_1 and S_2 which are connected to bit 6 and bit 7 of the same port, as shown in *Fig. 7*. Write a **6502** machine code program, starting at address 0200_{16}, which polls S_1 and S_2 and controls the motor and heater in the following manner:

S_1 open	:	heater	on
S_1 closed	:	heater	off
S_2 open	:	motor	on
S_2 closed	:	motor	off

Since current must flow in either the heater or the motor (or both) whenever either S_1 or S_2 is closed, a simple polling routine of the type shown in *Fig. 8* may be used. A suitable **6502** machine code routine to poil the switches and to operate the heater and the motor is as follows:

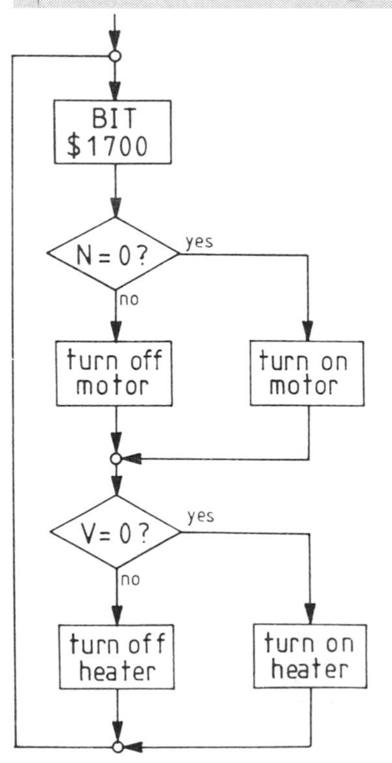

Fig. 8

```
                   PAD      = $1700
                   PADD     = $1701
                          * = $0200
                   ;
0200  A9 C0                 LDA #$03       ;configure PIA
0202  8D 01 17              STA PADD       ;b0,b1 are outputs, b6,b7 are inputs
0205  18                    CLC            ;create branch always for later
0206  A9 00                 LDA 0          ;clear accumulator
0208  8D 00 17     POLL     STA PAD        ;actuate motor/heater
020B  2C 00 17              BIT PAD        ;check switch positions
020E  10 04                 BPL MTRON      ;S2 closed so leave motor on
0210  29 FE                 AND #$FE       ;turn off motor control bit
0212  90 02                 BCC HTROFF     ;branch always in this program
0214  09 01        MTRON    ORA #$01       ;turn on motor control bit
0216  50 04                 BVC HTRON      ;S1 closed so leave heater on
0218  29 FD                 AND #$FD       ;turn off heater control bit
021A  90 EC                 BCC POLL       ;branch always in this program
021C  09 02        HTRON    ORA #$02       ;turn on heater control bit
021E  90 E8                 BCC POLL       ;branch always in this program
```

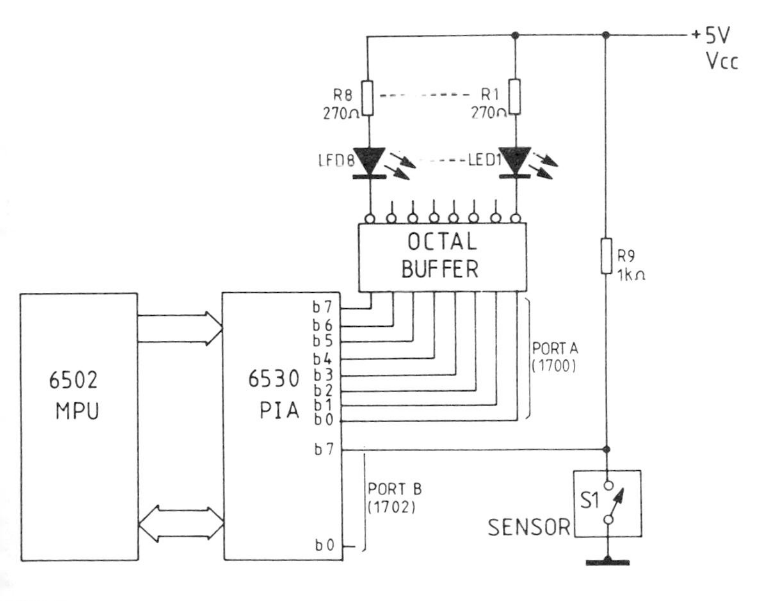

Fig. 9

Problem 4 A **6502** based microcomputer with a **6530 PIA** is used to monitor the number of objects passing along a conveyor belt. A photoelectric sensor (represented by S_1) is connected to bit 7 of port B of the PIA, and the presence of an object adjacent to this sensor breaks a beam and causes a logical 0 to be applied to bit 7 by the sensor. The total number of objects which pass this sensor are counted by the microcomputer, and are displayed, in binary, on eight LEDs which are connected to port A of the PIA. This arrangement is illustrated in *Fig. 9*.

Write a 6502 machine code program to poll S_1, and increment the display by one for each object counted. The program should start at address 0200_{16}.

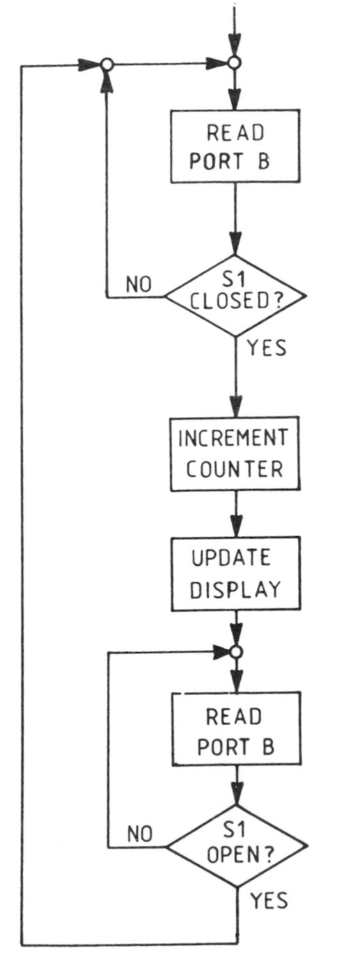

The polling routine required for this example is different to that used in *Problem 3*, since the counter must only be incremented once each time that an object breaks the beam of light to the photoelectric sensor. It is therefore necessary to poll the sensor to determine when the beam is broken so that the counter may be incremented, but then to poll the sensor again to determine when the beam is restored before repeating the process. Failure to carry out this procedure results in the counter being incremented many times while the beam is broken. A polling routine which satisfies these requirements is shown in *Fig. 10*.

Fig. 10

A suitable **6502** machine code program to poll the sensor and increment the displays is as follows:

```
                     PAD       =$1700
                     PADD      =$1701
                     PBD       =$1702
                     PBDD      =$1703
                             * =$0200
                     ;
0200  A9 FF                    LDA #$FF      ;configure port A
0202  8D 01 17                 STA PADD      ;all bits as outputs
0205  A2 00                    LDX #0        ;clear counter
0207  8E 03 17                 STX PBDD      ;all bits of port B are inputs
020A  8E 00 17       DISPLA    STX PAD       ;update displays
020D  2C 02 17       POLL0     BIT PBD       ;poll for S1 closure
0210  30 FB                    BMI POLL0     ;keep checking if not closed
0212  E8                       INX           ;increment object counter
0213  2C 02 17       POLL1     BIT PBD       ;poll for S1 open
0216  10 FB                    BPL POLL1     ;keep checking if not open
0218  30 F0                    BMI DISPLA    ;display new total and repeat
```

Problem 5 A **6502** based microcomputer with a **6530 PIA** is used to monitor the number of vehicles entering and leaving a car park. Sensors S_1 and S_2, sited at the entrance and exit of the car park are connected to bit 5 and bit 7 of port **B** of the **PIA** and are used to detect vehicles entering and leaving. Eight LEDs are connected to port A of the 6530 PIA and are used to display (in binary) the total number of vehicles currently in the car park. This arrangement is illustrated in *Fig. 11*.

Write a 6502 machine code program, starting at address 0200_{16}, to poll S_1 and S_2 and to increment the number displayed by one each time that a vehicle enters the car park, and to decrement the number displayed by one each time that a vehicle leaves (the sensors are arranged such that S_1 and S_2 close each time that a vehicle is detected).

Two polling routines of the type used in *Problem 4* could be implemented in series to detect closure of either S_1 or S_2. However, if S_1, for example, is closed for a long period of time, S_2 is 'locked out' and is prevented from being polled, because the program is looping around waiting for S_1 to be opened.

This problem may be resolved by polling S_1 and S_2, and setting a flag (one for each sensor) if a sensor switch is found to be closed. This flag is used to prevent further incrementing or decrementing of the counter until the appropriate switch is again in the open condition. Once the polling routine detects the open condition in either S_1 or S_2, the appropriate flag is reset. A polling routine of this type is illustrated by the flowchart shown in *Fig. 12*.

A **6502** machine code program to perform this function is as follows:

```
                        FLAG1   = $0000
                        FLAG2   = $0001
                        TEMP    = $0002
                        PAD     = $1700
                        PADD    = $1701
                        PBD     = $1702
                        PBDD    = $1703
                              * = $0200
                        ;
0200  A9 FF                     LDA #SFF
0202  8D 01 17                  STA PADD    ;port A all outputs
0205  A2 00                     LDX #$00
0207  8E 03 17                  STX PBDD    ;port B all inputs
020A  8E 00 17                  STX PAD     ;set display to zero
020D  86 00                     STX FLAG1   ;reset increment disable flag
020F  86 01                     STX FLAG2   ;reset decrement disable flag
0211  AD 02 17  POLL1           LDA PBD     ;read the sensors
0214  69 20                     ADC #$20    ;shift b5 to b6 position
0216  85 02                     STA TEMP
0218  24 02                     BIT TEMP    ;transfer b6 and b7 to V and N flags
021A  10 04                     BPL S1      ;S1 closed?
021C  86 00                     STX FLAG1   ;if not reset increment disable flag
021E  30 0B                     BMI S2      ;branch always in this case
0220  A5 00     S1              LDA FLAG1   ;get increment disable flag
0222  D0 07                     BNE POLL2   ;do not increment if flag is set
0224  EE 00 17                  INC PADD    ;increment display counter
0227  09 01                     ORA #$01    ;set increment disable flag to
0229  85 00                     STA FLAG1   ;inhibit further increments while S1 =0
022B  50 04     POLL2           BVC S2      ;S2 closed?
022D  86 01                     STX FLAG2   ;if not reset decrement disable flag
022F  70 E0                     BVS POLL1   ;branch always in this case
0231  A5 01     S2              LDA FLAG2   ;get decrement disable flag
0233  D0 DC                     BNE POLL1   ;do not decrement if flag is set
0235  CE 00 17                  DEC PAD     ;decrement display counter
0238  09 01                     ORA #$01    ;set decrement disable flag to
023A  85 01                     STA FLAG2   ;inhibit further decrements while S2 =0
023C  D0 D3                     BNE POLL1   ;keep polling S1 and S2
```

*Note: The pin on a **6530 PIA** to which bit 6 of port B would normally be connected is often used as a **chip select input**, thus making bit 6 of this port unavailable to the user. For this reason, bit 5 is used in this program, and is later shifted into bit 6 position.*

Problem 6 Describe the operation of the **reset** ($\overline{\text{RES}}$) input of a **6502** microprocessor.

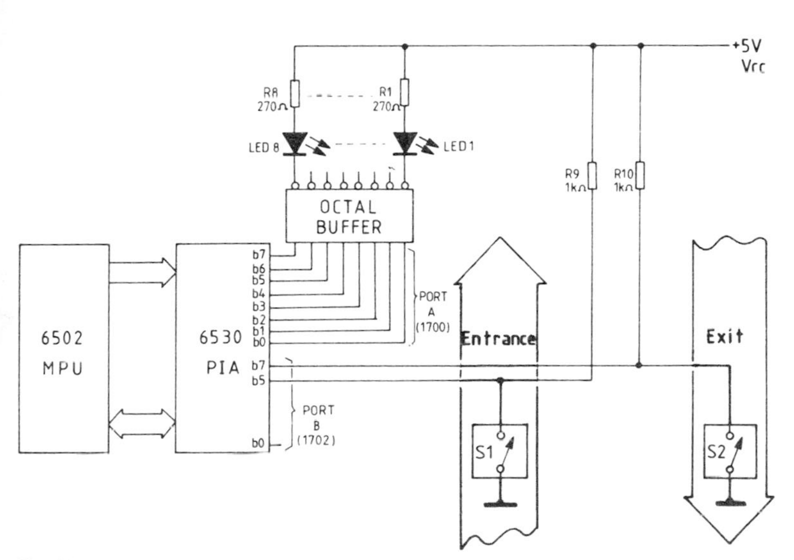

Fig. 11

This is an **interrupt input** which is used to **reset the microprocessor** (or to provide initial start-up from switch-on time), and it overrides all other inputs. While $\overline{\text{RES}}$ is held at logical 0, all microprocessor read and write operations are suspended. When a **positive edge** is detected on the $\overline{\text{RES}}$ input, the microprocessor immediately commences its restart sequence, and this consists of the following steps:

(a) a **delay** which is equivalent to **six clock cycle periods** occurs to allow time for the initialization of the microprocessor;
(b) the **interrupt mask flag is set**;
(c) addresses **$FFFC_{16}$** and **$FFFD_{16}$** are generated, in sequence, by the microprocessor, and **restart vector** is fetched from these locations; and
(d) the **program counter** is loaded with the **restart vector**, and control is handed over to the program whose starting address is defined by the contents of memory locations $FFFC_{16}$ and $FFFD_{16}$.

This process is shown in *Fig. 13*.

Problem 7 Describe the operation of the following interrupts on a **6502** microprocessor: (a) **interrupt request ($\overline{\text{IRQ}}$)**; (b) **non-maskable interrupt ($\overline{\text{NMI}}$)**.

(a) Interrupt request ($\overline{\text{IRQ}}$)
This **active low** input is used to initiate an interrupt sequence within the **6502**

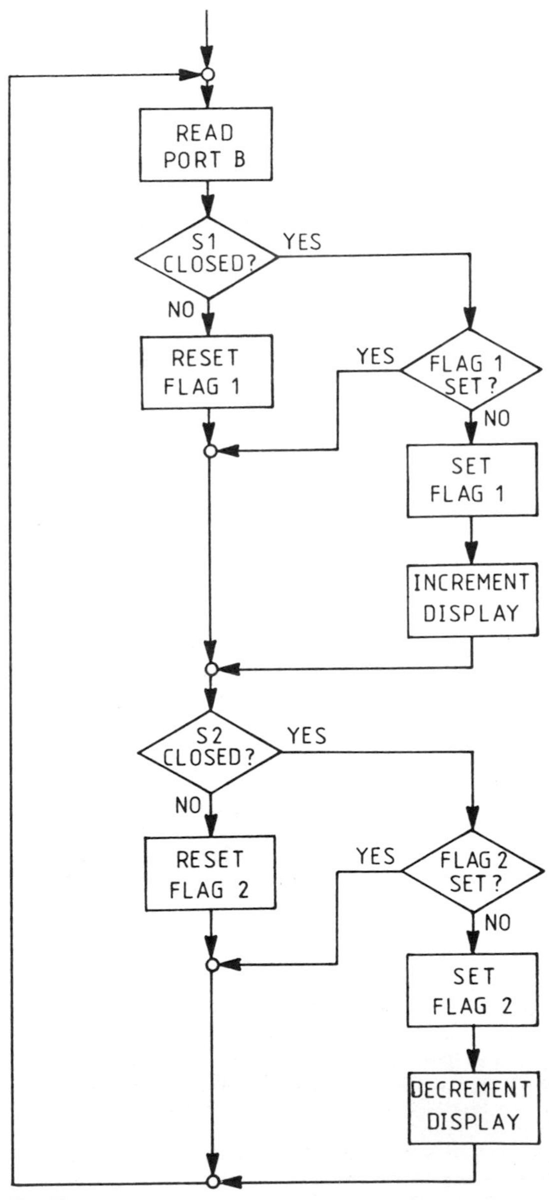
READ
PORT B
S1
CLOSED?
YES
NO
RESET
FLAG 1
YES
FLAG 1
SET?
NO
SET
FLAG 1
INCREMENT
DISPLAY
S2
CLOSED?
YES
NO
RESET
FLAG 2
YES
FLAG 2
SET?
NO
SET
FLAG 2
DECREMENT
DISPLAY

Fig. 12

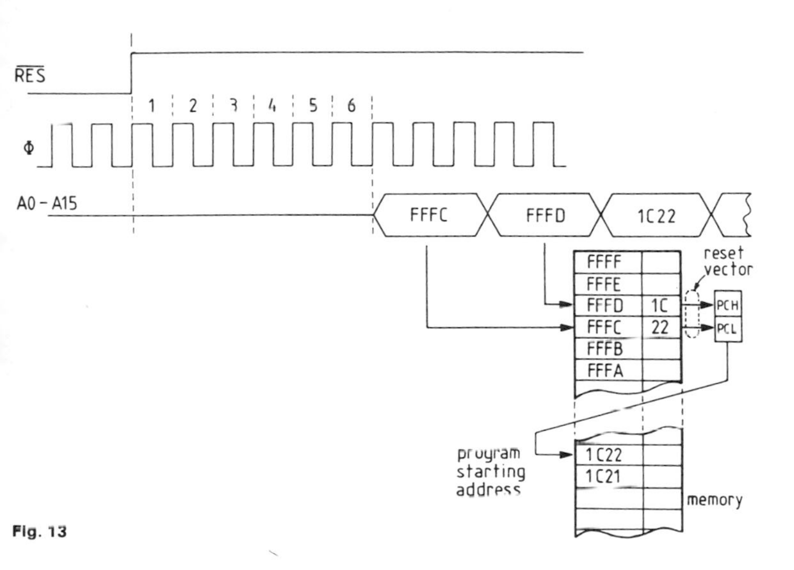

Fig. 13

microprocessor. The microprocessor completes the current instruction being executed before recognizing this request (the $\overline{IRQ}$ input is sampled during Φ2 and the interrupt sequence starts on the next Φ1). Next, the interrupt mask bit (I flag) in the status register is examined, and, provided that this flag is not set (i.e. I = 0), the microprocessor begins its interrupt sequence, which consists of the following steps:

(i) the **program counter** and **status register** contents are stored on the **stack**;
(ii) the **interrupt mask flag is set** (I = 1) to prevent the microprocessor from responding to further interrupt requests during this sequence;
(iii) addresses $\mathbf{FFFE_{16}}$ and $\mathbf{FFFF_{16}}$ are generated, in sequence, by the microprocessor, and the $\overline{\textbf{IRQ}}$ **vector** is fetched from these locations; and
(iv) the **program counter** is loaded with the $\overline{\textbf{IRQ}}$ **vector**, and control is handed over to the interrupt service routine (ISR) whose starting address is defined by the contents of memory locations $FFFE_{16}$ and $FFFF_{16}$.

This process is shown in *Fig. 14*.

(b) Non-maskable interrupt ($\overline{\textbf{NMI}}$)

This is an **edge triggered** input which is used to initiate a non-maskable interrupt sequence within the **6502** microprocessor. A negative going edge (i.e. logical 1 to logical 0 transition) is required to initiate the interrupt sequence, therefore this input must be returned to a logical 1 level before a further interrupt will be accepted. The $\overline{NMI}$ input may therefore be kept at a logical 0 level for as long as required after initiating an interrupt.

Following completion of the current instruction being executed, the sequence of operations described for $\overline{IRQ}$ in (a) is carried out, regardless of the state of the interrupt mask flag (i.e. it is non-maskable). The ISR starting address is,

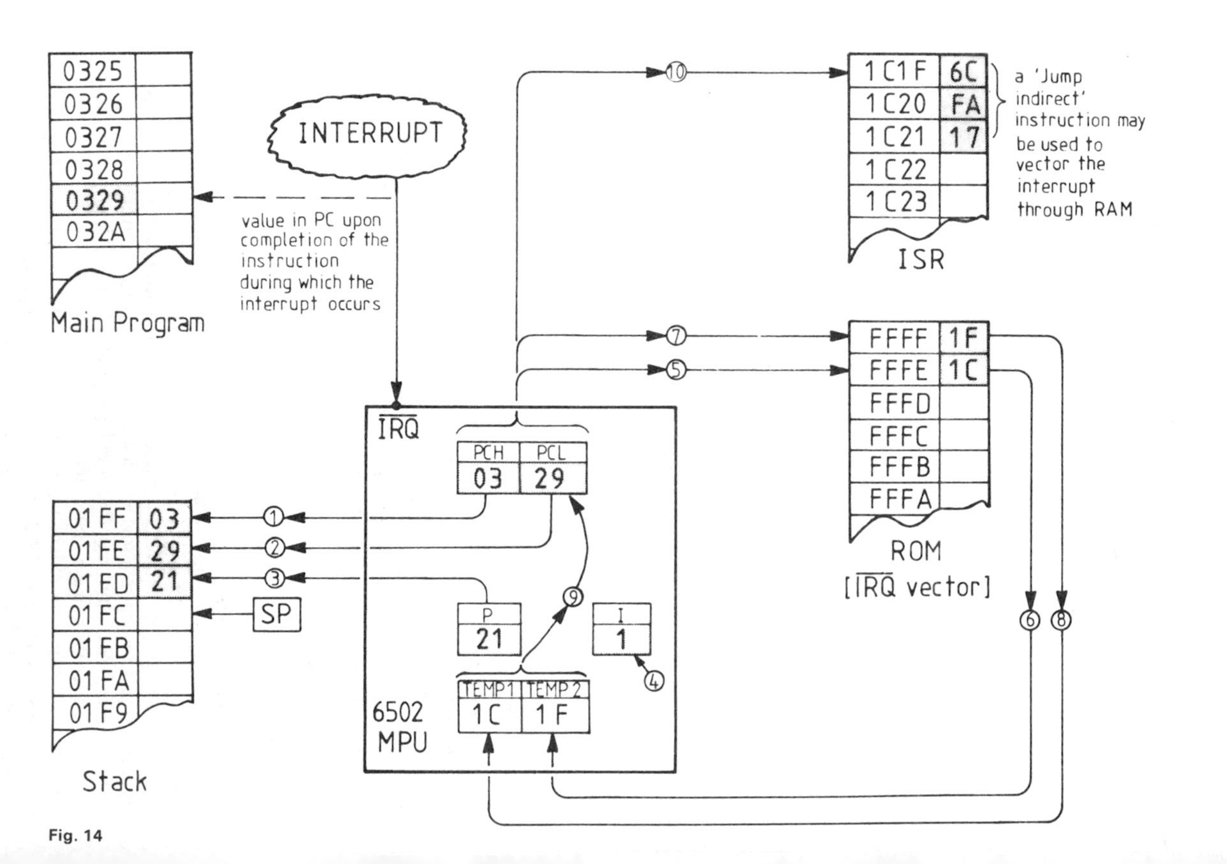
0325
0326
0327
0328
0329
032A
Main Program
INTERRUPT
value in PC upon completion of the instruction during which the interrupt occurs
IRQ
PCH
PCL
03
29
P
21
I
1
TEMP1
TEMP2
1C
1F
6502 MPU
01FF
03
01FE
29
01FD
21
01FC
01FB
01FA
01F9
SP
Stack
1C1F
6C
1C20
FA
1C21
17
1C22
1C23
ISR
a 'Jump indirect' instruction may be used to vector the interrupt through RAM
FFFF
1F
FFFE
1C
FFFD
FFFC
FFFB
FFFA
ROM
[IRQ vector]

Fig. 14

however, fetched from addresses **FFFA**$_{16}$ and **FFFB**$_{16}$ for a non-maskable interrupt.

Problem 8 A sensor, represented by S_1 is connected to bit 7 of port A of a **Z80** based microcomputer, as shown in *Fig. 15*. A polling routine is required which waits until S_1 is closed before continuing with the next section of the program.

With the aid of a flow chart, devise a **Z80** machine code polling routine to perform this task, assuming that port A is located at I/O address 4, and is configured to accept inputs.

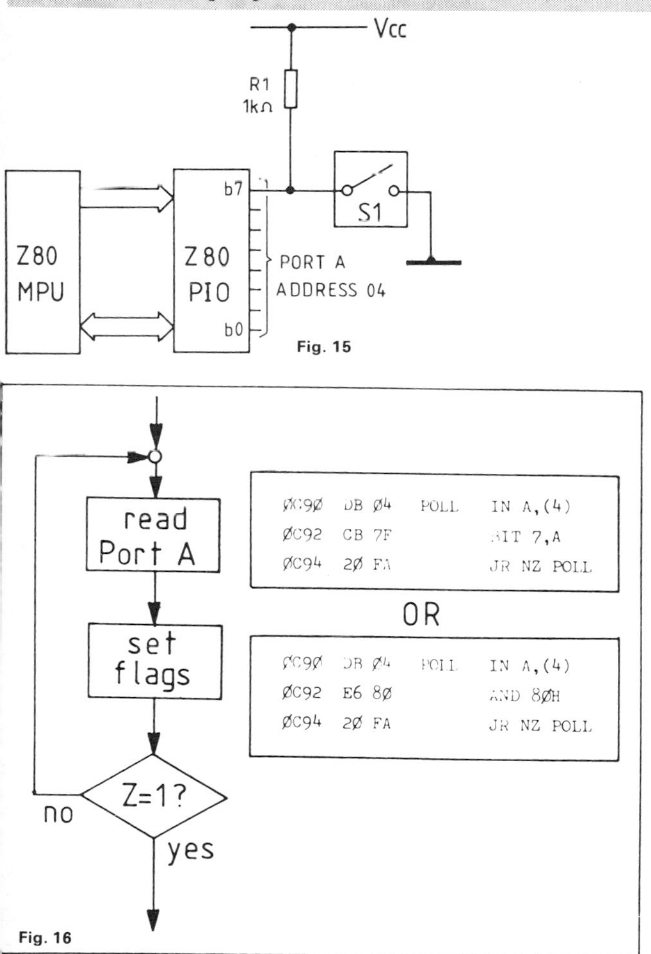

Fig. 15

Fig. 16

When data is read from port A into the accumulator, the flags in a Z80 flag register remain unchanged. Therefore, in order to detect the condition of S_1, a flag-setting instruction must follow each reading of the port. A logical AND or a bit instruction may be used for this purpose, and this results in the Z flag being set if S_1 is closed, or reset if S_1 is open.

A **jump relative if not zero** (JR NZ) instruction may then be used to determine what action to take, i.e. keep polling if $Z = 0$, or service the peripheral device if $Z = 1$. This action is shown in *Fig. 16*.

Problem 9 An industrial process which incorporates a heating element and an electric motor is controlled by a **Z80** based microcomputer with a **Z80 PIO**. The motor and heater circuits are connected to bit 0 and bit 1, respectively, of port A of the **Z80 PIO**, and are controlled by switches S_1 and S_2 which are connected to bit 6 and bit 7 of the same port, as shown in *Fig. 17*.

Write a **Z80** machine code program, starting at address $0C90_{16}$, which polls S_1 and S_2 and controls the heater and motor in the following manner:

S_1 open	:	heater	on
S_1 closed	:	heater	off
S_2 open	:	motor	on
S_2 closed	:	motor	off

Since current must flow in either the heater or the motor (or both) whenever either S_1 or S_2 is closed, a simple polling routine of the type shown in *Fig. 18* may be used. A suitable **Z80** machine code routine to poll the switches and to operate the heater and the motor is as follows:

```
                        ORG 0C90H
                        ;
0C90   3E CF            LD A,0CFH
0C92   D3 06            OUT (6),A      ;port A in bit mode
0C94   3E C0            LD A,0C0H
0C96   D3 06            OUT (6),A      ;bits 6 and 7 inputs, 0 and 1 outputs
0C98   AF               XOR A          ;clear accumulator
0C99   D3 04            OUT (4),A      ;motor and heater off initially
0C9B   DB 04    POLL    IN A,(4)       ;read the port
0C9D   CB 7F            BIT 7,A        ;test motor switch
0C9F   CB C7            SET 0,A        ;switch motor on
0CA1   28 02            JR Z,MTR       ;leave motor on if S2 is closed
0CA3   CB 87            RES 0,A        ;otherwise switch off motor
0CA5   CB 77    MTR     BIT 6,A        ;test heater switch
0CA7   CB CF            SET 1,A        ;switch heater on
0CA9   28 02            JR Z,HTR       ;leave heater on if S1 is closed
0CAB   CB 8F            RES 1,A        ;otherwise switch off heater
0CAD   D3 04    HTR     OUT (4),A      ;second control data to port
0CAF   18 EA            JR POLL        ;keep polling
```

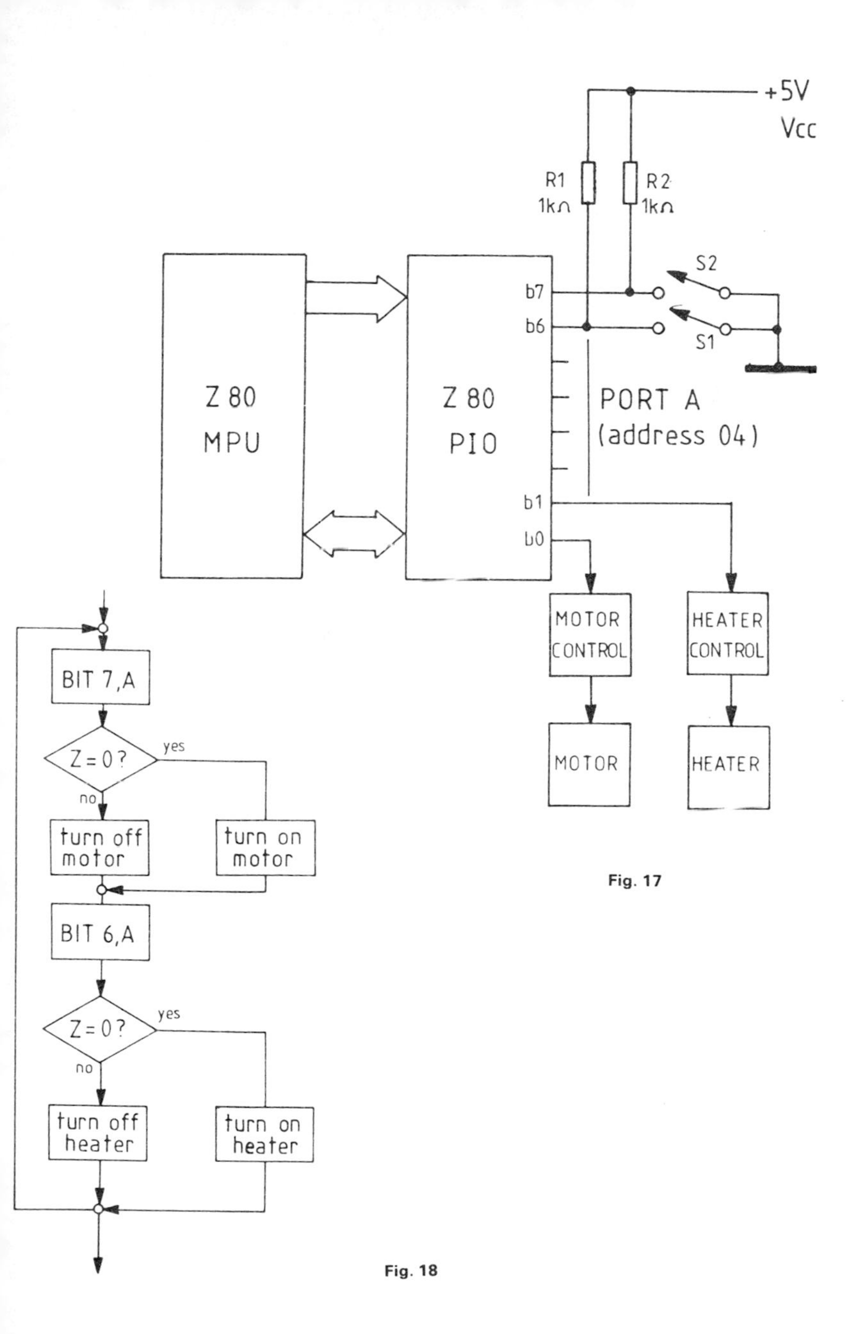
+5V
Vcc
R1
1kΩ
R2
1kΩ
S2
S1
b7
b6
b1
b0
Z 80
MPU
Z 80
PIO
PORT A
(address 04)
MOTOR CONTROL
HEATER CONTROL
MOTOR
HEATER
BIT 7,A
Z = 0 ?
yes
no
turn off motor
turn on motor
BIT 6,A
Z = 0 ?
yes
no
turn off heater
turn on heater

Fig. 17

Fig. 18

Problem 10 A **Z80** based microcomputer with a **Z80 PIO** is used to monitor the number of objects passing along a conveyor belt. A photoelectric sensor (represented by S_1) is connected to bit 7 of port B of the PIO, and the presence of an object adjacent to this sensor breaks a beam and causes a logical 0 to be applied to bit 7 by the sensor. The total number of objects which pass this sensor are counted by the microcomputer, and are displayed, in binary, on eight LEDs which are connected to port A of the PIO. This arrangement is illustrated in *Fig. 19*.

Write a Z80 machine code program to poll S_1, and increment the display by one for each object counted. The program should start at address $0C90_{16}$.

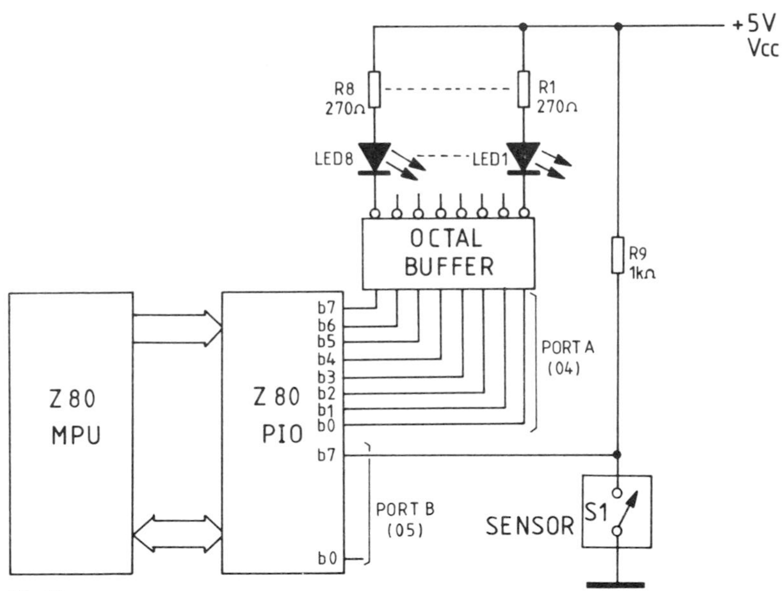

Fig. 19

The polling routine required for this example is different to that used in *Problem 9*, since the counter must only be incremented once each time that an object breaks the beam of light to the photoelectric sensor. It is therefore necessary to poll the sensor to determine when the beam is broken so that the counter may be incremented, but then to poll the sensor again to determine when the beam is restored before repeating the process. Failure to carry out this procedure results in the counter being incremented many times whilst the beam is broken. A polling routine which satisfies these requirements is shown in *Fig. 20*.

A suitable **Z80** machine code program to poll the sensor and increment the displays is as follows:

```
                        ORG 0C90H
                        ;
0C90  3E 0F             LD A,0FH
0C92  D3 06             OUT (6),A      ;port A output mode
0C94  3E 4F             LD A,4FH
0C96  D3 07             OUT (7),A      ;port B input mode
0C98  01 04 00          LD BC,0004 H   ;clear counter B, point C to port A
0C9B  ED 41    DISPLA   OUT (C),B      ;display counter
0C9D  DB 05    POLL0    IN A,(5)       ;read port B
0C9F  CB7F              BIT 7,A        ;and poll for S1 closure
0CA1  20 FA             JR NZ,POLL0    ;keep checking if not closed
0CA3  04                INC B          ;increment object counter
0CA4  DB 05    POLL1    IN A,(5)       ;read port B
0CA6  CB 7F             BIT 7,A        ;and poll for S1 open
0CA8  28 FA             JR Z,POLL1     ;keep checking if not open
0CAA  18 EF             JR DISPLA      ;display new total and repeat
```

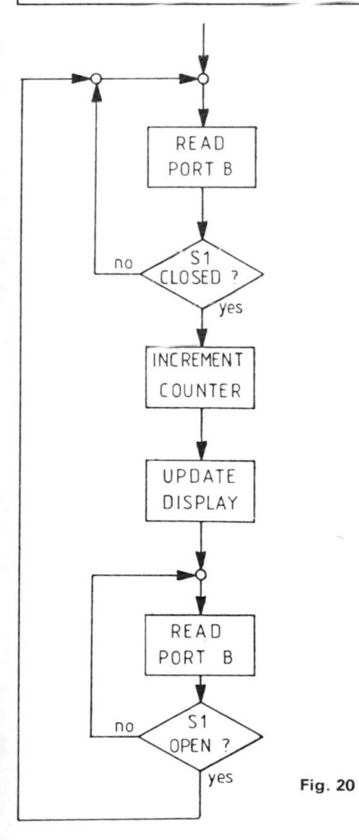

Fig. 20

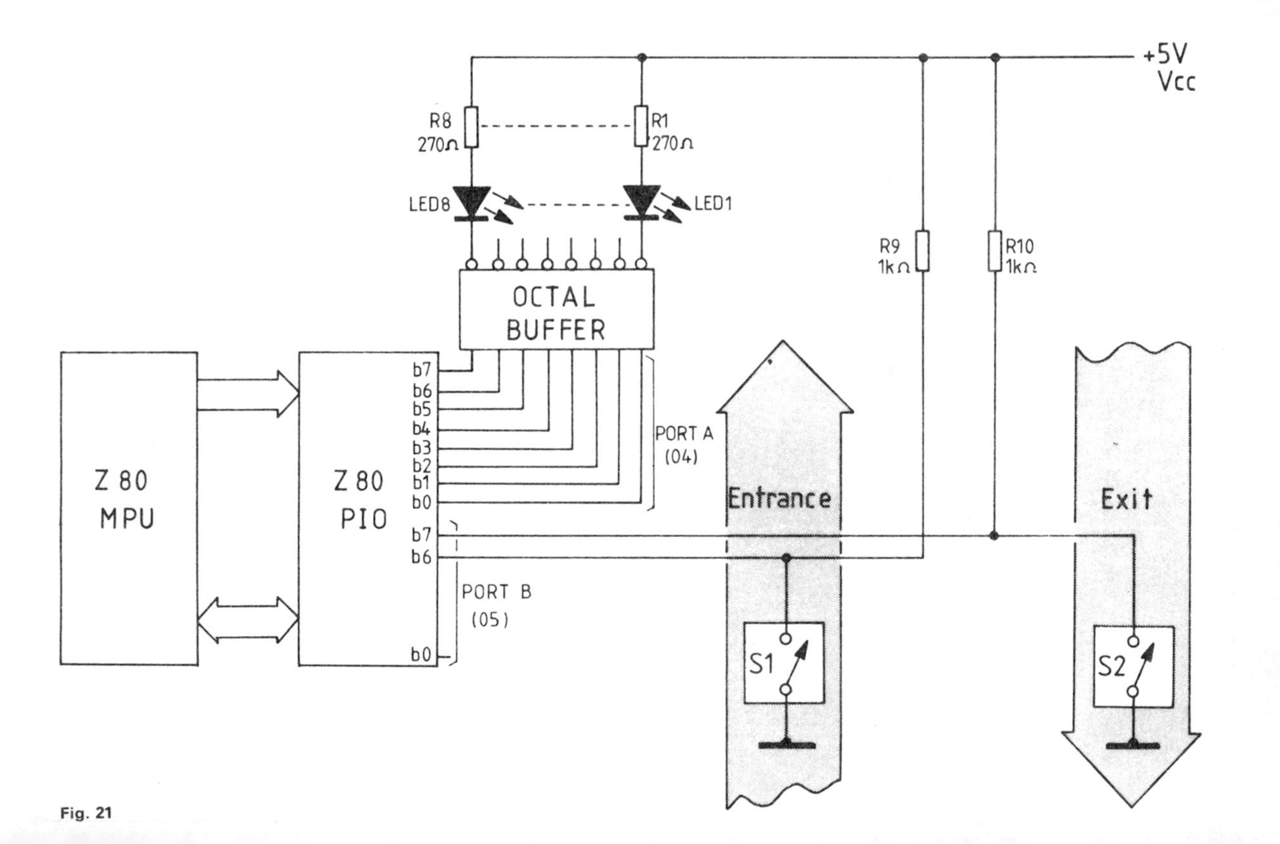
+5V
Vcc
R8
270Ω
R1
270Ω
LED8
LED1
R9
1kΩ
R10
1kΩ
OCTAL
BUFFER
Z 80
MPU
Z 80
PIO
b7
b6
b5
b4
b3
b2
b1
b0
PORT A
(04)
b7
b6
PORT B
(05)
b0
Entrance
Exit
S1
S2

Fig. 21

Problem 11 A **Z80** based microcomputer with a **Z80 PIO** is used to monitor the number of vehicles entering and leaving a car park. Sensors S_1 and S_2, sited at the entrance and exit of the car park are connected to bit 6 and bit 7 of port B of the PIO and are used to detect vehicles entering and leaving. Eight LEDs are connected to port A of the Z80 PIO and are used to display (in binary) the total number of vehicles currently in the car park. This arrangement is illustrated in *Fig. 21*.

Write a Z80 machine code program, starting at address $0C90_{16}$, to poll S_1 and S_2 and to increment the number displayed by one each time that a vehicle enters the car park, and to decrement the number displayed by one each time that a vehicle leaves (the sensors are arranged such that S_1 and S_2 close each time that a vehicle is detected).

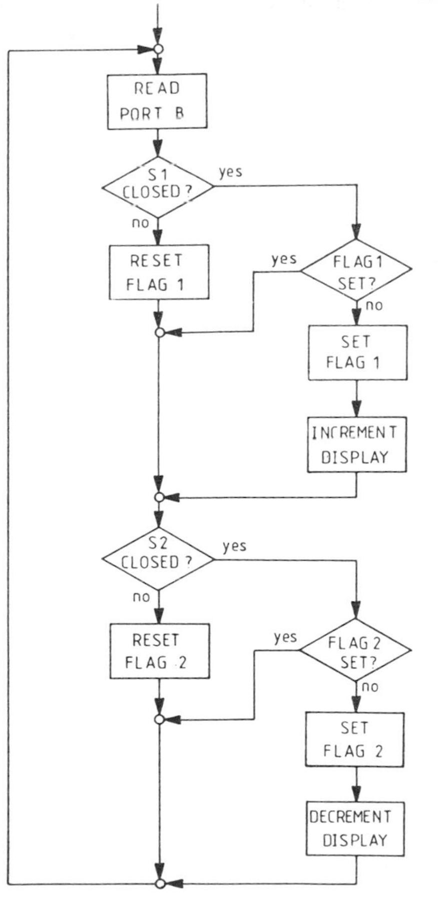

Fig. 22

Two polling routines of the type used in *Problem 10* could be implemented in series to detect closure of either S_1 or S_2. However, if S_1, for example, is closed for a long period of time, S_2 is 'locked out' and is prevented from being polled, because the program is looping around waiting for S_1 to be opened.

This problem may be resolved by polling S_1 and S_2, and setting a flag (one for each sensor) if a sensor switch is found to be closed. This flag is used to prevent further incrementing or decrementing of the counter until the switch is again in the open condition. Once the polling routine detects the open condition in either S_1 or S_2, the appropriate flag is reset. A polling routine of this type is illustrated by the flowchart shown in *Fig. 22*.

A **Z80** machine code program to perform this function is as follows:

```
                         ORG 0C90H
                ;
0C90   3E 0F             LD A,0FH
0C92   D3 06             OUT (6),A        ;byte output mode, port A
0C94   3E CF             LD A,0CFH
0C96   D3 07             OUT (7),A        ;bit mode, port B
0C98   3E C0             LD A,0C0H
0C9A   D3 07             OUT (7),A        ;b6 and b7 as inputs
0C9C   06 00             LD B,0           ;use B as a flag register
0C9E   48                LD C,B           ;use C as a binary counter
0C9F   79       DISPLA   LD A,C           ;move into A to display it
0CA0   D3 04             OUT (4),A        ;display counter value
0CA2   DB 05             IN A,(5)         ;read the sensors
0CA4   CB 77             BIT 6,A          ;test for S1 closure
0CA6   20 09             JR NZ,S1         ;do not increment counter if S1 open
0CA8   CB 40             BIT 0,B          ;check increment disable flag
0CAA   20 07             JR NZ,POLL2      ;do not increment if flag is set
0CAC   CB C0             SET 0,B          ;set increment disable flag
0CAE   0C                INC C            ;increment the counter register
0CAF   18 02             JR POLL2         ;skip the next instruction
0CB1   CB 80    S1       RES 0,B          ;clear the increment disable flag
0CB3   CB 7F    POLL2    BIT 7,A          ;test for S2 closure
0CB5   20 09             JR NZ,S2         ;do not decrement counter if S2 open
0CB7   CB 48             BIT 1,B          ;check decrement disable flag
0CB9   20 E4             JR NZ,DISPLA     ;do not decrement if flag is set
0CBB   CB C8             SET 1,B          ;set decrement disable flag
0CBD   0D                DEC C            ;decrement the counter register
0CBE   18 DF             JR DISPLA        ;update the display
0CC0   CB 88    S2       RES 1,B          ;clear the decrement disable flag
0CC2   18 DB             JR DISPLA        ;keep polling
```

Problem 12 Describe the operation of the $\overline{\text{RESET}}$ input of a **Z80** microprocessor.

This is an **interrupt input** which is used to **reset the microprocessor** (or to provide initial start-up from switch-on time), and it overrides all other inputs. The $\overline{\text{RESET}}$ input must be held low for at least 3 clock cycles in order to perform a reset. If $\overline{\text{RESET}}$ is then allowed to attain a logical 1 level, the MPU commences its **reset sequence**, and this consists of the following steps:

(a) the **interrupt enable flip-flop is reset**, and **interrupt mode 0 is selected**;
(b) the **I and R registers are reset** to zero;
(c) the **program counter is reset to zero**, and control is handed over to the program whose starting address is 0000_{16}.

Note that during the reset time, the address and data busses go to a high impedance state, and all control output signals go to the inactive state.

Problem 13 Describe the operation of the following interrupts on a **Z80** microprocessor:
(a) **interrupt request ($\overline{INT}$)**; (b) **non-maskable interrupt ($\overline{NMI}$)**.

(a) Interrupt request ($\overline{INT}$)
This **active low input** is used to initiate an interrupt sequence with the **Z80** microprocessor. The microprocessor completes the current instruction being executed before recognizing this request. Next the interrupt enable flip-flop (IFF) is checked, and, provided that this flip-flop is set (IFF = 1), the microprocessor begins its interrupt sequence. A Z80 MPU may respond to an interrupt request in one of three ways according to which mode is selected in the software. These three modes are called **mode 0**, **mode 1** and **mode 2**, and the required mode may be selected by the use of the appropriate instruction, **IM 0**, **IM 1** or **IM 2** (see Problem 14 for details for these interrupt modes).

(b) Non-maskable interrupt ($\overline{NMI}$)
This is an **edge triggered** input which is used to initiate a non-maskable interrupt sequence within the **Z80** microprocessor. A negative going edge (i.e. logical 1 to logical 0 transition) is required to initiate the interrupt sequence, therefore this input must be returned to a logical 1 level before a further interrupt will be accepted. The $\overline{NMI}$ input may therefore be kept at a logical 0 level for as long as required after initiating an interrupt. After completion of the current instruction being executed, the following sequence of operations is carried out regardless of the state of the interrupt enable flip-flop (IFF), i.e., this interrupt is non-maskable.

(i) the **program counter is saved on the stack**;
(ii) the **interrupt enable flip-flop (IFF) is reset** to prevent the MPU from responding to further interrupt requests during this sequence;
(iii) the **program counter** is loaded with address **0066_{16}**, and program execution continues from this address. It is most probable that a jump instruction will be located at this address to direct the program to the start of an interrupt service routine.

Problem 14 Describe the *three* different modes in which a **Z80** microprocessor may respond to an interrupt sequence ($\overline{INT}$).

A **Z80** microprocessor may be programmed to respond to a maskable interrupt request ($\overline{INT}$) in any one of the following three modes:

(i) Mode 0
This mode is entered by executing an **IM 0 instruction**. With this mode, the interrupting device can place any **instruction** on the **data bus** and this will be executed by the MPU. Thus the next instruction is supplied by the interrupting device rather than by the memory. Any instruction may be used for this purpose, although a single byte instruction such as a **'restart' instruction (RST)** may be easier to implement. The instruction is gated on to the data bus during the

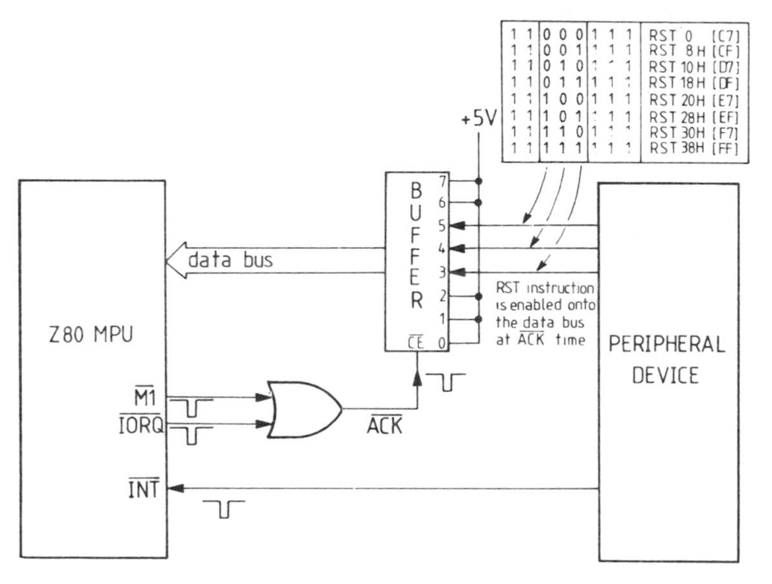

Fig. 23

interrupt acknowledge period when **both $\overline{\text{M1}}$ and $\overline{\text{IORQ}}$ are at logical 0**, as shown in *Fig. 23*.

(ii) Mode 1

This mode is entered by executing an **IM 1 instruction**. This mode of operation is similar to the response of a **Z80** to $\overline{\text{NMI}}$ except that the program counter is loaded with address **0038_{16}**, if the interrupt enable flip-flop is set. Program execution then continues from address 0038_{16} which most probably contains a jump instruction to direct the program to the start of the interrupt service routine. This action is identical to the response of a Z80 to an **RST 38H** instruction in a program.

(iii) Mode 2

This mode is entered by executing an **IM 2 instruction**. This mode is the most powerful interrupt response mode. An **8-bit vector** is supplied by the interrupting peripheral device, and this is combined with the contents of the **Z80 I register** to form a 16-bit memory pointer. This pointer is used to access a table of interrupt service routine starting addresses, and causes the program counter to be loaded with an address from this table. Program execution then continues from this address, i.e., the start of the selected interrupt service routine. This table of starting addresses may be changed by the programmer to allow different peripheral devices to be serviced. The operation of a mode 2 interrupt may be studied by reference to *Fig. 24*.

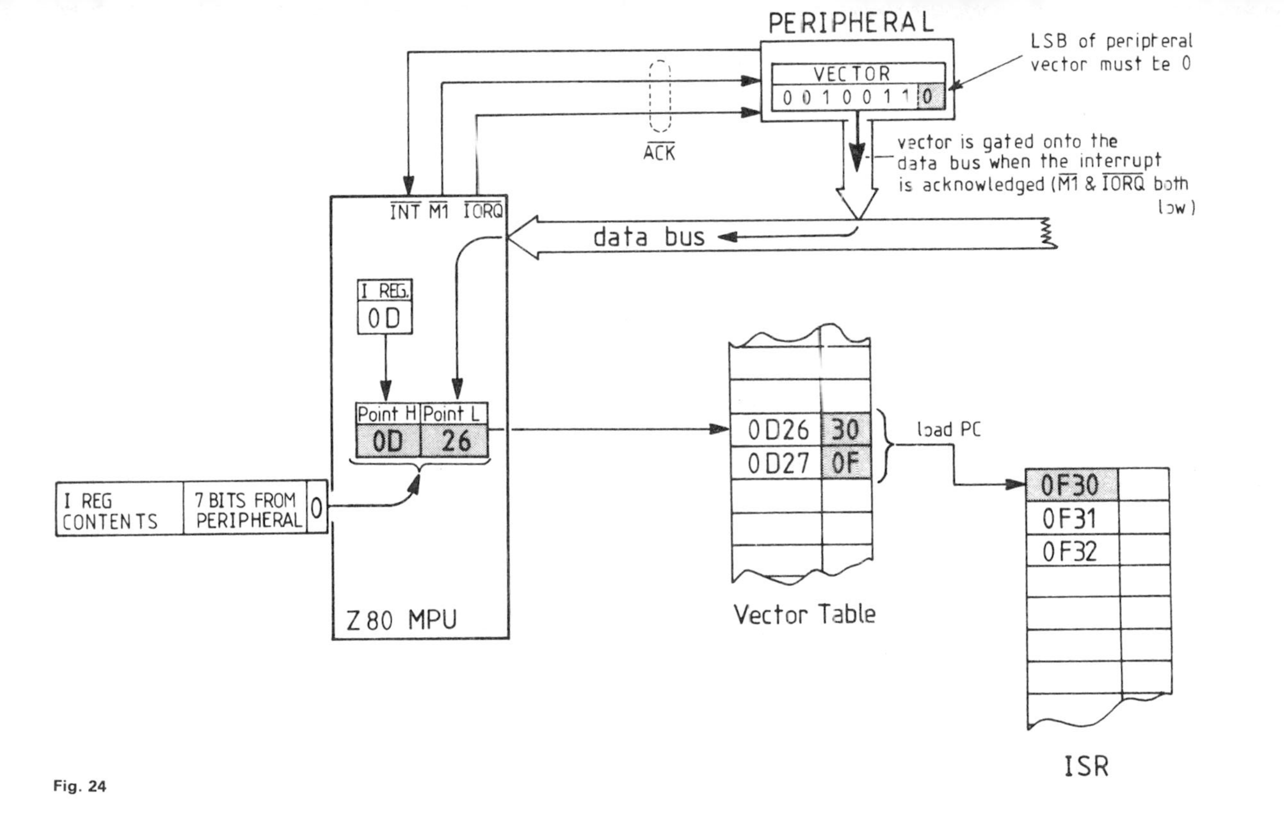
PERIPHERAL
VECTOR
0 0 1 0 0 1 1 0
LSB of peripheral vector must be 0
vector is gated onto the data bus when the interrupt is acknowledged (M1 & IORQ both low)
ACK
INT M1 IORQ
data bus
I REG.
0D
Point H
Point L
0D
26
I REG CONTENTS
7 BITS FROM PERIPHERAL
0
Z80 MPU
0D26
30
0D27
0F
load PC
Vector Table
0F30
0F31
0F32
ISR

Fig. 24

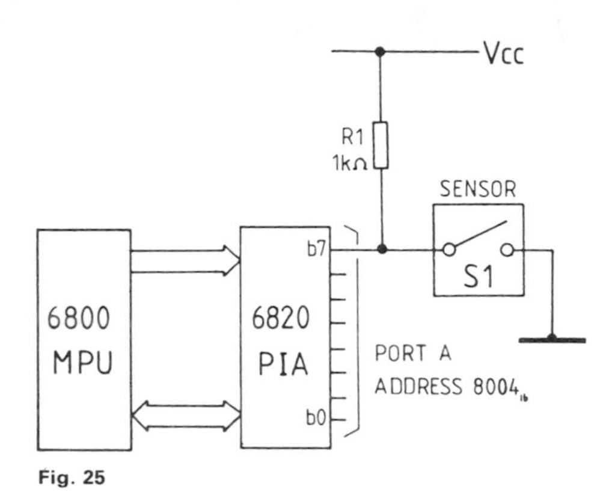

Fig. 25

Problem 15 A sensor, represented by S_1 is connected to bit 7 of port A of a **6800** microcomputer, as shown in *Fig. 25*. A polling routine is required which waits until S_1 is closed before continuing with the next section of the program.

With the aid of a flowchart, devise a **6800** machine code polling routine to perform this task, assuming that port A is configured to accept inputs at address 8004_{16}.

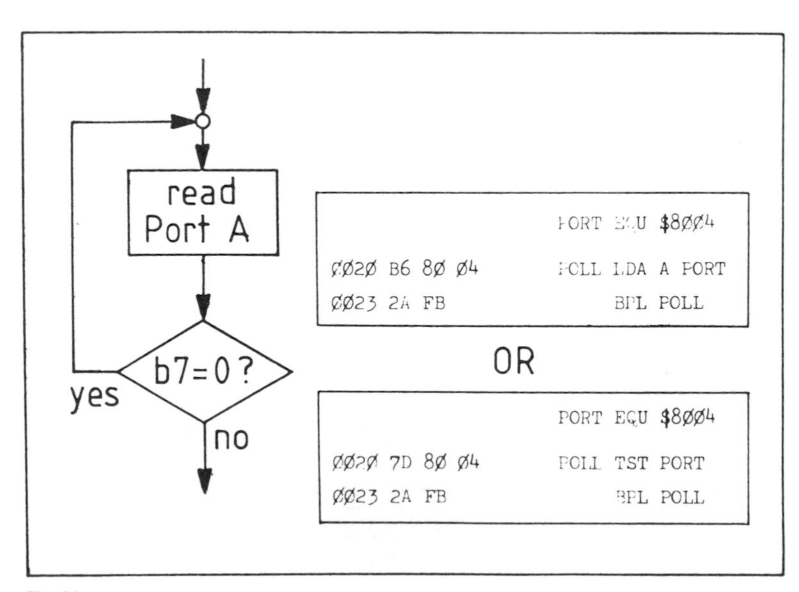

Fig. 26

If the data on port A is read into the accumulator, the N flag is set or reset according to the state of bit 7, since this action copies bit 7 into the N flag. When S_1 is open, bit 7 is at a logical 1 and the N flag is set, but if S_1 is closed, bit 7 is at logical 0 and the N flag is reset. Therefore a **branch if plus** (BPL) instruction may be used to determine what action must be taken, i.e., keep polling if $N = 1$; continue with the next section of program if $N = 0$. This action is shown in *Fig. 26*.

Alternatively, a TST instruction may be used to test the state of bit 7, and control the state of the N flag as before, and this may be used as an alternative to the LDA A instruction so that the test operation does not corrupt the contents of the accumulator.

Problem 16 An industrial process which incorporates a heating element and an electric motor is controlled by a **6800** based microcomputer with a **6820 PIA**. The motor and heater circuits are connected to bit 0 and bit 1, respectively, of port A of the **6820**, and are controlled by switches S_1 and S_2 which are connected to bit 6 and bit 7 of the same port, as shown in *Fig. 27*.

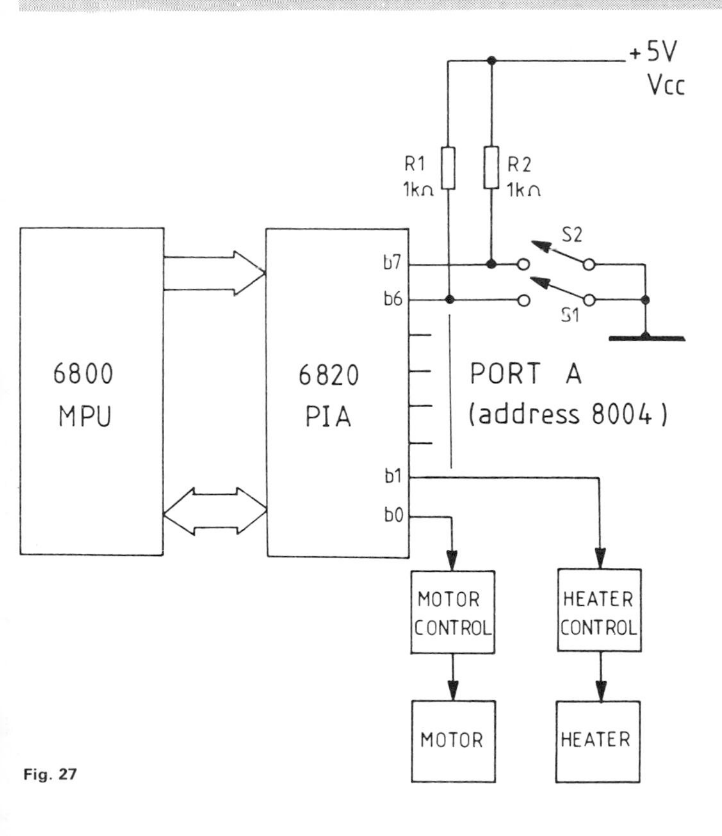

Fig. 27

Write a **6800** machine code program, starting at address 0100_{16}, which polls S_1 and S_2 and controls the motor and heater in the following manner:

S_1 open	:	heater	on
S_1 closed	:	heater	off
S_2 open	:	motor	on
S_2 closed	:	motor	off

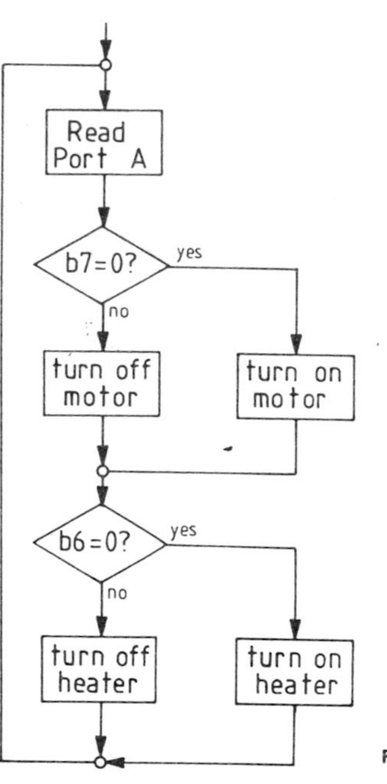

Fig. 28

Since current must flow in either the heater or the motor (or both) whenever either S_1 or S_2 is closed, a simple polling routine of the type shown in *Fig. 28* may be used.

A suitable **6800** machine code routine to poll the switches and to operate the heater and the motor is as follows:

```
                  DRA    EQU $8004
                  CRA    EQU $8005
                         ORG $0100
                  ;
0100  4F                 CLR A
0101  B7 80 05           STA A CRA     ;select data direction register A
0104  86 03              LDA A #03
0106  B7 80 04           STA A DRA     ;bits 0 and 1 outputs, rest inputs
0109  4C                 INC A         ;A = 04
010A  B7 80 05           STA A CRA     ;select data I/O register
```

```
010D  4F                   CLR A
010E  B7 80 04  POLL       STA A DRA      ;actuate motor/heater
0111  B6 80 04             LDA A DRA      ;read in the switch data
0114  2A 04                BPL MTRON      ;if S1 closed, switch on motor
0116  84 FE                AND A #$FE     ;turn off motor control bit
0118  20 02                BRA HTRTST     ;skip the next instruction
011A  8A 01     MTRON      ORA A #01      ;turn on motor control bit
011C  85 40     HTRTST     BIT A #$40     ;test S2 condition
011E  27 04                BEQ HTRON      ;if S2 closed, switch on heater
0120  84 FD                AND A #$FD     ;turn off heater control bit
0122  20 EA                BRA POLL       ;keep on polling
0124  8A 02     HRTON      ORA A #02      ;turn on heater control bit
0126  20 E6                BRA POLL       ;keep on polling
```

Problem 17 A **6800** based microcomputer with a **6820 PIA** is used to monitor the number of objects passing along a conveyor belt. A photoelectric sensor (represented by S_1) is connected to bit 7 of port B of the PIA, and the presence of an object adjacent to this sensor breaks a beam and causes a logical 0 to be applied to bit 7 by the sensor. The total number of objects which pass this sensor are counted by the microcomputer, and are displayed, in binary, on eight LEDs which are connected to port A of the PIA. This arrangement is illustrated in *Fig. 29*.

Write a 6800 machine code program to poll S_1, and increment the display by one for each object counted. The program should start at address 0100_{16}.

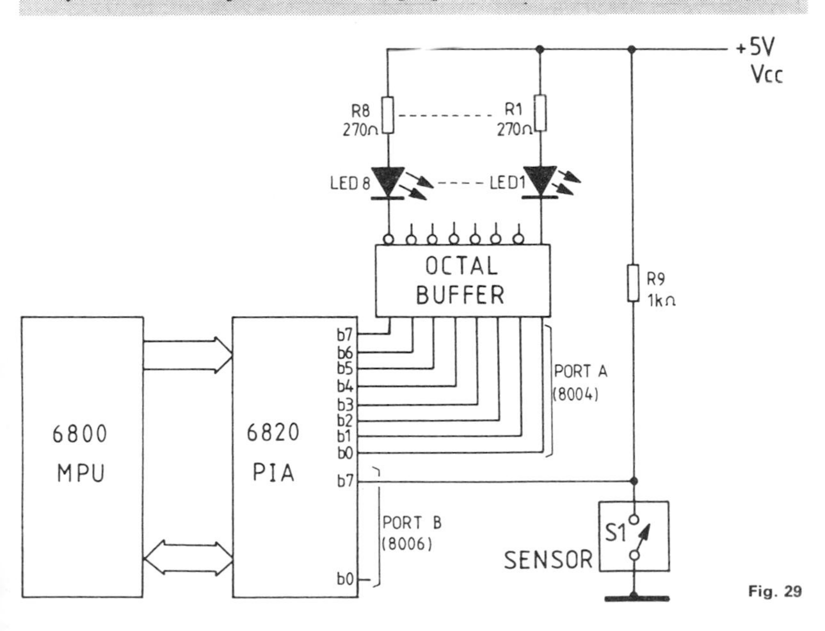

Fig. 29

The polling routine required for this example is different to that used in *Problem 16*, since the counter must only be incremented once each time that an object breaks the beam of light to the photoelectric sensor. It is therefore necessary to poll the sensor to determine when the beam is broken so that the counter may be incremented, but then to poll the sensor again to determine when the beam is restored before repeating the process.

Failure to carry out this procedure results in the counter being incremented many times whilst the beam is broken. A polling routine which satisfies these requirements is shown in *Fig. 30*.

A suitable **6800** machine code program to poll the sensor and increment the displays is as follows:

```
                     DRA      EQU $8004
                     CRA      EQU $8005
                     DRB      EQU $8006
                     CRB      EQU $8007
                              ORG $0100
                     ;
0100  4F                      CLR A
0101  B7 80 05                STA A CRA     ;select data direction register A
0104  B7 80 07                STA A CRB     ;select data direction register B
0107  B7 80 06                STA A DRB     ;port B all inputs
010A  43                      COM A
010B  B7 80 04                STA A DRA     ;port A all outputs
010E  86 04                   LDA A #04
0110  B7 80 05                STA A CRA     ;select data I/O register A
0113  B7 80 07                STA A CRB     ;select data I/O register B
0116  5F                      CLR B         ;reset counter B
0117  F7 80 04  DISPLA        STA B DRA     ;display counter on LEDs
011A  B6 80 06  POLL0         LDA A DRB     ;read the sensor
011D  2B FB                   BMI POLL0     :keep polling if S1 = 1
011F  5C                      INC B         ;increase total by one
0120  B6 80 06  POLL1         LDA A DRB     ;read the sensor again
0123  2A FB                   BPL POLL1     ;keep polling if S1 = 0
0125  20 F0                   BRA DISPLA    ;update displays and repeat
```

Problem 18 A **6800** based microcomputer with a **6820 PIA** is used to monitor the number of vehicles entering and leaving a car park. Sensors S_1 and S_2, sited at the entrance and exit of the car park are connected to bit 6 and bit 7 of port B of the PIA and are used to detect vehicles entering and leaving. Eight LEDs are connected to port A of the 6820 PIA and are used to display (in binary) the total number of vehicles currently in the car park. This arrangement is illustrated in *Fig. 31*.

Write a 6800 machine code program, starting at address 0100_{16}, to poll S_1 and S_2 and to increment the number displayed by one each time that a vehicle enters the car park, and to decrement the number displayed by one each time that a vehicle leaves (the sensors are arranged such that S_1 and S_2 close each time that a vehicle is detected).

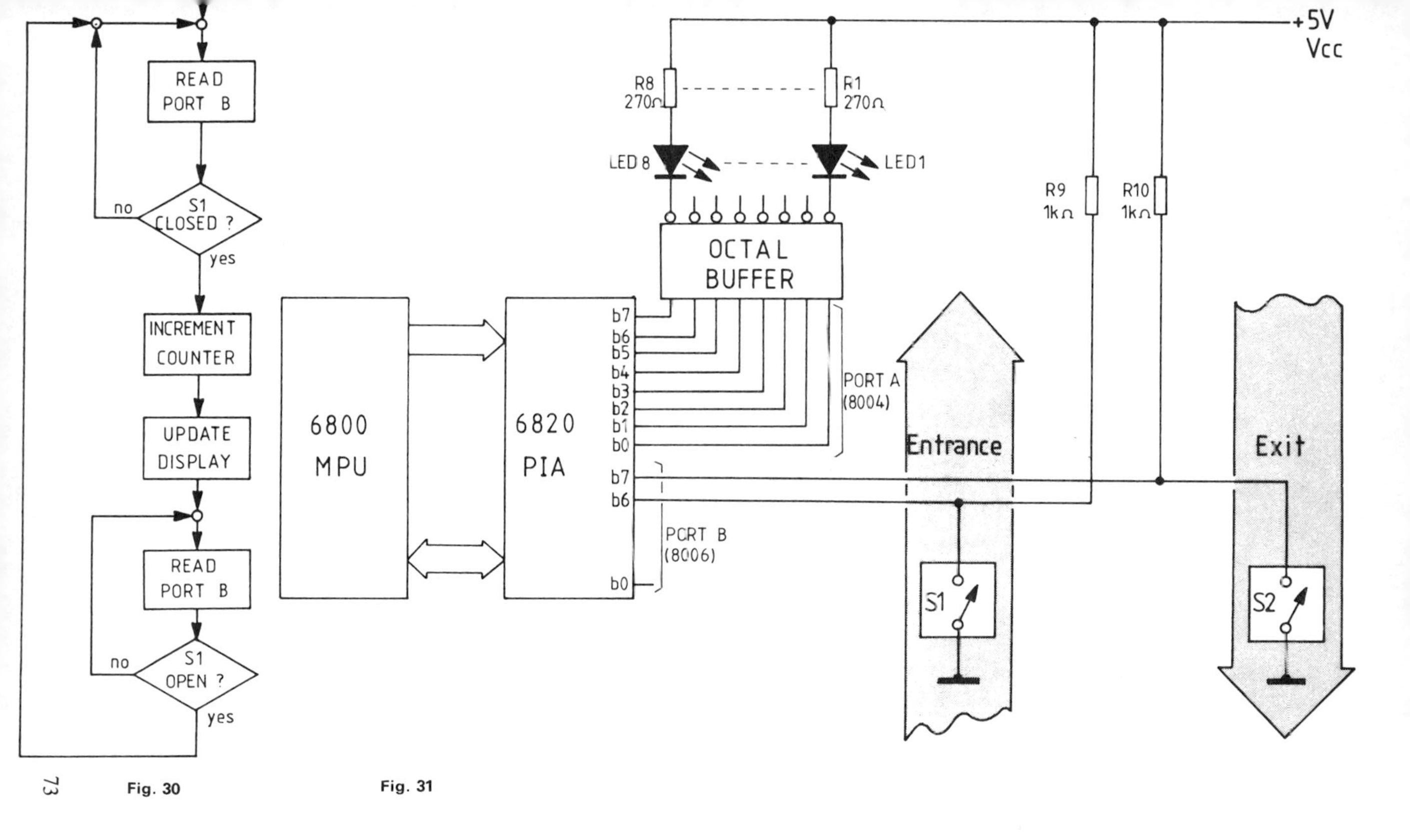
READ PORT B
S1 CLOSED ?
no
yes
INCREMENT COUNTER
UPDATE DISPLAY
READ PORT B
S1 OPEN ?
no
yes
6800 MPU
6820 PIA
b7
b6
b5
b4
b3
b2
b1
b0
b7
b6
b0
PORT A (8004)
PCRT B (8006)
OCTAL BUFFER
R8 270Ω
R1 270Ω
LED 8
LED1
R9 1kΩ
R10 1kΩ
+5V Vcc
Entrance
Exit
S1
S2

Fig. 30

Fig. 31

Two polling routines of the type used in *Problem 17* could be implemented in series to detect closure of either S_1 or S_2. However, if S_1, for example, is closed for a long period of time, S_2 is 'locked out' and is prevented from being polled, because the program is looping around waiting for S_1 to be opened.

This problem may be resolved by polling S_1 and S_2, and setting a flag (one for each sensor) if a sensor switch is found to be closed. This flag is used to prevent further incrementing or decrementing of the counter until the appropriate switch is again in the open condition. Once the polling routine detects the open condition in either S_1 or S_2, the appropriate flag is reset. A polling routine of this type is illustrated by the flow chart in *Fig. 32*.

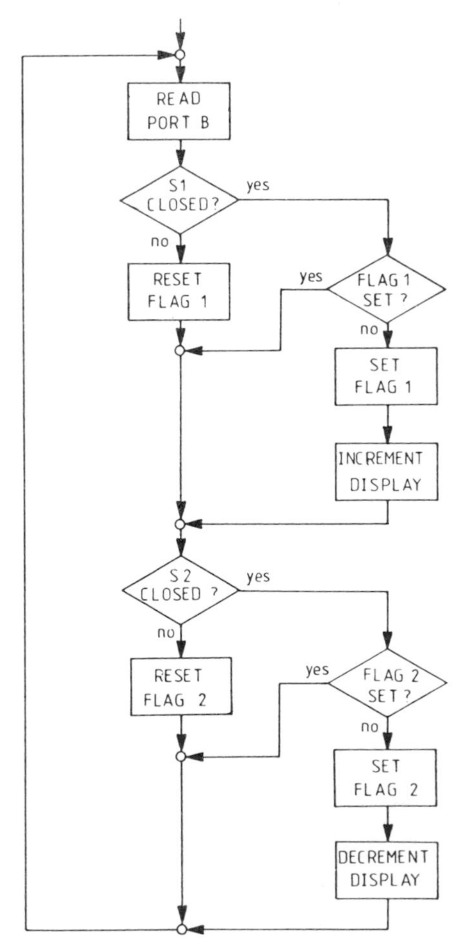

Fig. 32

A **6800** machine code program to perform this function is as follows:

```
                   FLAG1    EQU $0000
                   FLAG2    EQU $0001
                   DRA      EQU $8004
                   CRA      EQU $8005
                   DRB      EQU $8006
                   CRB      EQU $8007
                            ORG $0100
                   ;
0100  4F                    CLR A
0101  B7 80 05              STA A CRA     ;select data direction register A
0104  B7 80 07              STA A CRB     ;select data direction register B
0107  B7 80 06              STA A DRB     ;port B all inputs
010A  43                    COM A
010B  B7 80 04              STA A DRA     ;port A all outputs
010E  86 04                 LDA A #$04
0110  B7 80 05              STA A CRA     ;select data I/O register A
0113  B7 80 07              STA A CRB     ;select data I/O register B
0116  5F                    CLR B         ;
0117  D7 00                 STA B FLAG1   ;reset increment disable flag
0119  D7 01                 STA B FLAG2   ;reset decrement disable flag
011B  F7 80 04              STA B DRA     ;clear the displays
011E  B6 80 06     POLL1    LDA A DRB     ;read the sensors
0121  2A 04                 BPL S1        ;S1 closed?
0123  D7 00                 STA B FLAG1   ;if not, reset increment disable flag
0125  20 0B                 BRA POLL2     ;and check the other sensor
0127  96 00        S1       LDA A FLAG1   ;get increment disable flag
0129  26 07                 BNE POLL2     ;inhibit further increments while S1 = 0
012B  7C 80 04              INC DRA       ;increment display counter
012E  8A 01                 ORA A #$01    ;set increment disable flag bit
0130  97 00                 STA A FLAG1   ;and save in flag location
0132  B6 80 06     POLL2    LDA A DRB     ;read the sensors again
0135  85 40                 BIT A #$40    ;and test state of S2
0137  27 04                 BEQ S2        ;S2 closed?
0139  D7 01                 STA B FLAG2   ;if not, reset decrement disable flag
013B  20 E1                 BRA POLL1     ;and keep polling
013D  96 01        S2       LDA A FLAG2   ;get decrement disable flag
013F  26 DD                 BNE POLL1     ;inhibit further decrements while S2 = 0
0141  7A 80 04              DEC DRA       ;decrement display counter
0144  8A 01                 ORA A #$01    ;set decrement disable flag bit
0146  97 01                 STA A FLAG2   ;and save in flag location
0148  20 D4                 BRA POLL1     ;keep on polling
```

Problem 19 Describe the operation of the $\overline{\textbf{RESET}}$ input of a **6800** microprocessor.

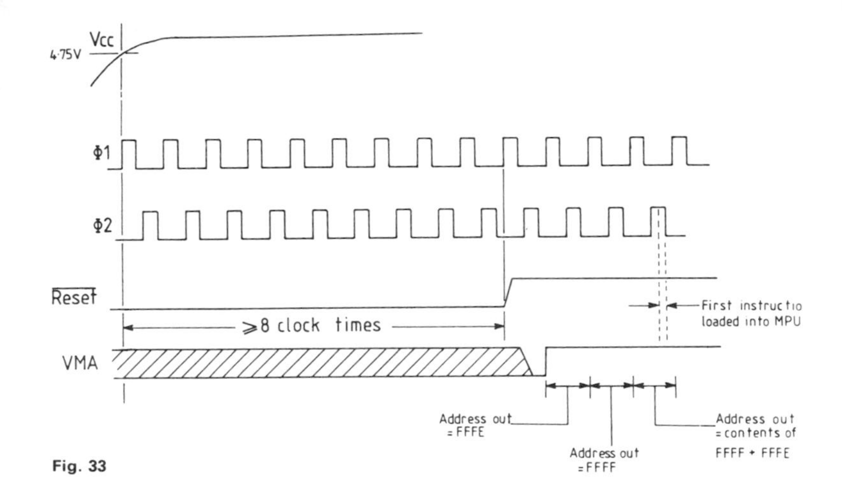

Fig. 33

This is an **interrupt input** which is used to **reset the microprocessor** (or to provide initial start-up from switch-on time), and it overrides all other inputs. The $\overline{\text{RESET}}$ input must be held low for **at least 8 clock periods** after a V_{cc} of 4.75 V minimum is established (3 clock periods if V_{cc} is already established), in order to perform a reset. If $\overline{\text{RESET}}$ is then allowed to attain a logical 1 level, the MPU commences its **reset sequence**, and this consists of the following steps:

(a) the **interrupt mask bit is set**,
(b) addresses **$FFFE_{16}$** and **$FFFF_{16}$** are generated, in sequence, and the **restart vector** is fetched from these locations,
(c) the **program counter** is loaded with the **restart vector**, and control is handed over to the program whose starting address is defined by the contents of memory locations $FFFE_{16}$ and $FFFF_{16}$.

This process is illustrated by *Fig. 33*.

Problem 20 Describe the operation of the following interrupts on a **6800** microprocessor: (a) **interrupt request ($\overline{\text{IRQ}}$)**; (b) **non-maskable interrupt ($\overline{\text{NMI}}$)**.

(a) Interrupt request ($\overline{\text{IRQ}}$)
This input is used to initiate an interrupt sequence within the 6800 microprocessor. The microprocessor completes the current instruction being executed before recognizing this request (the $\overline{\text{IRQ}}$ input is sampled during Φ2 and the interrupt sequence starts on the next Φ1). Next, the interrupt mask bit (I flag) in the condition codes register is examined, and, provided that this flag is not set (i.e., I = 0), the microprocessor begins its interrupt sequence, which consists of the following steps:

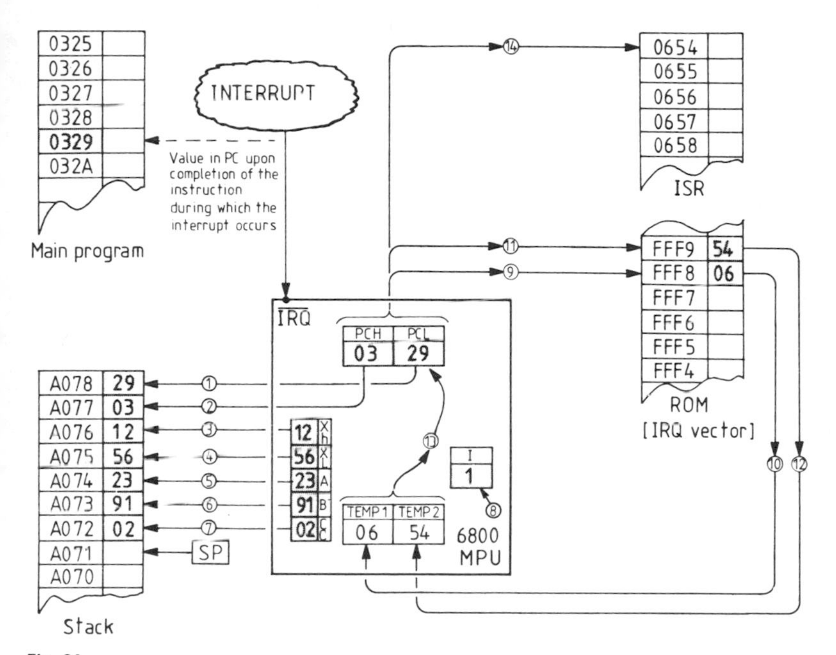

Fig. 34

(i) the **program counter**, **index register**, **accumulators** and **condition codes register are stored on the stack**;
(ii) the **interrupt mask bit is set** (I = 1) to prevent the microprocessor from responding to further interrupt requests during this sequence;
(iii) addresses **$FFF8_{16}$** and **$FFF9_{16}$** are generated, in sequence, by the microprocessor, and the **$\overline{IRQ}$ vector** is fetched from these locations; and
(iv) the **program counter** is loaded with the **$\overline{IRQ}$ vector**, and control is handed over to the interrupt service routine (ISR) whose starting address is defined by the contents of memory locations $FFF8_{16}$ and $FFF9_{16}$.

This process is illustrated in *Fig. 34*.

(b) **Non-maskable interrupt ($\overline{MNI}$)**
This is an **edge triggered** input which is used to initiate a non-maskable interrupt sequence within the **6800** microprocessor. A negative going edge (i.e., logical 1 to logical 0 transition) is required to initiate the interrupt sequence, therefore this input must be returned to a logical 1 level before a further interrupt will be accepted. The $\overline{NMI}$ input may therefore be kept at a logical 0 level for as long as required after initiating an interrupt.

Following completion of the current instruction being executed, the sequence of operations described for $\overline{IRQ}$ in (a) is carried out, regardless of the state of the interrupt mask bit (i.e. it is non-maskable). The ISR starting address is, however, fetched from addresses $\mathbf{FFFC_{16}}$ and $\mathbf{FFFD_{16}}$ for a non-maskable interrupt.

Problem 21 (a) Show how **multiple interrupts** may be implemented when using the $\overline{\text{IRQ}}$, (or $\overline{\text{INT}}$), input of a microprocessor. (b) Explain how it is possible to determine which device is responsible for initiating an interrupt in a multiple interrupt system.

(a) A **multiple interrupt system** is one in which many different peripheral devices may cause an interrupt to occur. Frequently a microprocessor has only one $\overline{\text{IRQ}}$ (or $\overline{\text{INT}}$) input, and it becomes necessary to combine signals from each interrupting device in the manner shown in *Fig. 35*.

(b) There are two possible methods for determining which device actually generated the interrupt in a multiple interrupt system, and these are:

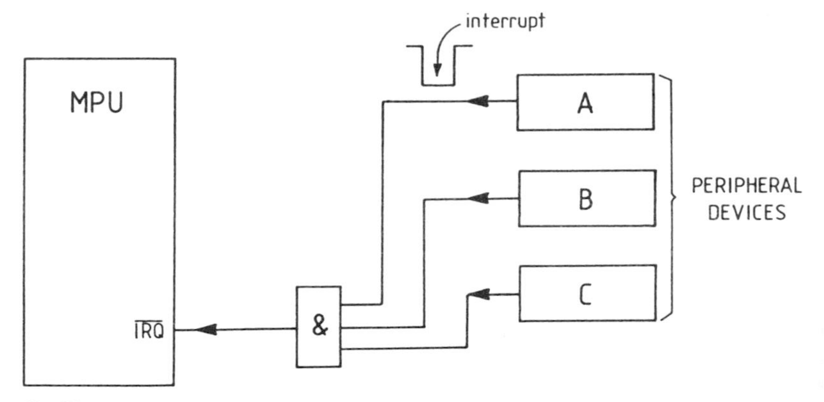

Fig. 35

(i) Polling

This method consists of interrogating each peripheral device, in turn, to determine which one caused the interrupt, and then selecting the interrupt service routine (ISR) appropriate for that device. Note, that unlike the use of polling in a non-interrupt system, at least one peripheral device does actually require attention when the polling routine is initiated. A multiple interrupt system of this type may be implemented as shown in *Fig. 36*. The interrupt lines are combined as shown in *Fig. 35* but are also separately connected to port A of a PIA (or PIO). Once an interrupt request is detected by the microprocessor, a polling routine is entered which tests each bit of port A in order to determine which ISR to use. This arrangement is shown in *Fig. 37*.

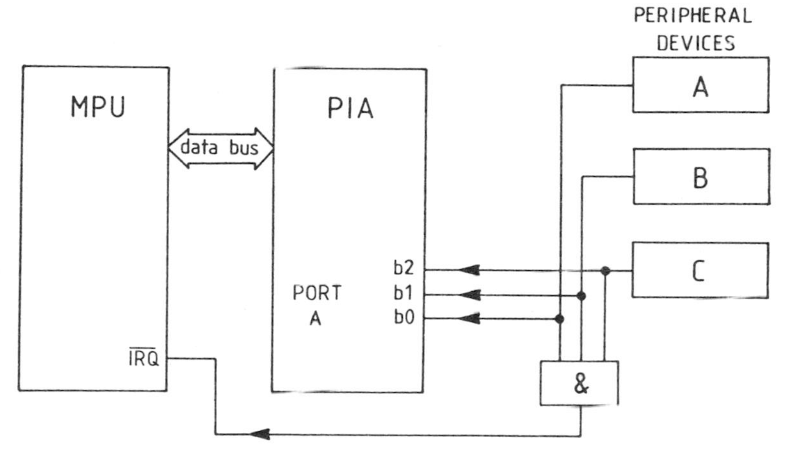
PERIPHERAL
DEVICES
A
B
C
MPU
PIA
data bus
PORT
A
b2
b1
b0
$\overline{IRQ}$
&

Fig. 36

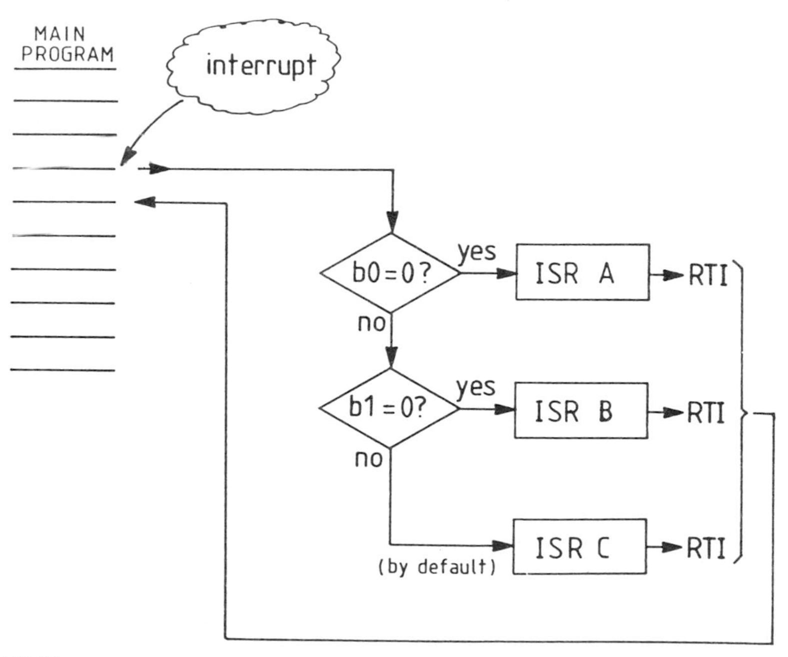
MAIN
PROGRAM
interrupt
b0 = 0 ?
yes
no
ISR A
RTI
b1 = 0?
yes
no
ISR B
RTI
ISR C
RTI
(by default)

Fig. 37

(ii) Vectoring
Using this technique, a peripheral device which causes an interrupt also **generates an address** or part of an address (i.e. **a vector**), which enables the correct ISR to be located. The ISR vector may be read into the microprocessor through an I/O port, or with certain microprocessors, it may be gated directly into the data bus at the appropriate time.

Problem 22 Explain the meaning of the term **priority** in connection with interrupts, and show how different priorities may be assigned to peripheral devices.

In a multiple interrupt system it is possible for two or more devices to interrupt simultaneously. If this situation arises, a decision must be made regarding which device should be serviced first, i.e., which device has the **greater priority**.

In a polled interrupt system, priority may be assigned to each peripheral device by organizing the software polling routine such that peripheral devices are polled in **descending order of priority**. This means that if two devices **simultaneously cause an interrupt**, the device with the higher priority will be serviced first. Upon completion of the ISR for the first device serviced, a further interrupt occurs (since the second device has not yet been serviced) and the program immediately enters the ISR for the second device. Note that it may be necessary to latch each interrupt so that it is kept active until the microprocessor enters its particular ISR.

In a vectored interrupt system, a **priority encoder** may be used to generate each vector. This type of encoder has *n* output lines on which are generated a unique binary code according to which of its input lines is made active. In addition, if two or more inputs are simultaneously made active, the **code for the most significant input is generated**. The characteristics of a typical priority encoder are shown in *Fig. 38*.

A priority encoder of this type may be used to generate eight interrupt vectors which may be used to determine the starting address of each ISR. The use of a priority encoder for this purpose is shown in *Fig. 39*.

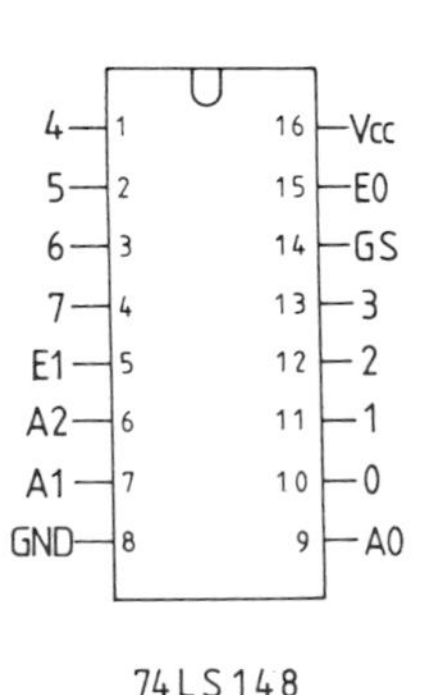

INPUTS									OUTPUTS				
E1	0	1	2	3	4	5	6	7	A2	A1	A0	GS	EO
1	X	X	X	X	X	X	X	X	1	1	1	1	1
0	1	1	1	1	1	1	1	1	1	1	1	1	0
0	X	X	X	X	X	X	X	0	0	0	0	0	1
0	X	X	X	X	X	X	0	1	0	0	1	0	1
0	X	X	X	X	X	0	1	1	0	1	0	0	1
0	X	X	X	X	0	1	1	1	0	1	1	0	1
0	X	X	X	0	1	1	1	1	1	0	0	0	1
0	X	X	0	1	1	1	1	1	1	0	1	0	1
0	X	0	1	1	1	1	1	1	1	1	0	0	1
0	0	1	1	1	1	1	1	1	1	1	1	0	1

Fig. 38

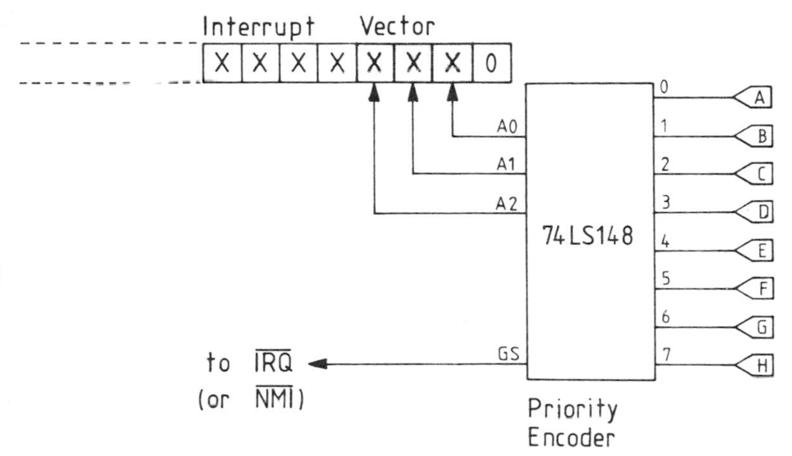

Fig. 39

Problem 23 A **6502** microprocessor with a **6530 PIA** is used to control an industrial process, as shown in *Fig. 40*. Eight LEDs are connected to port A of the PIA, and are arranged to slowly count up in binary to simulate the process control signals. A sensor is used to detect hazardous operating conditions in the process, e.g. overheating, loss of oil pressure, and is connected to $\overline{\text{NMI}}$ of the microprocessor. When such a condition occurs, the process is immediately terminated, and a warning tone is generated by an audio warning device (AWD) which is connected to bit 0 of port B. The warning tone is generated for a duration of approximately 5 seconds, after which further processing is suspended until the fault is cleared and the program is restarted.

Write suitable **6502** machine code routines for the main program and for the interrupt service routine, with the main program starting at address 0200_{16} and the ISR starting at 0250_{16}.

```
:MAIN PROGRAM
;
PAD     = $1700
PADD    = $1701
PBD     = $1702
PBDD    = $1703
DLY1    = $0000
DLY2    = $0001
FLAG    = $0002
IRQV    = $17FE
      * = 0200
```

```
0200  A2 FF                  LDX #$FF
0202  9A                     TXS            ;initialize the stack pointer
0203  A9 50                  LDA #$50
0205  8D FE 17               STA IRQV       ;set up IRQ vector (change to
0208  A9 02                  LDA #$02       ;suit user's system)
020A  8D FF 17               STA IRQV+1
020D  A9 00                  LDA #$00
020F  85 02                  STA FLAG       ;clear shut down flag
0211  A9 FF                  LDA #$FF
0213  8D 01 17               STA PADD       ;port A all outputs
0216  8D 03 17               STA PBDD       ;port B all outputs
0219  58                     CLI            ;enable interrupts
021A  CA         DELAY       DEX            ;time delay
021B  D0 FD                  BNE DELAY
021D  88                     DEY
021E  D0 FA                  BNE DELAY
0220  A5 02                  LDA FLAG       ;check state of shut down flag
0222  D0 FE      SDOWN       BNE SDOWN      ;end loop for shut down condition
0224  EE 00 17               INC PAD        ;simulate next part of process
0227  4C 1A 02               JMP DELAY      ;continue with the process
                 ;
                 ;INTERRUPT SERVICE ROUTINE
                             *=*+$2C
0250  A9 00                  LDA #$00
0252  8D 00 17               STA PAD        ;shut down the process
0255  A9 20                  LDA #$20
0257  85 00                  STA DLY1       ;set up tone duration constants
0259  85 01                  STA DLY2
025B  A2 90      TONE        LDX #$90       ;set up tone frequency constant
025D  CA         LOOP        DEX
025E  D0 FD                  BNE LOOP       ;half period time delay
0260  AD 02 17               LDA PBD
0263  49 01                  EOR #$01
0265  8D 02 17               STA PBD        ;toggle the AWD
0268  C6 01                  DEC DLY2       ;decrement tone duration counter 2
026A  D0 EF                  BNE TONE
026C  C6 00                  DEC DLY1       ;decrement tone duration counter 1
026E  D0 EB                  BNE TONE       ;end of tone?
0270  A9 01                  LDA #$01
0272  85 02                  STA FLAG       ;set shut down flag
0274  58                     CLI            ;re-enable interrupts
0275  40                     RTI
```

Problem 24 (a) Show how **priority vectored interrupts** may be implemented using a **6502** microprocessor. (b) Write a program, using **6502** machine code, to show how the appropriate interrupt service routine is selected when using the vectored interrupt system described in (a).

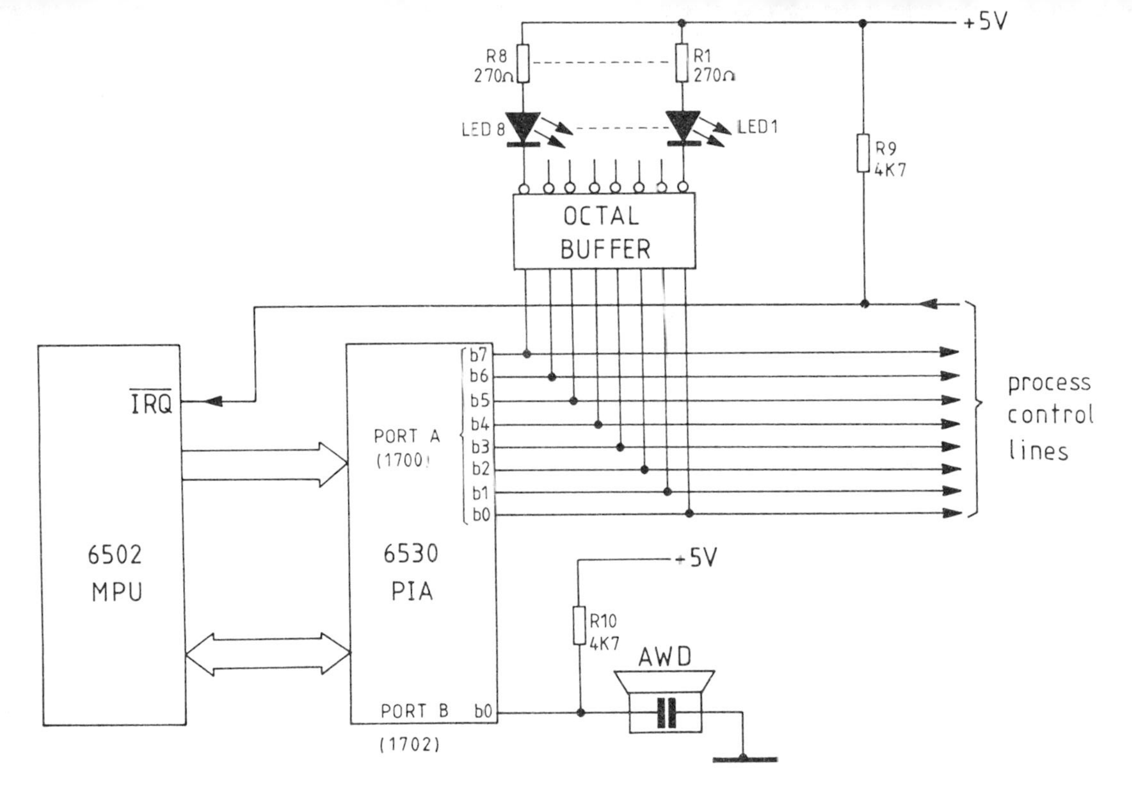
+5V
R8
270Ω
R1
270Ω
LED 8
LED1
R9
4K7
OCTAL
BUFFER
IRQ
PORT A
(1700)
b7
b6
b5
b4
b3
b2
b1
b0
process
control
lines
6502
MPU
6530
PIA
+5V
R10
4K7
AWD
PORT B
b0
(1702)

Fig. 40

(a) The **6502** microprocessor has no built-in facilities to directly handle vectored interrupts as, for example, a Z80 microprocessor can. Vectored interrupts may be implemented with a **6502** microprocessor, however, by reading in a vector from an I/O port and using this to create a **jump indirect** address to locate the appropriate interrupt service routine. The hardware required to implement such a vectored interrupt system is shown in *Fig. 41*.

(b) The 6502 machine code routine required to allow vectored interrupts to be implemented using the hardware shown in *Fig. 41* is as follows.

```
                    ;6502 VECTORED INTERRUPTS
                    ;
                    INVECT    =$0010
                    PBD       =$1702
                    PBDD      =$1703
                    NMIV      =$17FA
                            *=$0000
                    ;
                    ;ISR STARTING ADDRESS TABLE (may be changed
                    ;to suit user)
0000   70 02        TABLE     .WORD $0270       ;ISR0 (lowest priority)
0002   78 02                  .WORD $0278       ;ISR1
0004   80 02                  .WORD $0280       ;ISR2
0006   88 02                  .WORD $0288       ;ISR3
0008   90 02                  .WORD $0290       ;ISR4
000A   98 02                  .WORD $0298       ;ISR5
000C   A0 02                  .WORD $02A0       ;ISR6
000E   A8 02                  .WORD $02A8       ;ISR7 (highest priority)
                    ;
                      *=*+$01F0
                    ;MAIN PROGRAM
                    ;
0200   A2 FF                  LDX #$FF
0202   9A                     TXS               ;initialize the stack pointer
0203   A9 00                  LDA #$00
0205   8D 03 17               STA PBDD          ;port B all inputs
0208   A9 50                  LDA #$50
020A   8D FA 17               STA NMIV          ;set up NMI vector low
020D   A9 02                  LDA #$02
020F   8D FB 17               STA NMIV+1        ;set up NMI vector high
0212   4C 12 02     ENDLP     JMP ENDLP         ;endloop – replace with main prog
                    ;
                      *=*+$3B
                    ;INTERRUPT SERVICE ROUTINE
                    ;
0250   AD 02 17               LDA PBD           ;get vector from priority encoder
0253   29 0E                  AND #$0E          ;mask off unused bits
0255   AA                     TAX               ;use to index vector table
0256   B5 00                  LDA TABLE,X       ;get ISR starting address low
0258   85 10                  STA INVECT        ;save at indirect address
025A   E8                     INX
025B   B5 00                  LDA TABLE,X       ;get ISR starting address high
025D   85 11                  STA INVECT+1      ;save at indirect jump address
025F   6C 10 00               JMP (INVECT)      ;jump to selected ISR
```

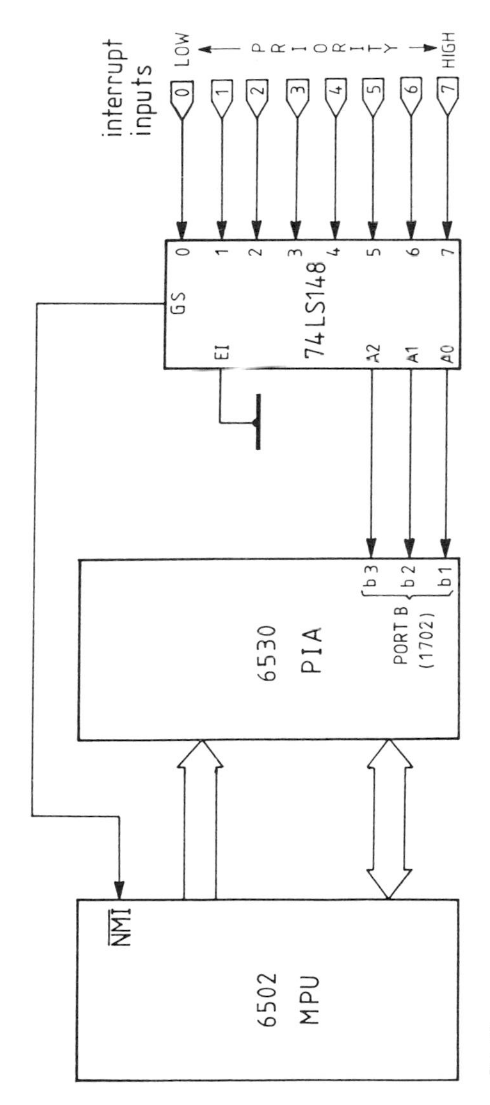

interrupt inputs
LOW
PRIORITY
HIGH
0
1
2
3
4
5
6
7
74LS148
GS
EI
A2
A1
A0
6530
PIA
PORT B
(1702)
b3
b2
b1
6502
MPU
$\overline{NMI}$

Fig. 41

Problem 25 A **Z80** microcomputer with a **Z80 PIO** is used to control an industrial process, as shown in *Fig. 42*. Eight LEDs are connected to port A of the PIO, and are arranged to count up slowly in binary to simulate the process control signals. A sensor is used to detect hazardous operating conditions in the process, e.g. overheating, loss of oil pressure, and is connected to $\overline{NMI}$ of the microprocessor. When such a condition occurs, the process is immediately terminated, and a warning tone is generated by an audio warning device (AWD) which is connected to bit 0 of port B. The warning tone is generated for approximately 5 seconds, after which further processing is suspended until the fault is cleared and the program is restarted.

Write suitable **Z80** machine code routines for the main program and for the interrupt service routine, with the main program starting at address $0C90_{16}$ and the ISR starting at $0D25_{16}$.

Note: In some microprocessor evaluation systems, $\overline{INT}$ or $\overline{NMI}$ may be used by the monitor for single step or for interrupt driven keyboard input. This should be checked by reference to the operating manual before attempting to implement an interrupt driven user program.

A **Z80** machine code main program and interrupt service routine to provide the required function is as follows:

```
                      ;MAIN PROGRAM
                      ;
                       NMIV    EQU 0C7EH
                               ORG 0C90H
                      ;
0C90   31 00 10                LD SP,1000H      ;initialize the stack pointer
0C93   21 25 0D                LD HL,0D25H      ;put ISR starting address in HL
0C96   22 7E 0C                LD (NMIV),HL     ;and load into NMI vector location
0C99   CB 82                   RES 0,D          ;use D as shut down flag and clear it
0C9B   3E 0F                   LD A,0FH
0C9D   D3 06                   OUT (6),A        ;port A byte output mode
0C9F   D3 07                   OUT (7),A        ;port B byte output mode
0CA1   0E 00                   LD C,0           ;use C as a binary counter and clear it
0CA3   79             DISPLA   LD A,C           ;transfer count to C
0CA4   D3 04                   OUT (4),A        ;and display it
0CA6   21 00 01                LD HL,0100H      ;set delay parameter in HL
0CA9   10 FE          DELAY    DJNZ DELAY       ;time delay
0CAB   2B                      DEC HL
0CAC   7C                      LD A,H
0CAD   B5                      OR L             ;set flags
0CAE   20 F9                   JR NZ,DELAY      ;keep looping if HL is not zero
0CB0   0C                      INC C            ;simulate next stage in the process
0CB1   CB 42                   BIT 0,D          ;check the state of the shut down flag
0CB3   20 FE          SDOWN    JR NZ,SDOWN      ;end loop for shut down condition
0CB5   18 EC                   JR DISPLA        ;continue with the process
```

```
                ;
                ;INTERRUPT SERVICE ROUTINE
                ;
                        ORG 0D25H
                ;
0D25  AF                XOR A
0D26  D3 04             OUT (4),A       ;shut down the process
0D28  CB C2             SET 0,D         ;set shut down flag
0D2A  21 80 30          LD HL,3080H     ;set up tone duration constant
0D2D  0E 00             LD C,0
0D2F  79        TONE    LD A,C
0D30  D3 05             OUT (5),A       ;output to AWD
0D32  06 38             LD B,38H        ;set up tone frequence constant
0D34  10 FE     LOOP    DJNZ LOOP       ;half period time delay
0D36  EE 01             XOR 01          ;toggle the AWD
0D38  4F                LD C,A          ;save A in C
0D39  2B                DEC HL
0D3A  7C                LD A,H
0D3B  B5                OR L            ;set flags
0D3C  20 F1             JR NZ,TONE      ;end of tone?
0D3E  ED 45             RETN            ;return from ISR
```

Problem 26 Show how the program developed in *Problem 25* may be modified so that a **maskable interrupt** may be initiated through the $\overline{STB}$ input of the PIO, using the **Z80 interrupt mode 2**.

In order to implement a **vector interrupt** via a $\overline{STB}$ input of the **Z80 PIO**, the initialization sequence of the main program must include the following:

(a) **initialize the I register of the Z80 CPU**;
(b) **load the peripheral vector into the Z80 PIO**;
(c) **store the ISR starting address in the RAM** addresses to which (a) and (b) point;
(d) **select interrupt mode 2**;
(e) **enable the CPU to accept interrupts**; and
(f) **enable the PIO to accept interrupts**.

In addition to this sequence, the CPU must be re-enabled to accept interrupts each time that the ISR is called, usually by including an EI instruction just before returning from the ISR.

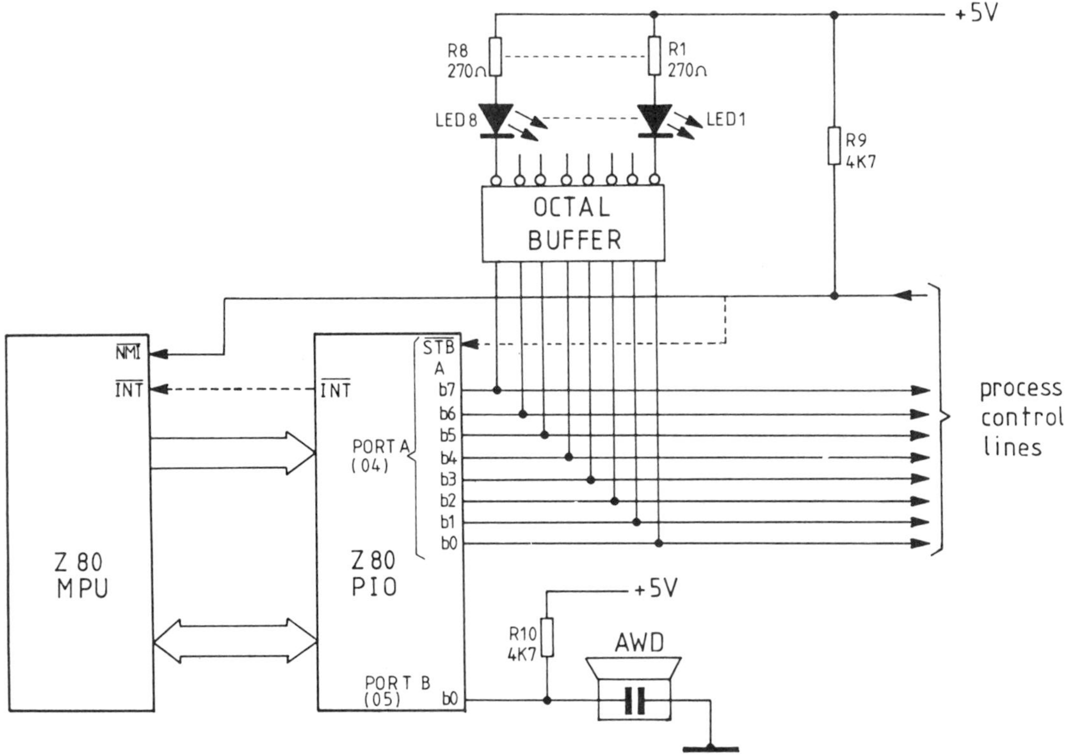

+5V

process
control
lines

R9
4K7

R1
270Ω

LED1

R8
270Ω

LED8

OCTAL
BUFFER

+5V

AWD

R10
4K7

$\overline{STB}$
A
b7
b6
b5
b4
b3
b2
b1
b0

PORT A
(04)

Z80
PIO

PORT B
(05)

b0

$\overline{INT}$

$\overline{NMI}$
$\overline{INT}$

Z80
MPU

Fig. 42

The sequence of instruction necessary to satisfy these requirements is as follows:

```
                              ;MAIN PROGRAM
                              ;
                                IRQV   EQU 0D20H
                                       ORG 0C90H
                              ;
     0C90   31 00 10                   LD SP,1000H      ;initialize the stack pointer
**   0C93   21 25 0D                   LD HL,0D25H      ;put ISR starting address in HL
**   0C96   22 20 0D                   LD (IRQV),HL     ;and load into IRQ vector location
     0C99   CB 82                      RES 0,D          ;use D as shut down flag and clear it
     0C9B   3E 0F                      LD A,0FH
     0C9D   D3 06                      OUT (6),A        ;port A byte output mode
     0C9F   D3 07                      OUT (7),A        ;port B byte output mode
**   0CA1   3E 0D                      LD A,0DH         ;high byte of ISR indirect address
**   0CA3   ED 47                      LD I,A           ;put it into the I register
**   0CA5   3E 20                      LD A,20H         ;low byte of ISR indirect address
**   0CA7   D3 06                      OUT (6),A        ;peripheral vector for port A
**   0CA9   ED 5E                      IM 2             ;CPU vectored interrupt mode
**   0CAB   FB                         EI               ;enable CPU to accept INT
**   0CAC   3E 87                      LD A,87H         ;interrupt control word
**   0CAE   D3 06                      OUT (6),A        ;enable PIO to generate INT
     0CB0   0E 00                      LD C,0           ;use C as a binary counter and clear it
     0CB2   79                DISPLA   LD A,C           ;transfer count to C
                                       .
                                       .
                                       .
                                       .
                                       .
                                       .
                                       .
                                       .
                                       .
                                       .
                                       .
     0D3B   B5                         OR L             ;set flags
     0D3C   20 F1                      JR NZ,TONE       ;end of tone?
**   0D3E   FB                         EI               ;re-enable interrupts
     0D3F   ED 45                      RETN             ;return from ISR
```

** indicates program lines which have been inserted or changed from the original.

Problem 27 (a) Show how **priority vectored interrupts** may be implemented using a **Z80** microprocessor. (b) Explain how the appropriate interrupt service routine is selected when using the vectored interrupt system described in (a).

(a) The hardware required to implement **priority vectored interrupts** with a **Z80** microprocessor is shown in *Fig. 43*.

(b) An interrupting device takes one of the **priority encoder** inputs (74LS148) to a logical 0 level. This action causes the priority encoder to generate a 3-bit binary code, whose value depends upon which input was activated, and which is used to form part of the peripheral vector. The priority encoder also generates a strobe output (GS) when any one of its inputs is activated, and this is used in the circuit shown to initiate an interrupt request.

The **Z80** acknowledges this request by making **both $\overline{\text{IORQ}}$ and $\overline{\text{MI}}$** to a logical 0 level. These signals are gated and used to enable an 81LS95 tri-state buffer, which then puts the peripheral vector onto the data bus where it may be read by the **Z80** microprocessor. The **Z80** microprocessor combines this vector with the contents of its I register and uses this as an indirect address to locate the start of the appropriate interrupt service routine.

Problem 28 (a) Explain the main disadvantage of the **priority vectored interrupt** system described in *Problem 27*. (b) Explain how the use of **Z80 PIOs** in a **daisy chain** configuration may be used to overcome the problem described in (a).

(a) When an interrupt is serviced using the system shown in *Fig. 43*, further interrupts must be inhibited until completion of the current ISR. This is necessary to prevent an ISR from being interrupted by a peripheral of lower priority than the one currently being serviced. Unfortunately, this also prevents an ISR from being interrupted by a peripheral device of higher priority than the one currently being serviced. Ideally, an interrupt system should allow higher priority peripherals to interrupt the ISR of a lower priority peripheral and resume service of the lower priority peripheral once the higher priority ISR is completed. A lower priority peripheral should not be allowed to interrupt the ISR of a higher priority peripheral, but should be queued for service at the next opportunity.

(b) The problem stated in (a) may be overcome by using a **daisy chain** interrupt system of the type shown in *Fig. 44*. The daisy chain is formed by linking together the **IEI (interrupt enable input)** and **IEO (interrupt enable output)** pins of each PIO as shown in *Fig. 44*. The PIO at the start of the chain has its IEI connected to a logical 1, and this assigns the highest priority to this device, and all subsequent PIOs connected in the chain have correspondingly lower priorities according to their position in the chain. The IEO and IEI of port A and port B of any one PIO are linked internally, and this accordingly assigns a higher priority to the A section. The IEI input of each PIO must be at a logical 1 to allow it to respond to an interrupt request and hence place its peripheral vector on the data bus when acknowledged by the MPU. A logical 0 on this input inhibits the PIO from generating an interrupt.

The IEO output of each PIO is held at a logical 1 only if its IEI pin is also at logical 1 and the MPU is not servicing an interrupt from this PIO.

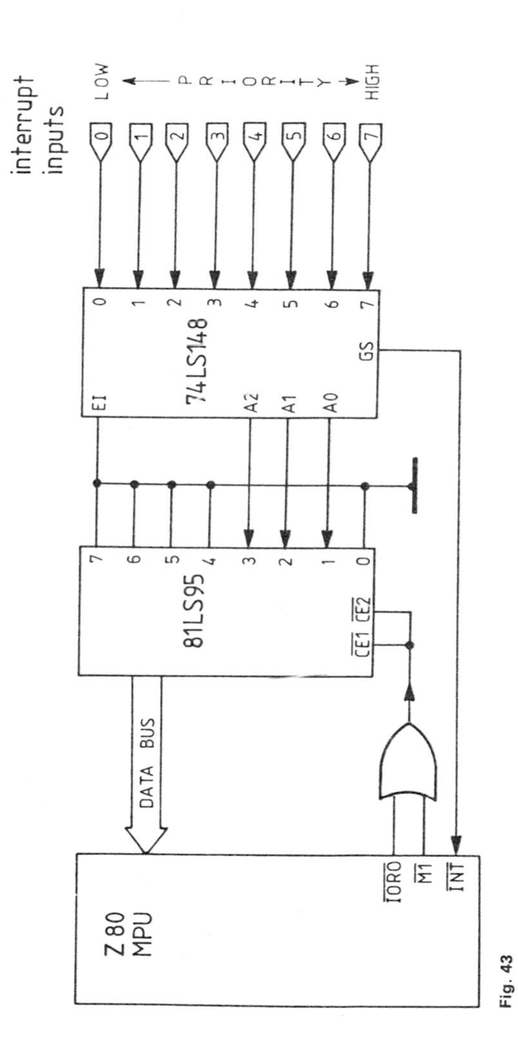
interrupt inputs
LOW
PRIORITY
HIGH
0
1
2
3
4
5
6
7
74LS148
EI
A2
A1
A0
GS
81LS95
CE1
CE2
DATA BUS
Z80 MPU
IORQ
M1
INT

Fig. 43

Therefore, if IEO on any PIO becomes a logical 0, all following IEIs and IEOs in the chain also become logical 0 and thus inhibit interrupts from that point in the chain. Up to four PIOs may be chained together in this manner without the need for extra logic circuits.

The behaviour of a daisy chain may be studied by reference to *Fig. 44*, which operates in the following manner:

(i) the main program is being run. All IEI and IEO points in the daisy chain are at logical 1, therefore all ports are ready to accept an interrupt;

(ii) an interrupt is initiated on port 2A (either by using the STB line or by means of I/O lines in mode 3 only). The IEO output of port 2A changes to a logical 0 and this causes all other IEI and IEO points further down the chain to also become logical 0 and inhibit interrupts from these ports. When acknowledged by the MPU ($\overline{IORQ}$ and $\overline{MI}$ both at logical 0), port 2A puts its peripheral vector out onto the data bus to be read by the MPU, which then jumps to the appropriate ISR for port 2A;

(iii) during the servicing of port 2A, an interrupt occurs on port 1B. This is accepted, since IEI and IEO of this port are both at logical 1. The service routine of port 2A is suspended, and port 1B puts its peripheral vector out onto the data bus. A jump to the start of the appropriate ISR for port 1B takes place;

(iv) when the service of port 1B is complete, the servicing of port 2A is resumed. The RETI instruction at the end of the ISR for port 1B causes IEO of the PIO under service to return to a logical 1 level to remake the chain (i.e., the IEO of a port which is at logical 0, but whose IEI is at logical 1);

(v) once the ISR for port 2A is complete, the daisy chain is completely restored and a return is made to the main program.

Problem 29 A **6800** microcomputer with a **6820 PIA** is used to control an industrial process, as shown in *Fig. 45*. Eight LEDs are connected to port A of the PIO, and are arranged to count up slowly in binary to simulate the process control signals. A sensor is used to detect hazardous operating conditions in the process, e.g. overheating, loss of oil pressure, and is connected to $\overline{IRQ}$ of the microprocessor. When such a condition occurs, the process is immediately terminated, and a warning tone is generated by an audio warning device (AWD) which is connected to bit 0 of port B. The warning tone is generated for approximately 5 seconds, after which further processing is suspended until the fault is cleared and the program is restarted.

Write suitable **6800** machine code routines for the main program and for the interrupt service routine, with the main program starting at address 0100_{16} and the ISR starting at 0180_{16}.

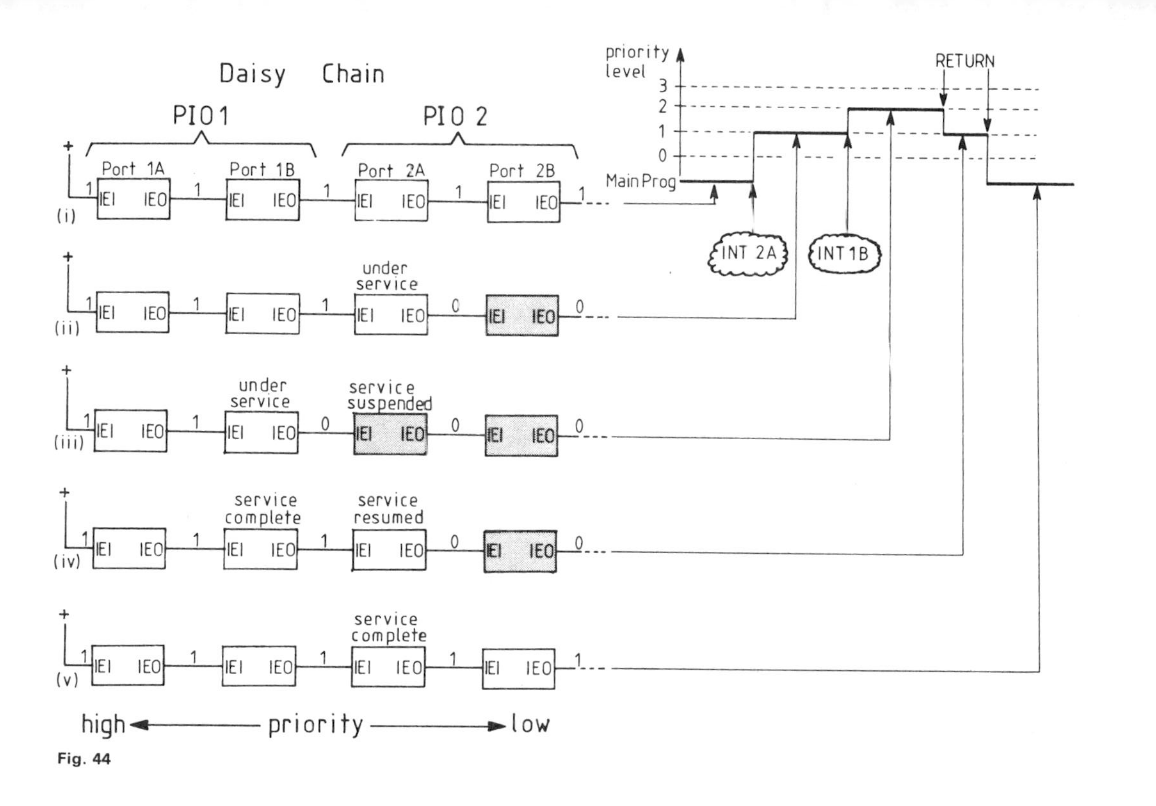

Daisy Chain
PIO1
PIO 2
Port 1A
Port 1B
Port 2A
Port 2B
IEI
IEO
under service
service suspended
service resumed
service complete
priority level
3
2
1
0
Main Prog
INT 2A
INT 1B
RETURN
(i)
(ii)
(iii)
(iv)
(v)
high
priority
low

Fig. 44

```
                    ;MAIN PROGRAM
                    ;
                     FLAG  EQU $0000
                     DRA   EQU $8004
                     CRA   EQU $8005
                     DRB   EQU $8006
                     CRB   EQU $8007
                     IRQV  EQU $A000
                           ORG $Ø1ØØ
                    ;
0100  8E A0 78             LDS #$A078    ;initialize stack pointer
0103  4F                   CLR A
0104  97 00                STA A FLAG    ;clear shut down flag
0106  B7 80 05             STA A CRA     ;select data direction register A
0109  B7 80 07             STA A CRB     ;select data direction register B
010C  43                   CPL A
010D  B7 80 04             STA A DRA     ;port A all outputs
0110  B7 80 06             STA A DRB     ;port B all outputs
0113  CE 01 80             LDX #$0180    ;put ISR starting address in X
0116  FF A0 00             STX IRQV      ;and save in vector location
0119  86 04                LDA A #$04    ;
011B  B7 80 07             STA A CRB     ;select I/O register B
011E  B7 80 05             STA A CRA     ;select I/O register A
0121  0E                   CLI           ;enable interrupts
0122  4F                   CLR A
0123  B7 80 04             STA A DRA     ;clear the displays
0126  CE 00 00 DELAY       LDX #$0000
0129  09       LOOP        DEX           ;time delay
012A  26 FD                BNE LOOP
012C  96 00                LDA A FLAG    ;check state of the shut down flag
012E  26 FE    SDOWN       BNE SDOWN     ;end loop for shut down condition
0130  7C 80 04             INC DRA       ;simulate next part of the process
0133  20 F1                BRA DELAY     ;continue with the process
                    ;
                    ;INTERRUPT SERVICE ROUTINE
                           ORG $0180
                    ;
0180  7F 80 04             CLR DRA       ;shut down the process
0183  CE 40 00             LDX #$4000    ;set up tone duration constant
0186  4F                   CLR A
0187  B7 80 06 TONE        STA A PBD     ;output to AWD
018A  C6 20                LDA B #$20    ;set up tone frequency constant
018C  5A       HDEL        DEC B
018D  26 FD                BNE HDEL      ;half period time delay
018F  88 01                EOR A #$01    ;toggle the AWD
0191  09                   DEX           ;decrement the tone duration counter
0192  26 F3                BNE TONE      ;end of tone?
0194  86 01                LDA A #$01
0196  97 00                STA A FLAG    ;set shut down flag
0198  0E                   CLI           ;re-enable interrupts
0199  3B                   RTI           ;return from interrupt routine
```

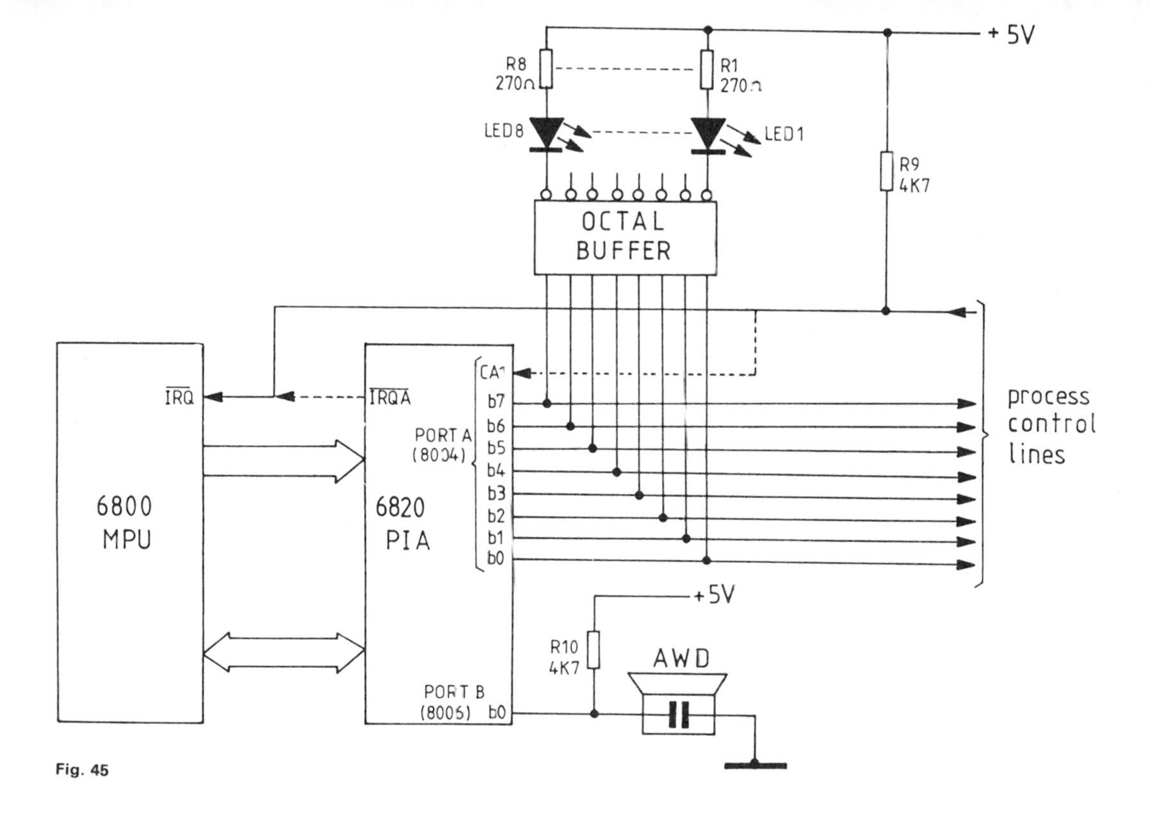
+5V
R8
270Ω
R1
270Ω
LED8
LED1
R9
4K7
OCTAL
BUFFER
IRQ
6800
MPU
IRQA
CA1
b7
b6
b5
b4
b3
b2
b1
b0
PORT A
(8004)
6820
PIA
process
control
lines
+5V
R10
4K7
AWD
PORT B
(8005)
b0

Fig. 45

Note: (i) The contents of addresses 0117_{16} and 0118_{16} may be changed if the equipment vectors the $\overline{IRQ}$ service routine through a different RAM address. Refer to the operating manual for the equipment in use.
(ii) The CLI instructions at addresses 0121_{16} and 0198_{16} may be removed if $\overline{NMI}$ is used instead of $\overline{IRQ}$.
(iii) The CLI instruction at address 0198_{16} is included so that the response of the system may be changed to allow several consecutive interrupts to be accepted, rather than the shut down at address $012E_{16}$.

Problem 30 Describe the changes which must be made to your answer to the previous problems so that the interrupt may be initiated **via CA1 of the PIA**.

Two changes must be made to the program in *Problem 29* to allow interrupts to be initiated via CA1 to the **6820 PIA**, and these are:

(a) Interrupts must be enabled through CA1 by storing appropriate bits in the **control register** of the PIA.
(b) The **interrupt flag** in the PIA must be **reset** after responding to the interrupt, and this may be accomplished by reading the appropriate data register (data register A for CA1).

The changes to the program given in *Problem 29* are as follows:

MAIN PROGRAM

```
                        .
                        .
                        .
  011B  B7 80 07        STA A CRB      ;select I/O register B
→ 011E  86 07           LDA A #S07     ;active edge low to high
  0120  B7 80 05        STA A CRA      ;select I/O register A
                        .
└─ insert this instruction
                        .
                        .

                        OR

                        .
                        .
                        .
  011B  B7 80 07        STA A CRB      ;select I/O register B
→ 011E  86 05           LDA A #S07     ;active edge high to low
  0120  B7 80 05        STA A CRA      ;select I/O register A
                        .
└─ insert this instruction
                        .
                        .
```

INTERRUPT SERVICE ROUTINE

```
                        .
                        .
                        .
  0196   97 00          STA A FLAG     ;set shut down flag
→ 0198   B6 80 04       LDA A DRA      ;read data register, clear I flag
  019B   0E             CLI            ;re-enable interrupts
  019C   3B             RTI            ;return from interrupt routine

└ insert this instruction
```

Problem 31 (a) Show how **vectored interrupts** may be implemented using a **6800** microprocessor. (b) Write a program, using **6800** machine code, to show how the appropriate interrupt service routine is selected when using the vectored interrupt system described in (a).

(a) The **6800** microprocessor has no built-in facilities to directly handle vectored interrupts as, for example, a Z80 microprocessor can. Vectored interrupts may be implemented with a **6800** microprocessor, however, by reading in a vector from an I/O port and using this data to implement a **jump indirect** to the address whose value is stored in index register X. Thus, the value read in from the I/O port is transferred to X and is used to select the appropriate interrupt service routine. The hardware required to implement such a vectored interrupt system is shown in *Fig. 46*.

The interrupt line from the 74LS148 **priority encoder** may go directly to the $\overline{IRQ}$ input of the **6800** microprocessor, or it may be connected to either CB1 or CB2 of the PIA which must then be configured to accept interrupts. The $\overline{IRQB}$ output of the PIA is then used to pass the interrupt request to $\overline{IRQ}$ of the microprocessor.

(b) The **6800** machine code routine required to allow vectored interrupts to be implemented using the hardware shown in *Fig. 46* is as follows:

```
                ;6800 VECTORED INTERRUPTS
                ;
                  DRB      EQU $8006
                  CRB      EQU $8007
                  IRQV     EQU $A000
                           ORG $0000
                ;
                ;ISR STARTING ADDRESS TABLE
                ;(may be changed to suit the user)
                ;
0000   01 90    TABLE      FDB $0190      ;ISR0 (lowest priority)
0002   01 98               FDB $0198      ;ISR1
0004   01 A0               FDB $01A0      ;ISR2
0006   01 A8               FDB $01A8      ;ISR3
```

```
0008   01 B0                 FDB $01B0        ;ISR4
000A   01 B8                 FDB $01B8        ;ISR5
000C   01 C0                 FDB $01C0        ;ISR6
000E   01 C8                 FDB $01C8        ;ISR7 (highest priority)
                  ;
                             ORG $0100
                  ;
                  ;MAIN PROGRAM
                  ;
0100   8E A0 78              LDS #$A078       initialize the stack pointer
0103   4F                    CLR A
0104   B7 80 07              STA A CRB        ;select data direction register B
0107   B7 80 06              STA A DRB        ;port B all inputs
010A   86 04                 LDA A #$04
010C   B7 80 07              STA A CRB        ;select I/O register B
010F   CE 01 80              LDX #$0180       ;put IRQ vector into X
0112   FF A0 00              STX IRQV         ;and store in vector location in RAM
0115   0E                    CLI              ;enable interrupts
0116   20 FE      ENDLP      BRA ENDLP        ;replace end loop with main program
                  ;
                             ORG $0180
                  ;
                  ;INTERRUPT SERVICE ROUTINE
                  ;
0180   CE FF FF              LDX #$FFFF       ;use X as ISR table pointer
0183   F6 80 04              LDA B DRB        ;get vector from priority encoder
0186   08         POINT      INX              ;advance X up the ISR table
0187   5A                    DEC B
0188   2A FC                 BPL POINT        ;stop X at the right one
018A   EE 00                 LDX TABLE,X      ;put ISR starting address in X
018C   6E 00                 JMP X            ;and jump to it
```

C FURTHER PROBLEMS ON INTERRUPTS

(a) SHORT ANSWER PROBLEMS

1 A technique which interrogates each peripheral device, in turn, to determine whether it requires attention is known as ....

2 A technique which allows a peripheral device to cause a break in normal program execution when it requires attention is called an ....

3 The main reason why special techniques are required when transferring data between a microcomputer and its peripheral devices is that they do not operate ....

4 Events which occur at totally random times, but which require immediate attention when they do occur, are best dealt with by using ....

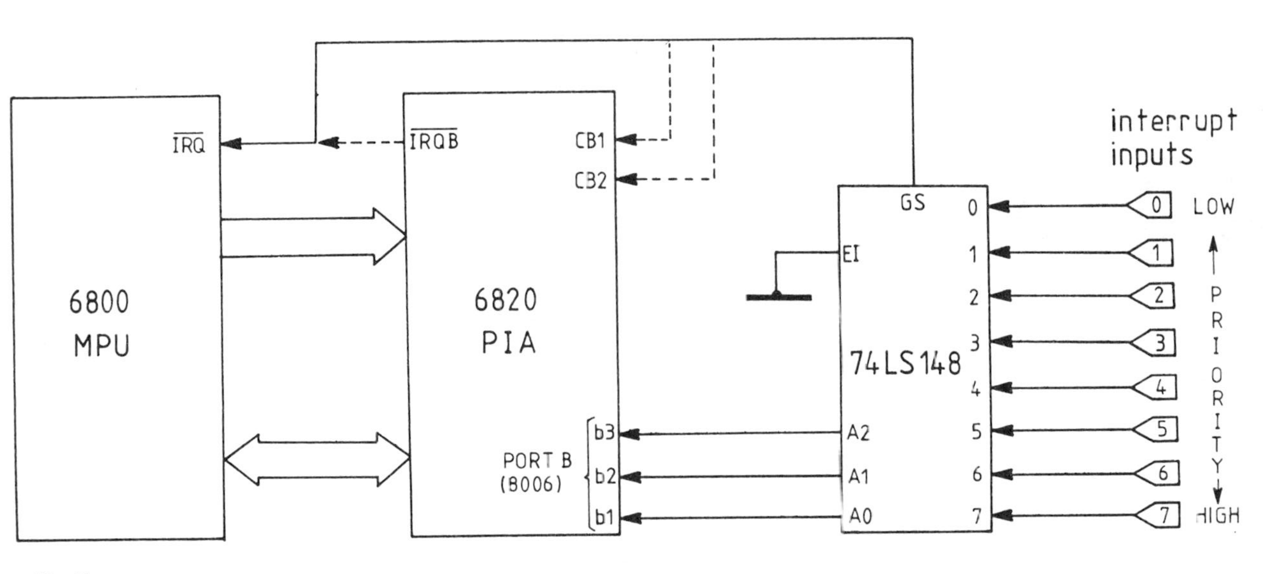
interrupt inputs
LOW
PRIORITY
HIGH
0
1
2
3
4
5
6
7
74LS148
GS
EI
A2
A1
A0
6820
PIA
CB1
CB2
IRQB
PORT B
(8006)
b3
b2
b1
6800
MPU
IRQ

Fig. 46

5 Events which occur regularly, at reasonably predictable times, may be dealt with by using . . . . techniques.

6 A major disadvantage of using a technique of repeatedly interrogating peripheral devices to determine whether they require attention is that it . . . .

7 One precaution which must be taken if a peripheral device is allowed to break into a main program is that all of the important . . . .

8 If it is important that a peripheral device must be capable of breaking into a main program at all times, it should be connected to the . . . . pin of the MPU.

9 If a peripheral device must be prevented from breaking into a main program at times selected by the main program, it should be connected to the . . . . pin of the MPU.

10 When it is possible for several different peripheral devices to break into a main program, a . . . . system is in use.

11 If several different peripheral devices may cause a break in execution of a main program, but each peripheral device is assigned a different level of importance, a . . . . system is in use.

12 A program which is initiated by external means and which causes a peripheral device to receive the required attention is known as an . . . .

13 A technique in which a peripheral device supplies part of an address which enables one of several different peripheral servicing routines to be selected is known as a . . . . system.

14 If the NMI pin on a 6502 microprocessor is taken to a logical 0, addresses . . . . and . . . . are put out into the address bus.

15 If the NMI pin on a Z80 microprocessor is taken to a logical 0, address . . . . is put out onto the address bus.

16 If the NMI pin on a 6800 microprocessor is taken to a logical 0, addresses . . . . and . . . . are put out onto the address bus.

17 After a reset, a 6502 microprocessor starts to execute the program whose starting address is stored in addresses . . . . and . . . .

18 After a reset, a Z80 microprocessor starts to execute the program whose starting address is . . . .

19 After a reset, a 6800 microprocessor starts to execute the program whose starting address is stored in addresses . . . . and . . . .

20 With regard to the $\overline{\text{RESET}}$, $\overline{\text{NMI}}$ and $\overline{\text{INT}}$ (or $\overline{\text{IRQ}}$) inputs of a microprocessor, $\overline{\text{RESET}}$ has the greatest . . . .

(b) CONVENTIONAL PROBLEMS

1 Rewrite *Worked problem 3* to include an interrupt driven protection arrangement which operates in the following manner. When an overload condition is detected in the process, a non-maskable interrupt is generated which causes both the motor and heater to be switched off, and which then causes an LED connected to bit 2 of port A to flash at a 1 Hz rate until the system is reset.

2 Rewrite *Worked problem 4*, using the sensor to generate a non-maskable interrupt in order to cause the object counter to be incremented.

3 Rewrite *Worked problem 5*, using a multiple interrupt system connected to $\overline{NMI}$ of the 6502 microprocessor (use a circuit similar to that shown in *Fig. 41*) in order to increment or decrement the vehicle counter.

4 A 6502 based microcomputer with a 6530 PIA is used to construct a digital frequency meter (frequency range 0 to 99 Hz. The frequency to be measured is suitably processed to convert it to TTL levels, and is connected to bit 0 of port B. A polling routine is used to sample bit 0 of port B and to increment a counter by one for each input cycle detected. An accurate external 2 Hz pulse generator is connected to the $\overline{NMI}$ input of the 6502 MPU, and each interrupt causes the contents of the counter to be converted to hertz (multiplied by 2), and to be displayed on two 7-segment LEDs connected to port A. Draw a circuit to show how such a frequency counter may be constructed, and write a 6502 machine code main program and interrupt service routine to operate the circuit.

5 Rewrite *Worked problem 9* to include an interrupt driven protection arrangement which operates in the following manner. When an overload condition is detected in the process, a non-maskable interrupt is generated which causes both the motor and heater to be switched off, and which then causes an LED connected to bit 2 of port A to flash at a 1 Hz rate until the system is reset.

6 Rewrite *Worked problem 10*, using the sensor to generate a non-maskable interrupt in order to cause the object counter to be incremented.

7 Rewrite *Worked problem 11*, using a multiple interrupt system connected to $\overline{NMI}$ of the Z80 microprocessor (use a circuit similar to that shown in *Fig. 43*) in order to increment or decrement the vehicle counter.

8 A Z80 based microcomputer with a Z80 PIO is used to construct a digital frequency meter (frequency range 0 to 99 Hz). The frequency to be measured is suitably processed to convert it to TTL levels, and is connected to bit 0 of port B. A polling routine is used to sample bit 0 of port B and to increment a counter by one for each input cycle detected. An accurate external 2 Hz pulse generator is connected to the $\overline{NMI}$ input of the Z80 MPU, and each interrupt causes the contents of the counter to be converted to hertz (multiplied by 2), and to be displayed on two 7-segment LEDs connected to port A. Draw a circuit to show how such a frequency counter may be constructed, and write a Z80 machine code main program and interrupt service routine to operate the circuit.

9 Rewrite *Worked problem 16* to include an interrupt driven protection arrangement which operates in the following manner. When an overload condition is detected in the process, a non-maskable interrupt is generated which causes both the motor and heater to be switched off, and which then causes an LED connected to bit 2 of port A to flash at a 1 Hz rate until the system is reset.

10 Rewrite *Worked problem 17*, using the sensor to generate a non-maskable interrupt in order to cause the object counter to be incremented.

11 Rewrite *Worked problem 18*, using a multiple interrupt system connected to $\overline{NMI}$ of the 6800 microprocessor (use a circuit similar to that shown in *Fig. 46*) in order to increment or decrement the vehicle counter.

12 A 6800 based microcomputer with a 6820 PIA is used to construct a digital frequency meter (frequency range 0 to 99 Hz). The frequency to be measured is suitably processed to convert it to TTL levels, and is connected to bit 0 of port B. A polling routine is used to sample bit 0 of port B and to increment a counter by one for each input cycle detected. An accurate external 2 Hz pulse generator is connected to the $\overline{\text{NMI}}$ input of the 6800 MPU, and each interrupt causes the contents of the counter to be converted to hertz (multiplied by 2), and then to be displayed on two 7-segment LEDs connected to port A. Draw a circuit to show how such a frequency meter may be constructed, and write a 6800 machine code main program and interrupt service routine to operate the circuit.

3 Logic families

A MAIN POINTS CONCERNED WITH LOGIC FAMILIES

1 Many different electronic circuit configurations are available for the construction of logic circuits. A complete set of logic functions which use a common circuit configuration is known as a **logic family**. Many different logic families are available in integrated form, and those most commonly used are:
(a) **transistor-transistor logic (TTL)**,
(b) **complementary metal oxide semiconductor (CMOS)**, and
(c) **emitter-coupled logic (ECL)**.
Logic families may be of the saturating or non-saturating type, and selection of a particular family depends upon the application being considered.

2 The most widely used type of saturating logic is **transistor-transistor logic (TTL)**, which uses special multi-emitter transistors for its fabrication. A wide range of circuit functions from simple gates to complex counters and registers are available in the TTL family. The circuit of a basic positive NAND gate is shown in *Fig. 1*.
This circuit operates in the following manner:
(a) A or B (or both) at logical 0
When one (or both) of the input terminals *A* or *B* is held at logical 0 (0 V–0.4 V), TR1 base-emitter junction becomes forward biased, and base current I_B flows. As a result, TR1 becomes saturated and its collector potential falls to almost 0 V. Therefore TR2 is cut off since it receives insufficient base bias, and this in turn, causes TR3 to be cut off. The collector potential of TR3 rises towards $+V_{CC}$ (5 V) and the gate output is therefore at logical 1.
(b) A and B both at logical 1
When both input terminals *A* and *B* are held at logical 1 (2.4 V–5 V), TR1 base potential rises towards $+V_{CC}$. As a result, the base-collector junction of TR1 becomes forward biased, and base current I_B is diverted into TR2 base-emitter circuit, thus turning TR2 on. This, in turn, causes TR3 to conduct heavily and its collector potential falls to almost 0 V. The gate output is therefore at logical 0.
(See Problems 1 to 10, 14, 15, and 20.)

3 Where low power consumption and compact assemblies are required, e.g. battery-operated portable equipment, a very wide range of circuits are available which use **complementary metal oxide semiconductor (CMOS)** construction. CMOS circuits are constructed from field effect transistors (FETs), using a mixture of *n*-channel and *p*-channel devices. A typical CMOS NAND gate circuit is shown in *Fig. 2*.

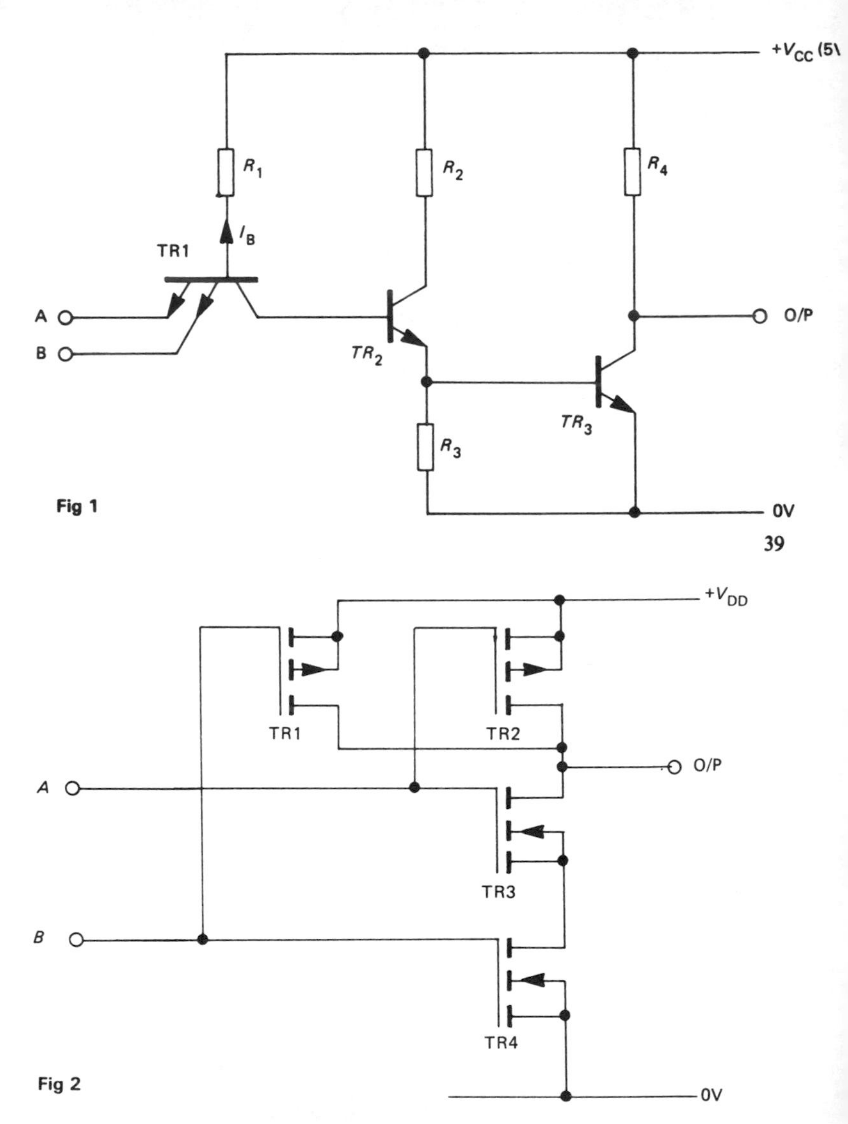

This circuit operates in the following manner:

(a) A or B (or both) at logical 0

When one (or both) of the input terminals *A* or *B* is held at logical 0 (0 V), TR1 or TR2 (or both) are biased into conduction since they are both *n*-channel devices. TR3 or TR4 (or both), however, are in the non-conducting state with this input, since they are both p-channel devices. Therefore, the output potential rises to $+V_{DD}$ and the gate output is at logical 1, as illustrated by the equivalent switching diagram in *Fig. 3(a)*.

(b) A and B both at logical 1

When both input terminals A and B are held at logical 1 (3 V–15 V), TR1 and TR2 are biased into the non-conducting state since they are both n-channel devices. TR3 and TR4, however, are biased into conduction with this input, since they are both p-channel devices. Therefore, the output potential falls to 0 V and the gate output is at logical 0, as illustrated by the equivalent switching diagram, *Fig. 3(b)*. (See Problems 11 to 16 and 20.)

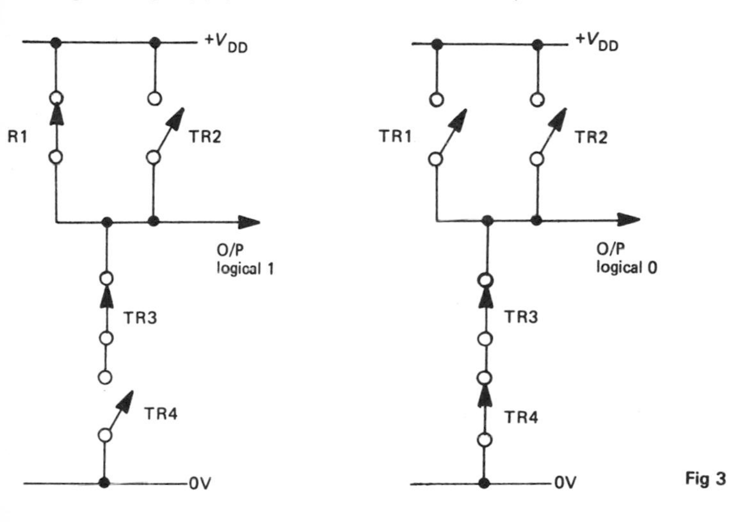

Fig 3

4 For applications that require very fast switching speeds (>200 MHz), a non-saturating type of logic, known as **emitter-coupled logic (ECL)** may be used. The basic circuit for this type of logic consists of two transistors with their emitter circuits coupled – hence the term 'emitter coupled logic' – and a circuit of this type is shown in *Fig. 4*.

The operation of this circuit may be studied by considering TR1 and TR2 operating under the following conditions:

(a) TR1 and TR2 conducting equally,

$$I_1 = I_2 = \tfrac{1}{2}I_S \quad \text{and} \quad V_1 = V_2$$

(b) TR1 conducting very much more than TR2

$$I_1 \approx I_S, \qquad I_2 \approx 0,$$
$$\text{and} \qquad V_1 \approx -(I_S \times R_L), \qquad V_2 \approx 0$$

(c) TR1 conducting very much less than TR2

$$I_1 \approx 0, \qquad I_2 \approx I_S$$
$$\text{and} \qquad V_1 \approx 0, \qquad V_2 \approx -(I_S \times R_L)$$

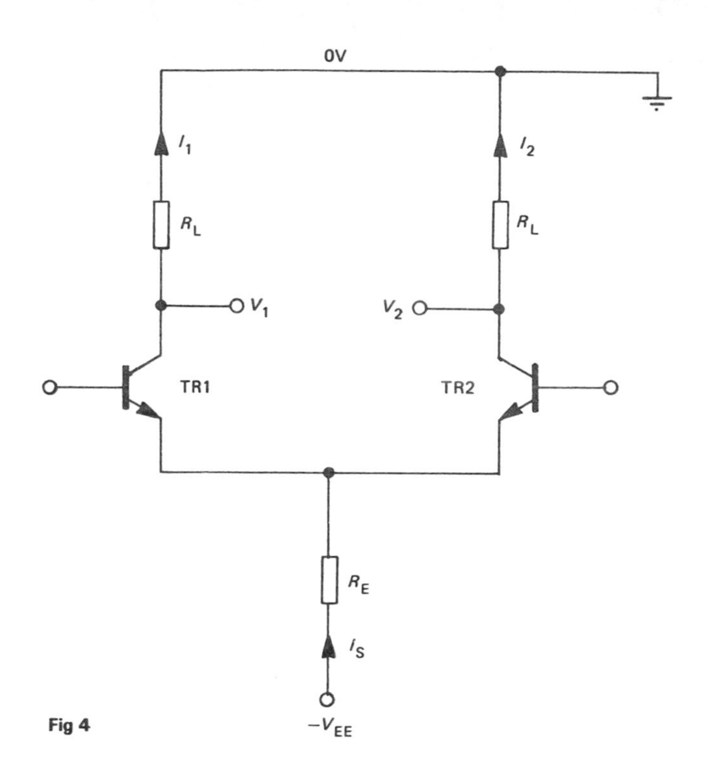

Fig 4

Thus, current I_S switched between TR1 and TR2 according to the conduction of TR1 relative to TR2, and for this reason, ECL is often known as **current mode logic (CML)**.

Note, for positive logic, 0 V corresponds to logical 1 and $-(I_S \times R_L)$ V corresponds to logical 0.

The circuit of a basic OR/NOR gate using ECL technology is shown in *Fig. 5*.

This circuit operates in the following manner:

(a) One or more inputs A, B or C at logical 1

When one of the input terminals *A*, *B* or *C* is held at logical 1 (−0.75 V), TR1, TR2 or TR3 conduct and current I_1 increases, causing an increase in p.d. across R_3. Due to the emitter coupling, TR4 emitter potential rises towards its fixed base potential ($-V_{BB}$) and causes this transistor to cut off, with the result that I_2 falls to zero. Therefore, V_1 falls to logical 0 ($-(V_{BB}+0.6$ V)) and V_2 rises to logical 1 (0 V).

(b) All inputs A, B and C at logical 0

When all input terminals, *A*, *B* and *C* are held at logical 0 (−1.55 V), TR1, TR2 and TR3 are all in a non-conducting state, and I_1 falls to zero, causing the p.d. across R_3 to fall also. Due to the emitter coupling, TR4 emitter potential falls relative to its fixed base potential ($-V_{BB}$) and causes this transistor to conduct

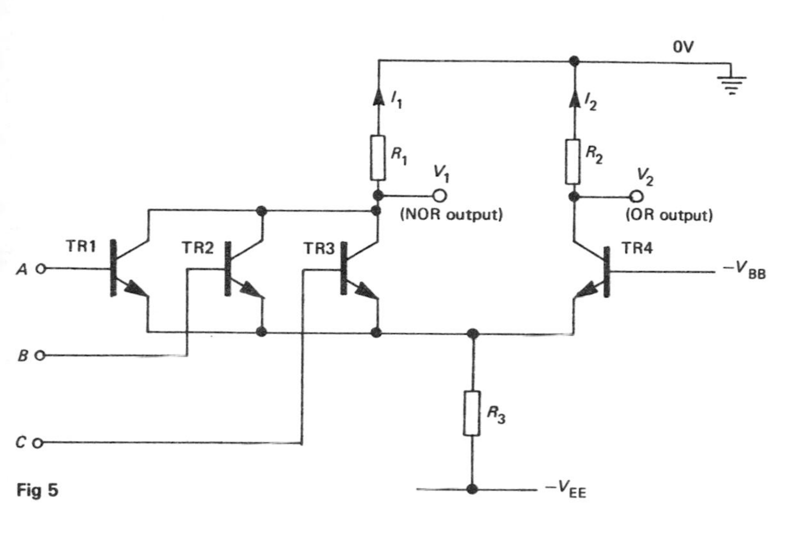

Fig 5

more heavily. Therefore, I_2 increases and all of the current is diverted through TR4 with the result that V_2 falls to logical 0 ($-(V_{BB}+0.6\text{ V})$) and V_1 rises to logical 1 (0 V).

From this description, it can be seen that V_1 acts logically as the result of NORing inputs A, B and C, and V_2 acts logically as the result of ORing inputs, A, B and C. (See Problems 17 to 20.)

B WORKED PROBLEMS ON LOGIC FAMILIES

Problem 1 Draw the circuit diagram of a dual input TTL NAND-gate.

A typical dual input TTL NAND-gate is the type 7400, and its circuit diagram is shown in *Fig. 6*.

Problem 2 Explain the reason for using an **active load** in a TTL gate output circuit.

One of the factors which limit the switching speed of a gate is capacitance across its output. This capacitance (formed by junction capacitance of the transistors in the gate circuit and stray wiring capacitance) increases greatly as the number of gate inputs connected to an output is increased.

The circuit of a gate output stage which uses a passive load is shown in *Fig. 7*.

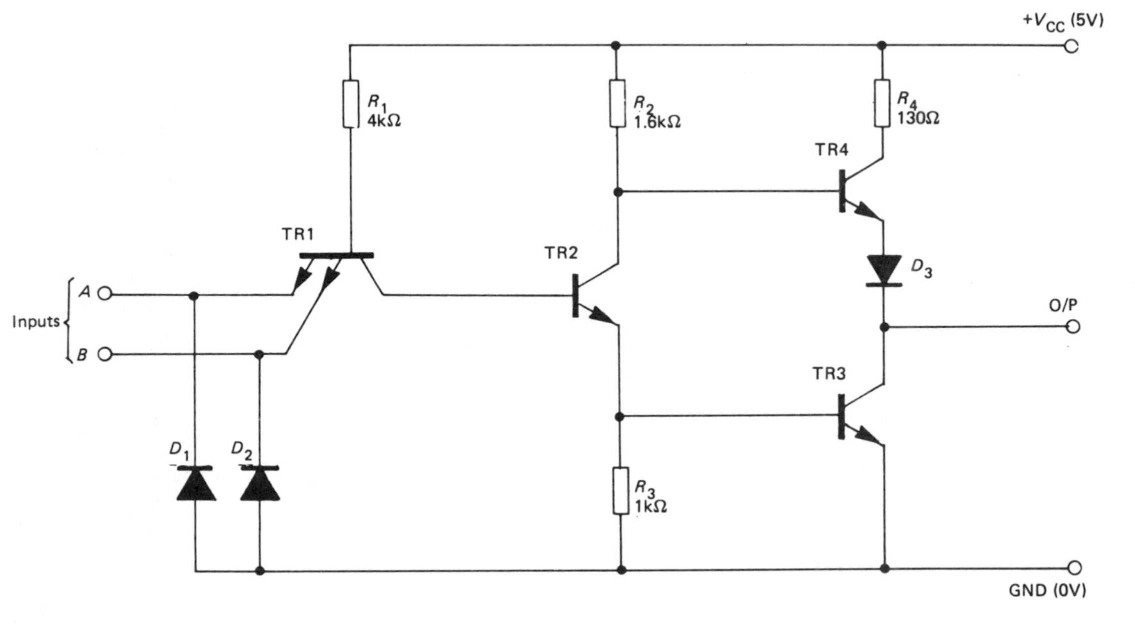
+V_{CC} (5V)
R_1
4kΩ
R_2
1.6kΩ
R_4
130Ω
TR4
TR1
TR2
D_3
A
Inputs
B
O/P
TR3
D_1
D_2
R_3
1kΩ
GND (0V)

Fig 6

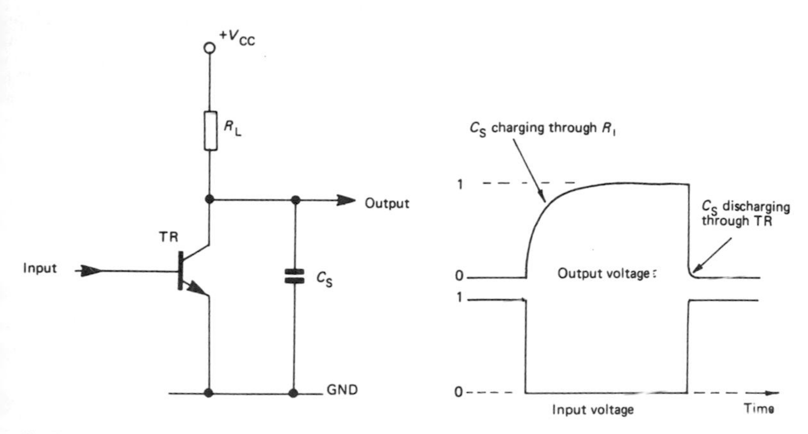

Fig 7

The total circuit capacitances are considered to be lumped together across TR outputs and depicted as C_S in *Fig. 7*.

When a low-to-high transition occurs at the output of this circuit, the output potential is prevented from changing instantaneously, since C_S must first be charged through R_L.

On a high-to-low transition, however, this problem does not occur, since TR is saturated and its low resistance is able to discharge C_S rapidly.

One possible solution to speed up the low-to-high transition is to reduce the value of R_L and thus reduce the charging time for C_S. This greatly increases the power dissipation in the circuit when TR is saturated, which is undesirable. The ideal solution is to have a variable value for R_L, such that it has a high resistance for high-to-low output transitions (TR saturated), but a low resistance for low-to-high output transitions (TR cut off). Such a solution may be implemented by replacing R_L with a transitor, known as an **active load**, and vary its conduction according to circuit requirements. A circuit of this type is shown in *Fig. 8*, and is frequently called a **totem pole** output stage.

The circuit is arranged so that when TR2 conducts for a low-to-high output transition, TR1 is non-conducting (high resistance) and C_S charges (*Fig. 8(a)*). When TR2 is cut off for a high-to-low output transition, TR1 is saturated (low resistance) so that C_S is rapidly discharged (*Fig. 8(b)*). Driving signals V_1 and V_2 must be at opposite logic levels at any instant in time to achieve correct operation.

Problem 3 A practical **totem pole** output stage of a TTL gate is shown in *Fig. 6*. Explain the function of: (a) R_4 and (b) D_3.

(a) Due to charge storage effects, it takes longer for a transistor to switch from saturation to cut off than it does to switch from cut off to saturation. Therefore, when there is a change in logic state at the gate output, both

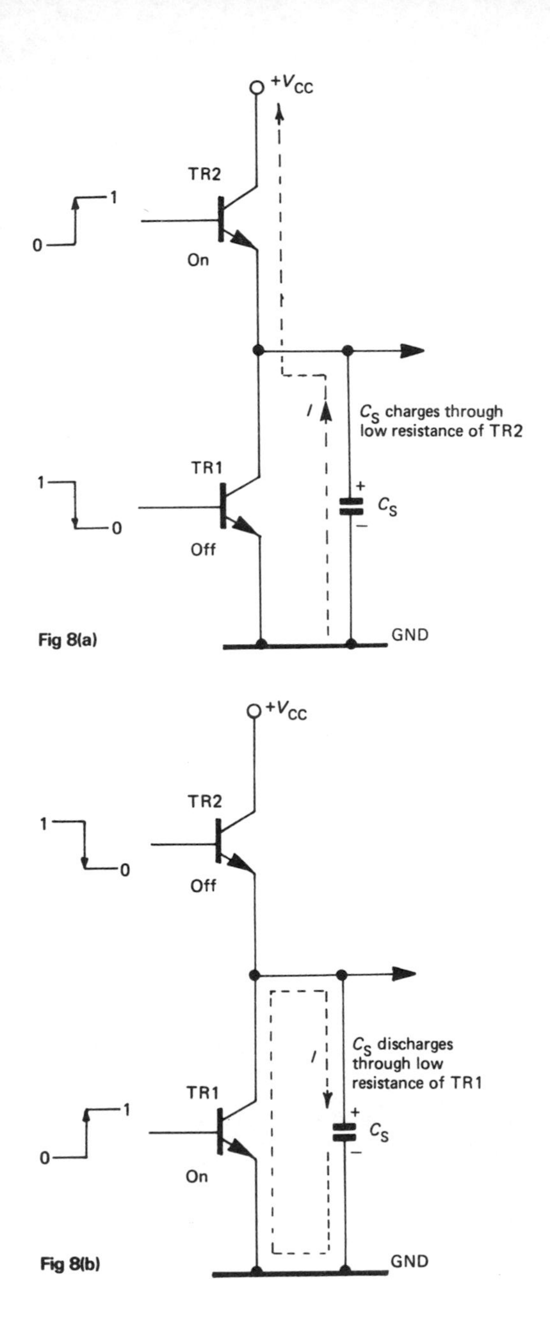
+V_{CC}
TR2
1
0
On
C_S charges through low resistance of TR2
I
TR1
1
0
Off
+
C_S
−
GND
Fig 8(a)
+V_{CC}
TR2
1
0
Off
C_S discharges through low resistance of TR1
I
TR1
1
0
On
+
C_S
−
GND
Fig 8(b)

transistors TR3 and TR4 simultaneously saturate for a short period of time. The function of R_4 is to limit the supply current during the period when both TR3 and TR4 are saturated to prevent excessive disturbance of the supply voltage V_{CC}.

(b) When the gate output is at logical 0, TR2 and TR3 are saturated (on) and TR4 is cut off (off). These conditions are illustrated in *Fig. 9*.

From *Fig. 9* it can be seen that:

(i) TR4 base potential = TR2 collector potential = $V_{CE_{SAT}} + V_{BE_{SAT}}$

$$= 0.3\text{ V} + 0.7\text{ V} = \mathbf{1.0\text{ V}}$$

and (ii) if D_3 is considered to be out of circuit,

TR4 emitter potential = TR3 collector potential = $V_{CE_{SAT}} = \mathbf{0.3\text{ V}}$

Therefore the potential difference between TR4 base and emitter (0.7 V) is sufficient to cause it to conduct. This is unacceptable, since TR3 is also conducting and a low resistance path across the supply results.

If D_3 is included in series with TR4 emitter, an additional 0.6 V is required on TR4 base to cause conduction, i.e., 1.6 V total. Since TR2 collector potential is only 1.0 V, this is insufficient to cause TR4 to conduct.

When the gate output is at logical 1, D_3 has little effect since the gate output is required to source only a very small current (≈40 μA).

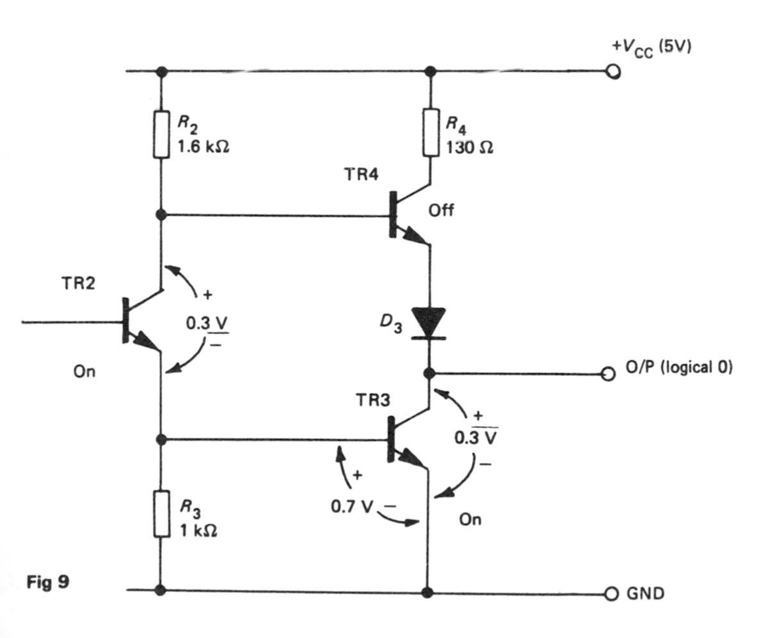

Fig 9

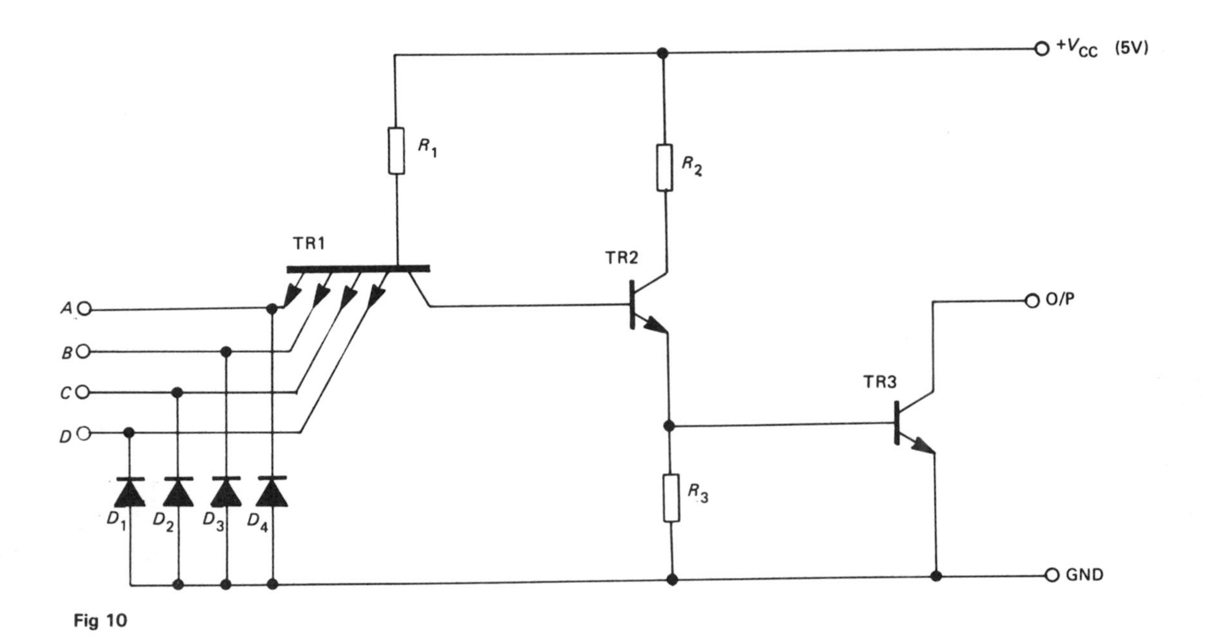

Fig 10

Problem 4 Draw the circuit diagram of a quadruple input TTL NAND gate with open collector output.

A typical quadruple TTL NAND gate with open collector output is the type 74LS22, and the circuit diagram of this gate is shown in *Fig. 10*.

Problem 5 Explain the reason for using logic gates with open collector outputs.

On occasions it may be necessary to connect the outputs of two (or more) TTL gates in parallel. Two ordinary TTL outputs must not be connected in parallel, since they may try to force each other to opposite logic levels and consequently cause damage. This situation is illustrated in *Fig. 11*.

Logic gates *A* and *B* have their outputs connected in parallel. If the input conditions on gate *A* are such as to cause its output to become logical 1, TR1 is saturated and TR2 is cut off. If the input conditions on gate *B* cause its output to become logical 0, TR3 is cut off and TR4 is saturated. When these conditions occur simultaneously, an excessively large current flows through TR1 and TR4 and this may lead to damage in either or both gates.

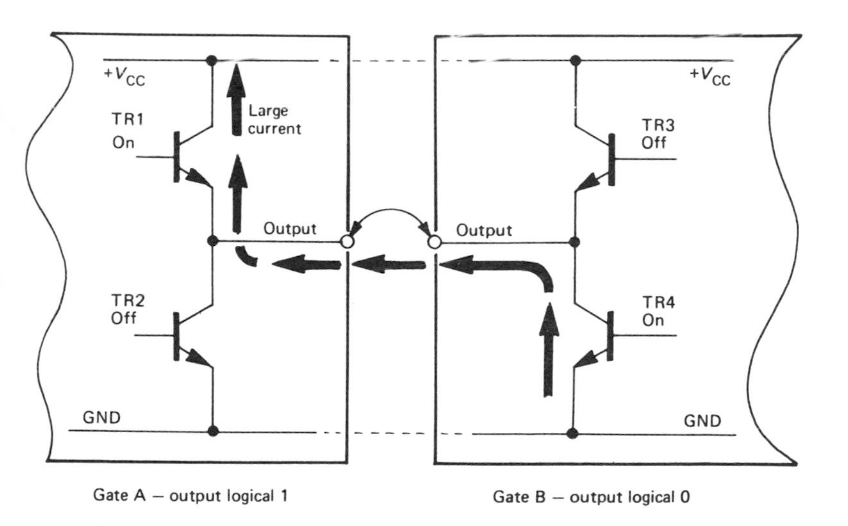

Fig 11

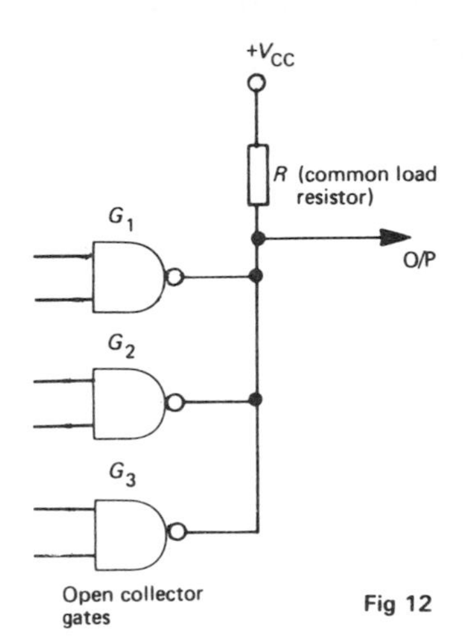

Fig 12

Open collector gates are available in which the load resistor of the output stage is missing and must be provided external to the gate IC by its user. The outputs of these gates may be connected in parallel and supplied by a common load resistor (R) (see *Fig. 12*).

With this configuration, the output is at logical 0 if any one of the outputs of G_1, G_2 or G_3 is at logical 0 level. This arrangement is frequently known as **wired-OR** output, although this term is somewhat misleading, since logically this does not give the same result as connecting G_1, G_2 and G_3 outputs to a conventional OR gate.

Problem 6 Describe how the logic levels used in practical TTL circuits are derived.

TTL circuits operate with a supply potential (V_{CC}) of $+5$ V, and for this reason it is often assumed that the logic levels used are $+5$ V for logical 1 and 0 V for logical 0.

This may well be the case for externally derived signals, but when considering interconnected TTL circuits a different situation exists. Logic levels are defined as bands of voltage levels rather than fixed values, and the reason for this is that it is not practicable to be precise because of supply variations and circuit loading effects.

When a TTL output is used to source an external load (see *Fig. 13(a)*), the logic 1 output voltage level (V_{OH}) may be expressed as:

(a) $V_{OH} = V_{CC} - (I_B \times R_2) - V_{BE} - V_{D3}$

Under most conditions the term $I_B \times R_2$ is very small and may be ignored. Terms V_{BE} and V_{D3} represent volt drops across two forward biased diodes in the output structure, therefore equation (a) may be simplified to:

(b) $V_{OH} = V_{CC} - 2V_{BE}$

The value of V_{BE} used depends upon the output load current value (I_L) and the ambient temperature. Typical values for V_{BE} are quoted in *Table 1*.

When a TTL output is used to sink current from an external load, the logic 0 output voltage level (V_{OL}) may be expressed as:

$V_{OL} = V_{CE_{SAT}}$ (see *Fig. 13(b)*)

The value of $V_{CE_{SAT}}$ varies from 0.2 V (small load current) to 0.4 V (max. load

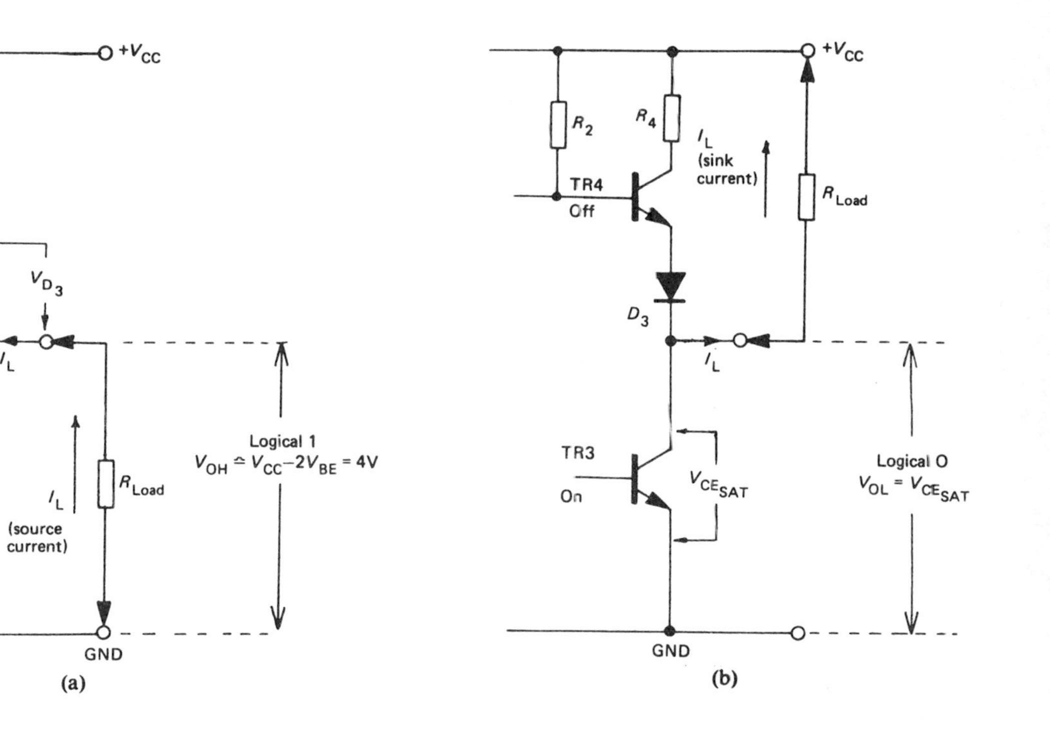

$+V_{CC}$
I_B
$I_B \times R_2$
R_2
R_4
TR4
On
V_{BE}
V_{D_3}
D_3
I_L
TR3
Off
I_L
(source current)
R_{Load}
Logical 1
$V_{OH} \simeq V_{CC} - 2V_{BE} = 4V$
GND
(a)
$+V_{CC}$
R_2
R_4
I_L
(sink current)
R_{Load}
TR4
Off
D_3
I_L
TR3
On
$V_{CE_{SAT}}$
Logical 0
$V_{OL} = V_{CE_{SAT}}$
GND
(b)

Fig 13

Table 1

Load current	V_{BE} *at 0°C*	V_{BE} *at 25°C*
10^{-2} mA	450 mV	500 mV
10^{-1} mA	500 mV	550 mV
1 mA	550 mV	600 mV
10 mA	620 mV	670 mV
100 mA	720 mV	770 mV

current of 16 mA). The voltage levels required at a gate input for a particular logic output level may be determined by reference to the TTL NAND gate circuit shown in *Fig. 6*.

To obtain an output of logical 0 from this gate, TR2 and TR3 must be saturated, and TR1 must be cut off with its collector/base junction forward biased. A minimum input potential of 2.4 V is required on both inputs *A* and *B* to maintain this condition.

To obtain an output of logical 1, TR2 and TR3 must be cut off and TR1 must be saturated. A maximum input potential of 0.8 V is required on either input *A* or *B* to maintain this condition.

A safety margin of 0.4 V is adopted, so that the maximum logical 0 level at a gate output is specified as 0.4 V less than the maximum logical 0 level permitted at a gate input. Similarly, the minimum logical 1 level at a gate output is specified as 0.4 V more than the minimum logical 1 level permitted at a gate input. These levels are illustrated in *Fig. 14*.

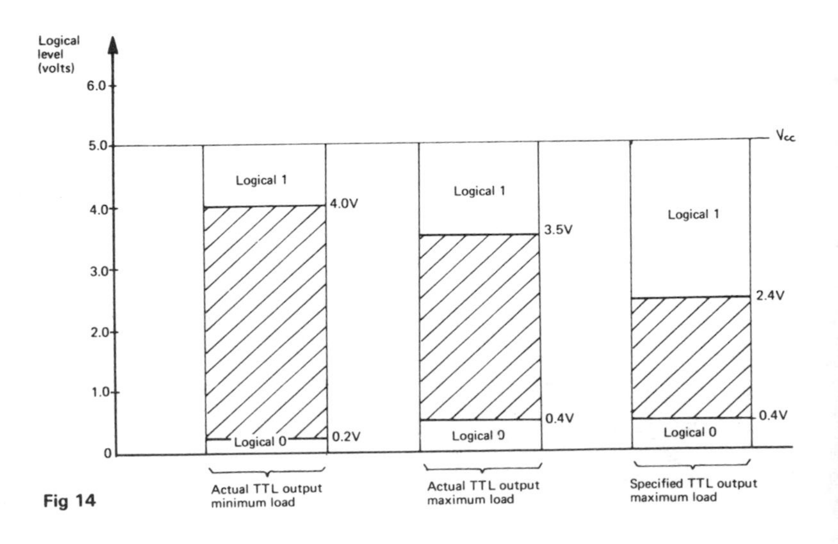

Fig 14

Problem 7 With regard to TLL gate circuits, explain: (a) why unused inputs must not be left 'floating', and (b) how unused inputs may be dealt with in practical logic circuits.

(a) Current pulses due to switching in one part of a logic circuit may cause voltages to be induced in other parts of the circuit. Such induced voltages are known as **noise**, and any noise pulse of sufficient amplitude may cause an unwanted change in logic level if it is allowed to reach the input of a logic circuit. This may cause circuit malfunction.

An unconnected or floating TTL input has a potential of approximately 1.6 V, and although this normally acts as a logical 1, it is an undefined level (see *Fig. 14*) and is therefore not consistently predictable. Small changes in input voltage may cause a logical input in the undefined region to swing to either logical 0 or logical 1. Therefore, unconnected TTL inputs are more likely to be affected by noise pulses than those inputs which are connected to well-defined logic levels.

The amplitude of a noise pulse necessary to cause a logical 0 or a logical 1 input to be driven into the undefined region is known as the circuit **noise margin**, and this is typically 400 mV (0.4 V) for TTL circuits. Unused TTL inputs should therefore not be left unconnected if noise problems are to be avoided.

(b) An unused TTL input may be dealt with in one of the following ways:
   (i) connect it to a valid logical 0 or logical 1 level, as appropriate, so that normal circuit operation with the used inputs is maintained (see *Fig. 15(a)*),
   (ii) connect it in parallel with a driven input, providing the fan out of the driving gate is not exceeded (see *Fig. 15(b)*), or
   (iii) connect it to another input (or output) which is permanently held at the appropriate logic level (see *Fig. 15(c)*).

Problem 8 Draw the circuit diagram of a TTL NOR gate and briefly explain its operation.

A typical dual input TTL NOR gate is the type 7402, and its circuit diagram is shown in *Fig. 16*.

This circuit operates in the following manner:

(a) Inputs *A* and *B* both at logical 0

TR1 and TR4 are both saturated and their collector potentials fall to almost 0 V. This causes TR2 and TR3 to both be cut off and their collector potential rises to $+V_{CC}$ thus biasing TR5 into conduction. TR2/TR3 emitter potential falls to 0 V, thus causing TR6 to be cut off. Therefore the gate output potential rises towards $+V_{CC}$, i.e. logical 1.

(b) Input *A* or input *B* (or both) at logical 1

The base/collector junction of either TR1 or TR4 (or both) becomes forward biased, and a large base current flows in either TR2 or TR3 (or both). Therefore TR2 or TR3 (or both) conduct heavily and TR2/3 emitter potential rises, biasing TR6 into conduction. TR2/3 collector potential falls, and this biases TR5 to cut off. Therefore the gate output potential falls towards 0 V, i.e. logical 0.

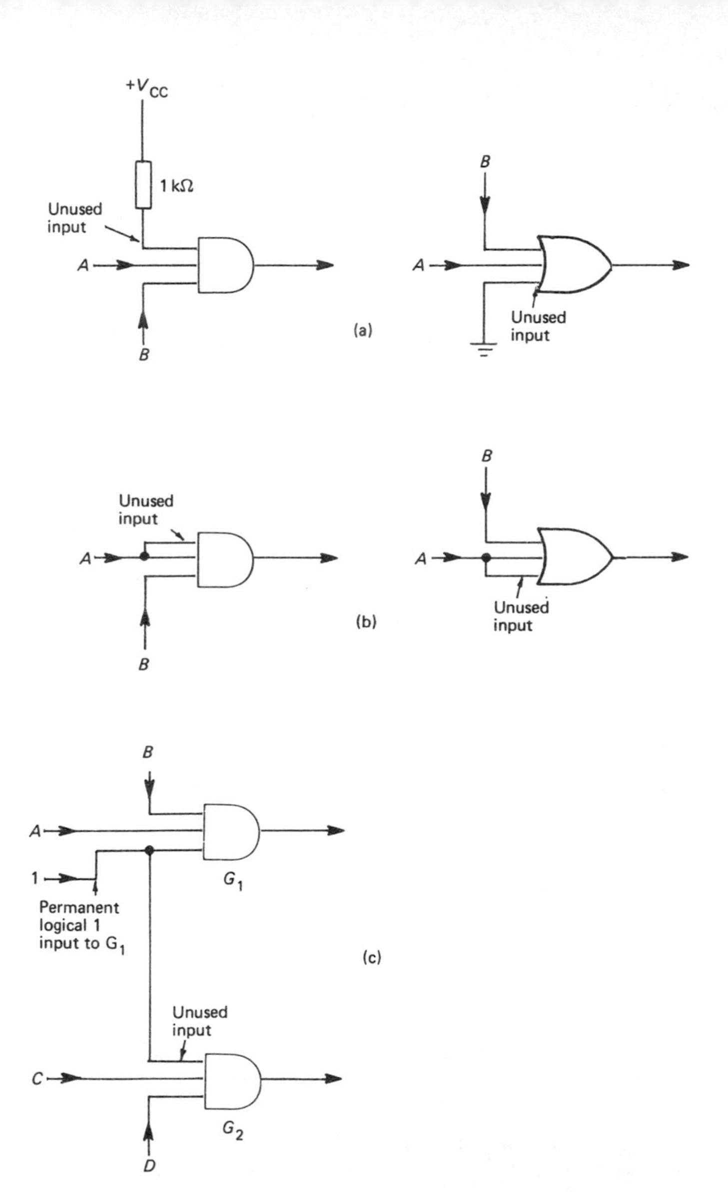
+V_{CC}
1 kΩ
Unused input
A
B
(a)
B
A
Unused input
Unused input
A
B
(b)
B
A
Unused input
B
A
1
G_1
Permanent logical 1 input to G_1
(c)
Unused input
C
G_2
D

Fig 15

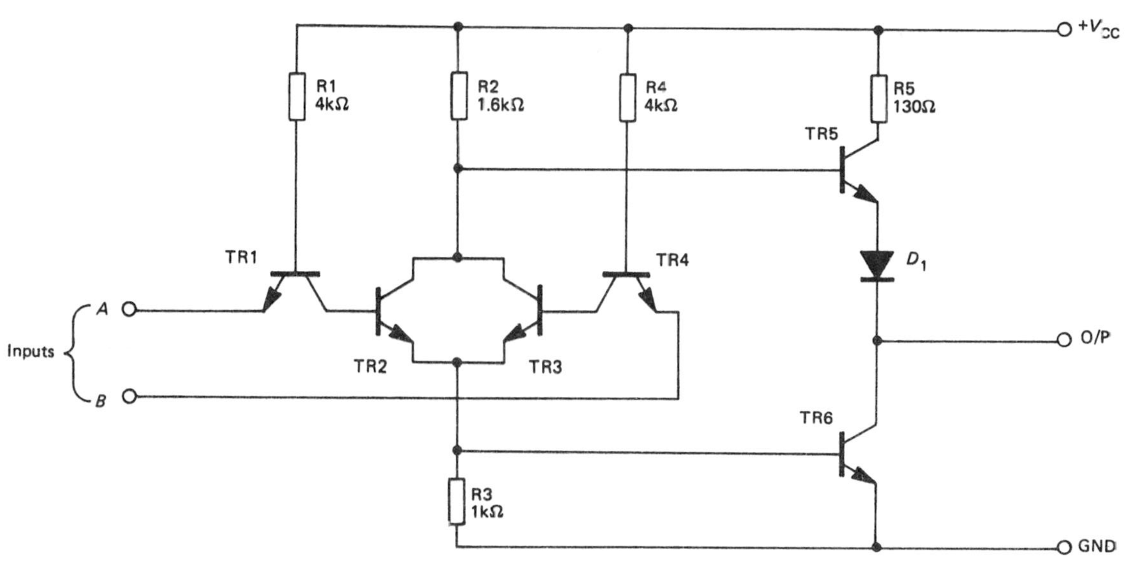
R1
4kΩ
R2
1.6kΩ
R4
4kΩ
R5
130Ω
+V_{CC}
TR5
TR1
TR4
D_1
A
Inputs
B
TR2
TR3
O/P
TR6
R3
1kΩ
GND

Fig 16

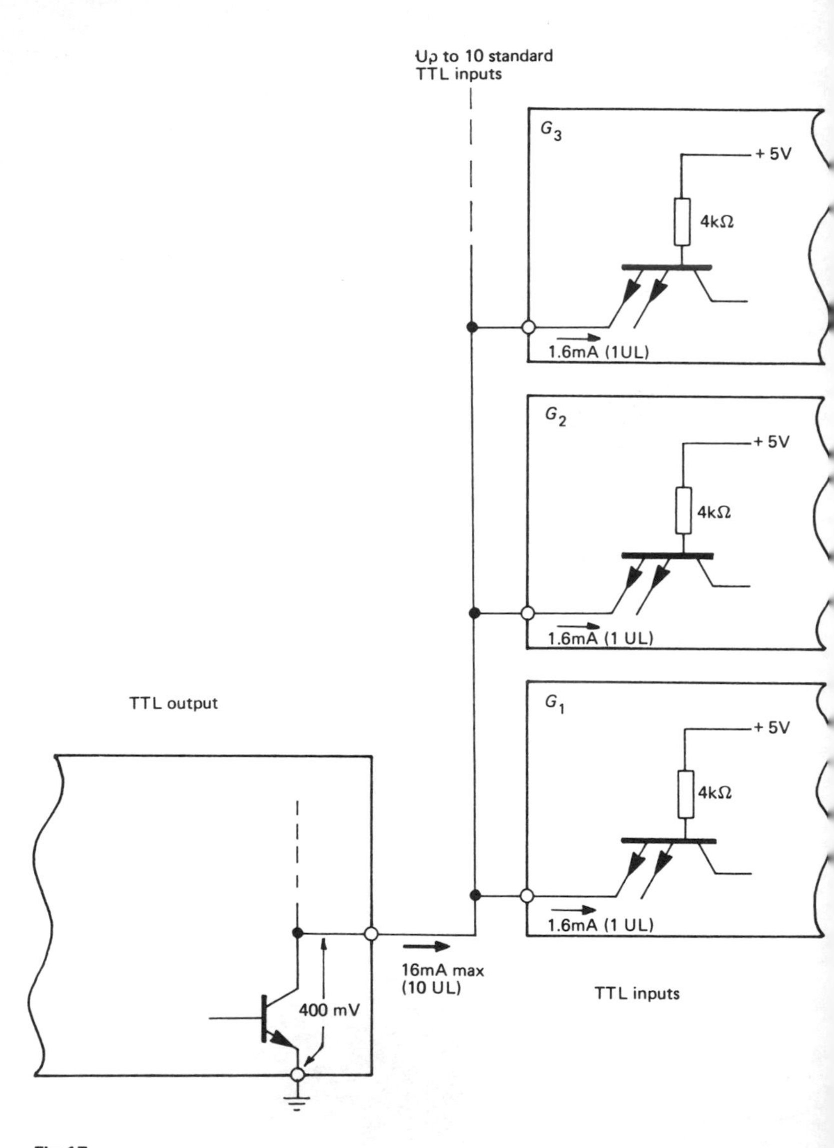
Up to 10 standard TTL inputs
G3
+ 5V
4kΩ
1.6mA (1UL)
G2
+ 5V
4kΩ
1.6mA (1 UL)
TTL output
G1
+ 5V
4kΩ
1.6mA (1 UL)
400 mV
16mA max (10 UL)
TTL inputs

Fig 17

Problem 9 Explain the meanings of the term **unit load** and **fan out** in relation to TTL circuits.

The load imposed by a single, standard TTL input on the output of a circuit to which it is connected is known as a **unit load (UL)**. For a standard TTL input, this represents a load current of 1.6 mA at a logical 0 level of 400 mV. The load imposed by other circuits may be expressed in terms of unit loads.

The **fan out** of a TTL circuit refers to the number of TTL gate inputs that may be connected to a single TTL output. A TTL output transistor is required to sink a current of 1.6 mA for each standard TTL input connected to it when in the logical 0 state (see *Fig. 17*). An increase in the current through this output transistor causes a rise in $V_{CE_{SAT}}$, therefore there is a limit to the number of inputs that may be connected to a TTL output, otherwise the logical 0 level may increase to a value which puts it into the undefined region.

For a standard TTL circuit, the maximum number of unit loads that may safely be connected to a single TTL output is 10, therefore this type of circuit has a fan out of 10. The fan out of open collector TTL circuits is usually 30. The loads imposed by various inputs are shown in *Fig. 18*.

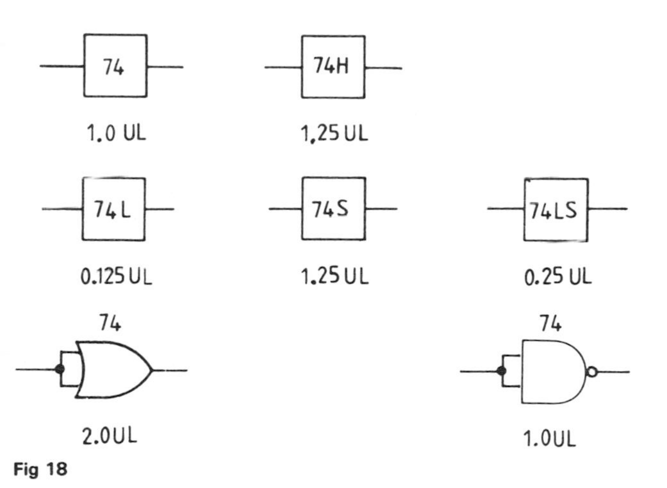

Fig 18

Problem 10 Briefly explain the main differences between the following types ot TTL devices: (a) 74XX; (b) 74HXX; (c) 74LXX; (d) 74SXX; (e) 74LSXX.

In an ideal logic family, the switching speeds should be as high as possible, and the power consumption as low as possible.

In a TTL circuit, use is made of the various resistors in the circuit to charge and discharge various transistor and stray capacitances, and the time constants so formed ($C \times R$) determine the maximum switching speed. Also, since one or more transistors are conducting at any given time in a TTL circuit, current paths are provided through the circuit resistors which keep the power consumption of such circuits high.

The various TTL families available offer variations on the conflicting requirements of low power consumption and fast switching times.

(a) **74XX series** This is the **standard TTL** family, and represents a compromise between low power consumption and high switching speed. Typical power consumption is 10 mW per gate and switching time is 10 ns.

(b) **74HXX series** This is a **high speed TTL** family, and the faster switching speed is obtained by using lower resistance values throughout to shorten the internal *CR* time constants. This inevitably results in higher power consumption. Typical power consumption is 22 mW per gate and a switching time of 6 ns.

(c) **74LXX series** This is a **lower power consumption TTL** family. The reduction in power consumption is obtained by increasing the resistor values throughout. This inevitably results in slower switching speeds, since the internal *CR* time constants are increased. Typical power consumption is 1 mW per gate and switching time is 33 ns.

(d) **74SXX (Schottky) series** The speed of switching of heavily saturated bipolar transistors is limited due to the problem of excess stored charge carriers. If transistors are operated on the edge of saturation, however, the number of excess stored charge carriers is considerably reduced and faster switching times are possible. This mode of operation may be implemented with the aid of **Schottky barrier diodes (SBD)**.

A Schottky diode consists of a semiconductor to metal interface rather than a *p-n* junction and therefore only one type of charge carrier is involved. This means that Schottky diodes do not suffer from charge storage effects. In addition, Schottky diodes can be manufactured with a lower forward volt-drop (400 mV) than that encountered with *p-n* junction diodes (700 mV).

A typical switching circuit with the addition of a Schottky diode is shown in *Fig. 19*. In this circuit, when TR1 is saturated, the volt-drop across the base-collector junction is limited to 400 mV by a forward biased Schottky diode (SBD). The Schottky diode has the effect of diverting the surplus base current, thus holding TR1 at the edge of saturation and preventing the build up of excess stored charge carriers in the transistor. As a result of this, and due to the fact that a Schottky diode does not suffer from storage effects, the switching speed is greatly increased.

The circuit depicted in *Fig. 19* may be implemented using Schottky transistors, and the simplified construction of such a transistor is shown in *Fig. 20*. The Schottky diode is formed by extending the base contact over the collector region to form a semiconductor to metal interface. This type of transistor may be used to construct Schottky TTL circuits which have the advantage of higher switching speeds without the penalty of excessively increased power consumption. Typical power consumption of a Schottky TTL circuit is 20 mW per gate with a switching time of 3 ns.

(e) **74LSXX series** This is a low power Schottky TTL family in which Schottky transistors are used to compensate for the reduced switching speeds that

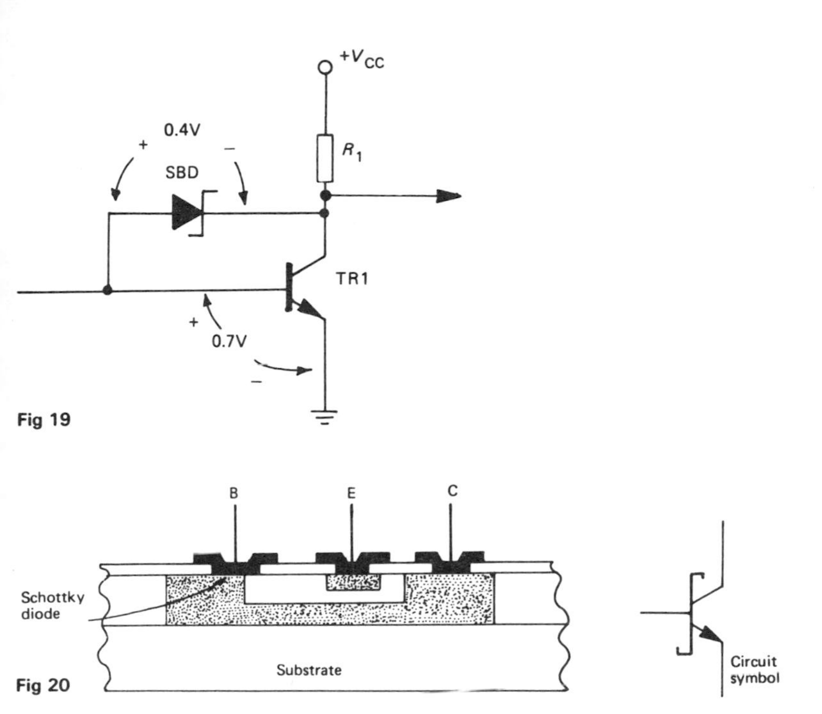

Fig 19

Fig 20

would otherwise occur when increasing resistor values throughout to obtain low power consumption. This logic family therefore has switching speeds similar to standard TTL, but with power consumption similar to that of the 74LXX series. Typical power consumption is 2 mW per gate with a switching time of 10 ns.

Problem 11 Drawing the circuit diagram of a CMOS inverter and explain its operation.

A CMOS inverter consists of a *p*-channel MOSFET and an *n*-channel MOSFET connected in series across a supply, as shown in *Fig. 21*. The *n*-channel MOSFET (TR1) acts as a switch with the *p*-channel MOSFET (TR2) as its active load.

When the input to this circuit is held at logical 1, TR1 is biased into conduction and TR2 is biased to cut off, therefore the output is at logical 0 (0 V).

When the input is held at logical 0, TR1 is biased to cut off and TR2 is biased into conduction, therefore the output is at logical 1 ($+V_{DD}$).

Since FETs do not have base-emitter or saturation volt drops, the output voltage swing of an unloaded CMOS inverter is equal to the power supply

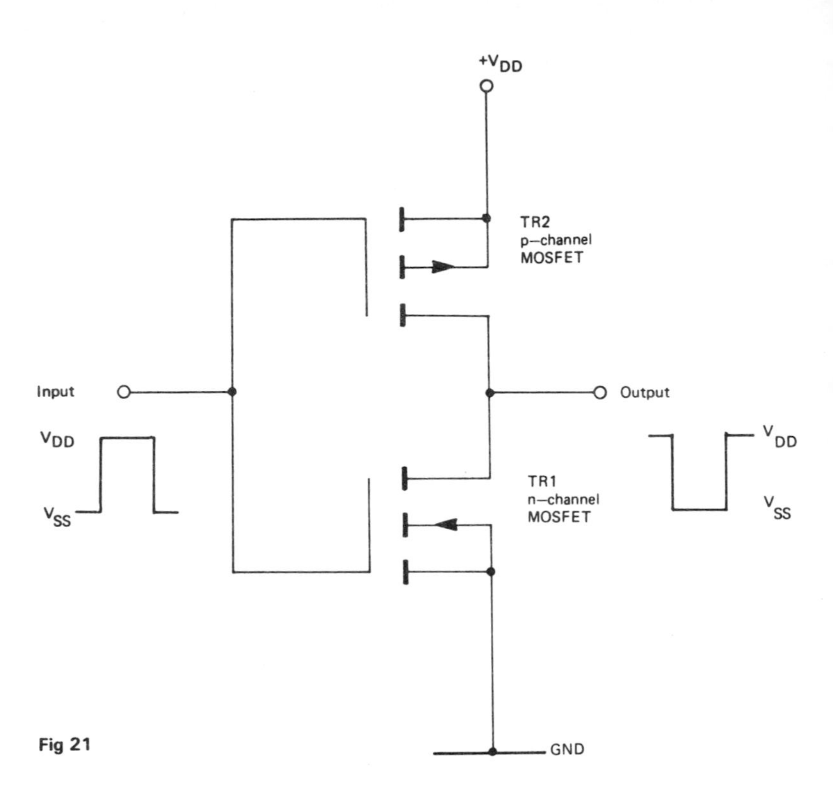

Fig 21

voltage. This is illustrated by the CMOS inverter transfer characteristic shown in *Fig. 22.*

It can be seen from this characteristic that changes in logic output level of a CMOS inverter take place at an input voltage of around 50 per cent of the power supply potential.

Problem 12 Draw the circuit diagram of a dual input CMOS NOR gate and explain its operation.

The circuit diagram of a CMOS NOR gate is illustrated in *Fig. 23*. In this circuit, parallel connected transistors TR3 and TR4 act as the switching transistors with series connected TR1 and TR2 as their active load.

It can be seen from *Fig. 23* that it is only possible to obtain an output of logical 1 if TR3 and TR4 are both cut off. An input of 0 V (logical 0) is required on both inputs *A* and *B* to achieve this condition, and this also cause TR1 and TR2 to conduct, thus pulling the output to $+V_{DD}$ (logical 1). All other input combinations cause TR3 or TR4 (or both) to conduct and TR1 or TR2 (or both) to cut off, thus giving an output of 0 V (logical 0).

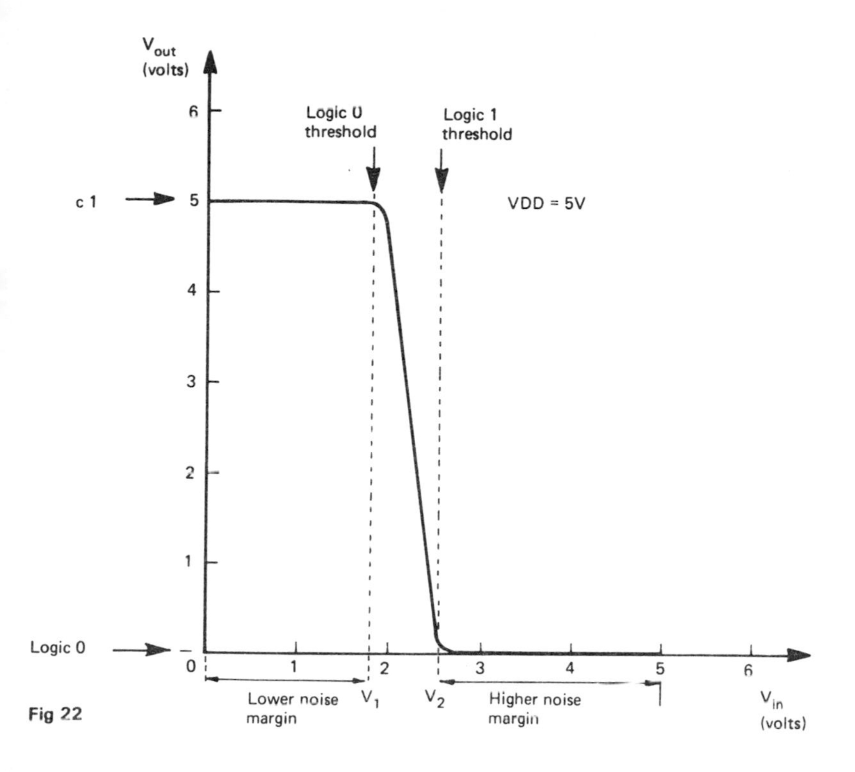

Fig 22

Problem 13 Explain why care must be exercised in handling CMOS ICs, and list the main precautions to be observed.

The inputs of CMOS logic circuits are connected to the gates of MOSFETs and are insulated from the remainder of the circuit by a very thin layer of silicon dioxide. Breakdown of this silicon dioxide layer and consequent destruction of the circuit occurs if a gate potential is allowed to rise above 50 V. Static charges of several thousand volts may build up on objects insulated from ground, and if this is allowed to happen with CMOS inputs, damage is inevitable.

CMOS inputs are usually protected against the build-up of static charges by the use of protection networks similar to that shown in *Fig. 24*. Despite this, it is still advisable to observe the following precautions when handling CMOS ICs:

(a) always store CMOS ICs with their pins shorted together by embedding them in conductive foam or aluminium foil or use a special aluminium IC carrier,
(b) do not remove ICs from their protective storage until required for insertion into a circuit,

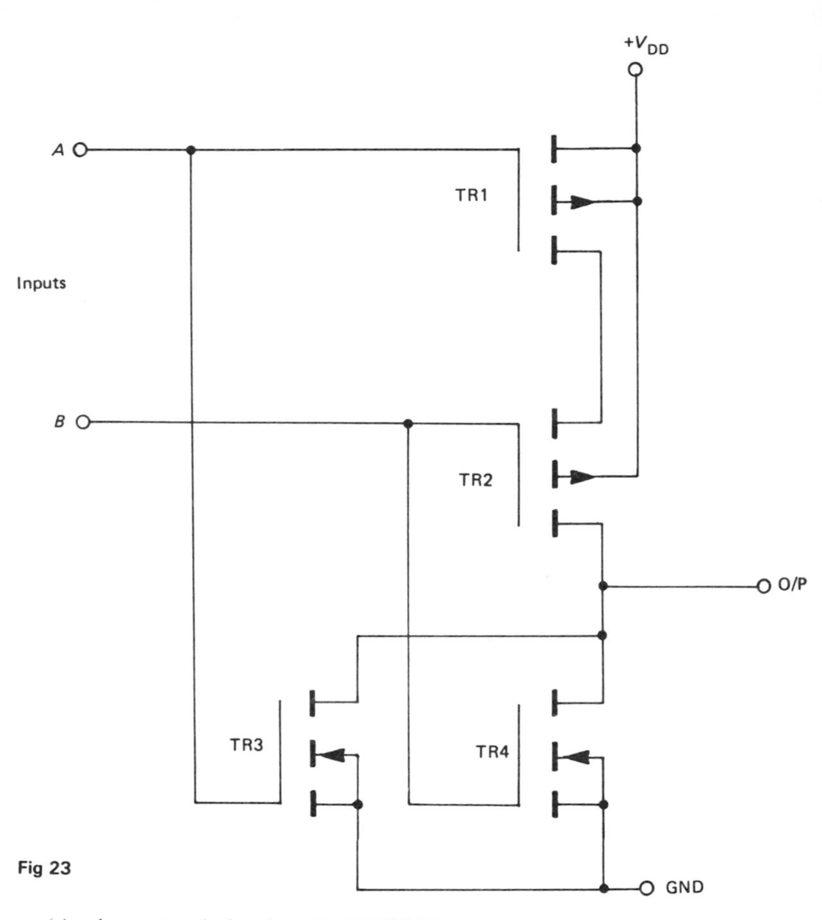

Fig 23

(c) do not touch the pins of a CMOS IC,
(d) do not install or remove a CMOS IC from circuit whilst it is powered,
(e) use a low leakage earthed soldering iron when installing CMOS ICs (better still, use IC sockets),
(f) do not apply input signals to CMOS ICs when the power is disconnected,
(g) do not leave CMOS IC inputs floating, and
(h) avoid the use of materials which encourage the build-up of static electricity, e.g. nylon for clothing or working surfaces.

Problem 14 (i) Explain why CMOS ICs have low power consumption. (ii) Show how power consumption in CMOS ICs varies with operating frequency. (iii) Compare the power consumption of a standard TTL gate with that of a CMOS gate.

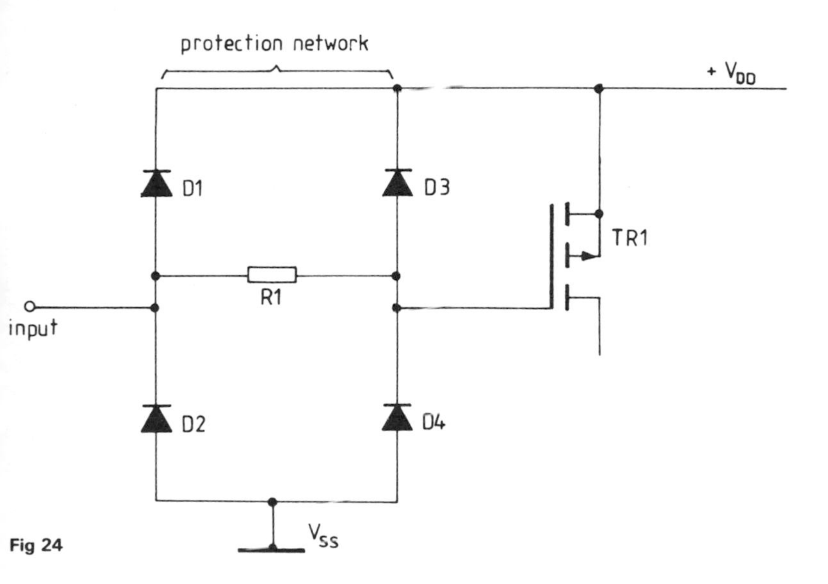

Fig 24

(a) The reason for CMOS circuits having low power consumption may be determined by considering a basic CMOS inverter circuit which forms the basis of all CMOS logic gates (see *Fig. 25*). Under steady conditions, either TR1 or TR2 is non-conducting, depending upon the output logic level. Since MOSFETs have a very high resistance between source and drain when biased off, virtually no current flows in this circuit regardless of output logic

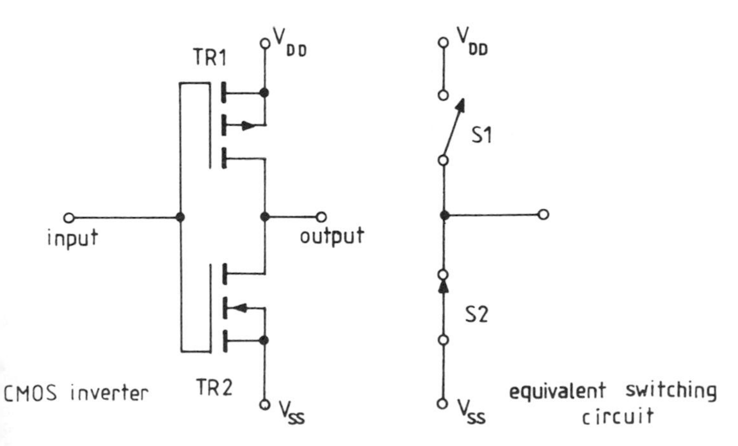

Fig 25

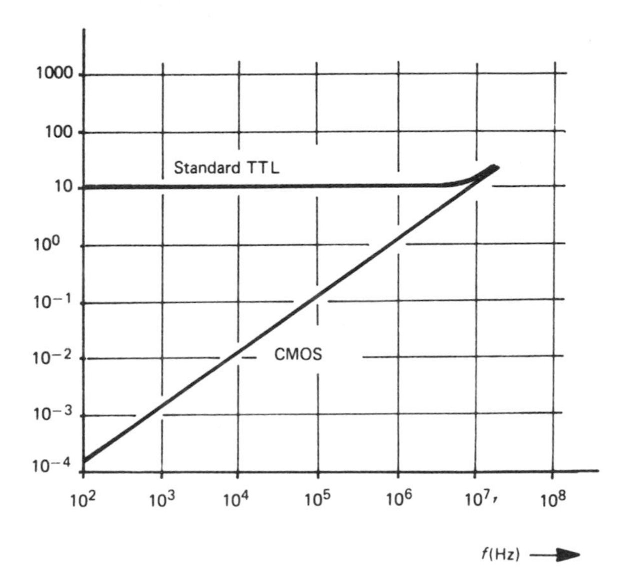

Fig 26

level. The power consumption is therefore very low ($\approx 0.5\ \mu W$ per gate at 5 V, V_{DD}).

(b) During an output transition from logical 0 to logical 1 (or vice versa), TR1 and TR2 are momentarily both conducting, and power consumption during this period is high. At low switching rates, the average power dissipated in a CMOS circuit is still very low, but as switching rates are increased power consumption rises, as shown in *Fig. 26*.

(c) In equivalent TTL circuits, at least one transistor is conducting at any instant in time, and therefore provides a current path across the supply at all times. For this reason, power consumption is relatively high under static conditions, and is largely independent of switching rate (see *Fig. 26*), except when operating close to its maximum operating speed.

Problem 15 With the aid of a diagram, explain how: (a) TTL outputs may be interfaced to CMOS inputs, and (b) CMOS outputs may be interfaced to TTL inputs.

It is sometimes necessary to use a mixture of TTL and CMOS ICs for a given application. Interfacing TTL and CMOS circuits may be achieved by the following methods:

(a) TTL output to CMOS input

No problem exists in driving CMOS inputs directly from TTL outputs since they have greater output drive capabilities than CMOS outputs. The logical

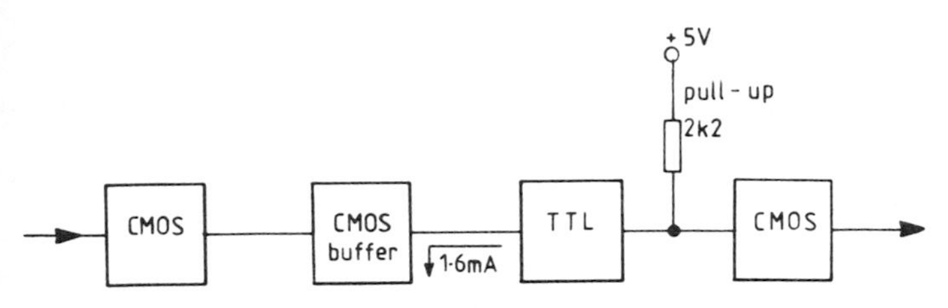

Fig 27

1 output from a TTL circuit may not be high enough to drive a CMOS input, however, but this problem may be overcome by the use of a 2K2 'pull-up' resistor between the TTL output and $+V_{CC}$ (see *Fig. 27*). This resistor ensures that the TTL logical 1 output rises to $+V_{CC}$ (5 V).

(b) CMOS output to TTL input

The output circuit of a normal CMOS IC cannot sink a current of more than 0.5 mA. A standard TTL input requires that the circuit driving it is capable of sinking 1.6 mA. Therefore it is necessary to use a CMOS buffer with this drive capability between a standard CMOS output and a standard TTL input. An example of such a buffer is the type 4049 which is capable of driving two TTL loads.

Low power (74LXX) and low power Schottky (74LSXX) TTL circuits have much lower input current requirements (see *Problem 10*) and these may be driven directly by normal CMOS outputs without further buffering.

Problem 16 Explain the difference between unbuffered (A series) and buffered (B series) CMOS circuits.

The circuit of an unbuffered CMOS NOR circuit is shown in *Fig. 23*, and the state of each MOSFET in this circuit for all possible input conditions is shown in *Table 2*.

When conducting fully, MOSFETs do not behave as perfect switches, but have a resistance of several hundred ohms. Therefore, a CMOS gate has an

Table 2

Inputs		*State of each transistor*				
B	*A*	TR1	TR2	TR3	TR4	*Output resistance*
0	0	on	on	off	off	TR1 and TR2 in series
0	1	off	on	on	off	TR3
1	0	on	off	off	on	TR4
1	1	off	off	on	on	TR3 and TR4 in parallel

output resistance of several hundred ohms, and as can be seen from *Table 2*, this varies in value according to its input conditions.

Each CMOS gate has an input capacitance of approximately 5 pF which is formed by the capacitance between its gate electrode and the substrate. When two CMOS gates are interconnected, the output resistance of one gate and the input capacitance of the following gate behave as a low-pass filter which limits the maximum usable switching rate. Therefore, with unbuffered CMOS gates, the rise and fall times of the switching signals vary according to the particular input conditions present.

The circuit of a buffered CMOS NOR gate is shown in *Fig. 28*. An inverter circuit, consisting of TR9 and TR10 is used to isolate or buffer the output

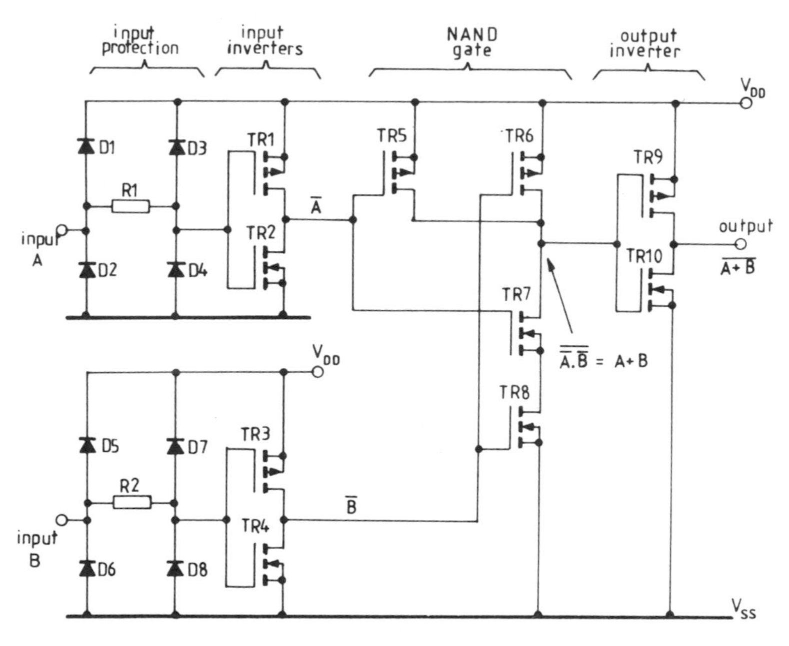

Fig 28

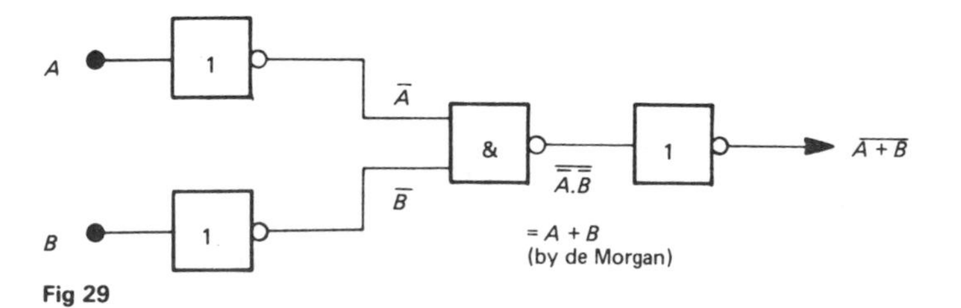

Fig 29

terminal from the actual gate circuit output. This has the advantage of providing a constant output resistance, equivalent to the source to drain resistance of whichever MOSFET is conducting. Therefore the output resistance of a buffered CMOS gate does not vary according to its input conditions. Note that since the buffer causes inversion of the output logic level, the logic element used in *Fig. 28* is a NAND gate with inverted inputs, i.e., a direct application of de Morgan's theorem is used (see *Fig. 29*).

Problem 17 Draw the circuit diagram of emitter-coupled logic (ECL) OR/NOR gate.

The circuit diagram of an ECL OR/NOR gate is shown in *Fig. 30*.

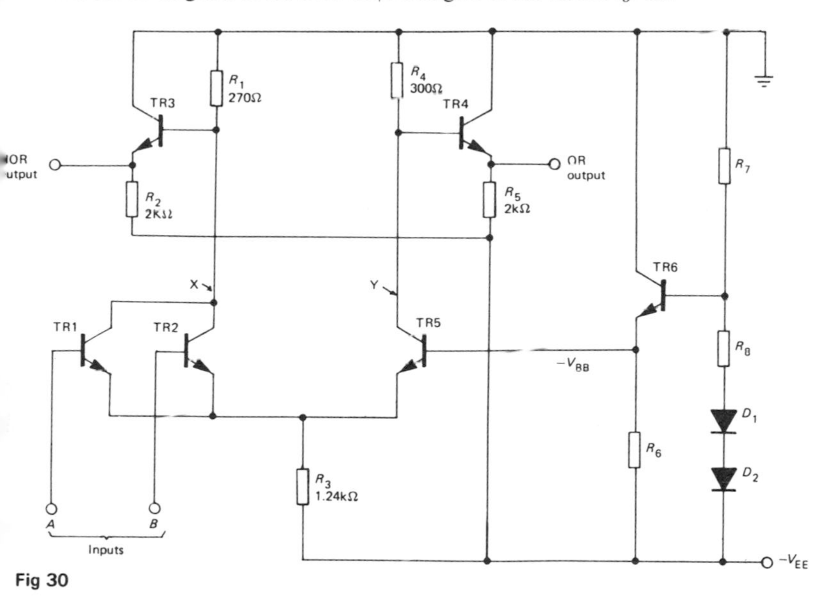

Fig 30

Problem 18 Describe the action of each transistor stage in *Fig. 30*.

TR1, TR2 and TR3 form a basic ECL OR/NOR gate. When inputs A and B are both held at logical 0 ($-V_{EE}$), TR1 and TR2 are cut off and point X rises towards 0 V (logical 1). TR5 base-emitter junction is forward biased by $-V_{BB}$, therefore TR5 conducts and point Y falls towards $-V_{EE}$ (logical 0). A p.d. is developed across R_3 due to the current through TR5, and this is used to bias the emitters of TR1 and TR2.

If either TR1 or TR2 base (or both) is increased in potential towards 0 V (logical 1), collector current flows and causes point X to fall towards $-V_{EE}$ (logical 0). The current through TR1 and TR2 (or both) causes extra current to flow through R_3 and the p.d. across this resistor increases. TR5 emitter potential rises towards $-V_{BB}$ and causes a reduction in its base-emitter potential. This causes TR5 to conduct less heavily and point Y rises towards 0 V (logical 1). From this description it may be seen that point X behaves as a NOR output, and point Y behaves as an OR output.

TR3 and TR4 behave as output emitter followers. A problem arises if the ECL gate shown in *Fig. 30* is used to drive the inputs of a similar ECL gate directly. The collectors of the driving transistors (points X and Y) are connected to the bases of similar circuits, which is clearly unacceptable. Emitter follower transistors TR3 and TR4 are used to shift the output voltage level to a value which is suitable for driving other ECL gates. The use of emitter followers also results in a low output impedance which improves the fan out capabilities of an ECL gate.

TR6 and its associated components form a bias compensating circuit. The noise margin of an ECL circuit may vary widely due to temperature variations, causing changes in circuit parameters. The effect can be minimized by using a temperature compensated supply for TR5 base ($-V_{BB}$).

An increase in temperature causes TR5 to conduct more heavily and changes the logic thresholds at inputs A and B. An increase in temperature also causes D_1 and D_2 to have a lower forward resistance, and this causes a decrease (more negative) in potential in TR6 base. This, in turn, causes a decrease in TR6 emitter potential ($-V_{BB}$), which offsets the change in conditions in the ECL circuit caused by the temperature increase. A temperature decrease has the opposite effect.

Problem 19 Explain the logic elements used in practical ECL circuits.

Transistors in ECL circuits do not saturate, but in order to obtain the maximum output voltage swing, it is desirable to operate between cut off and the edge of saturation (i.e. up to the point just before the collector-base junction becomes forward biased).

Typical voltages encountered in ECL circuits are:

$$V_{CC} = 0\text{ V},$$
$$V_{EE} = -5.2\text{ V}$$
$$V_{BB} = -1.15\text{ V}$$
and $$V_{BE_{SAT}} = 0.75\text{ V}$$

The circuit diagram of part of an ECL gate circuit is shown in *Fig. 31*. Using this circuit, various parameters may be calculated to determine the actual logic levels at its output terminal.

The logical 0 output condition, when TR5 is conducting heavily may first be

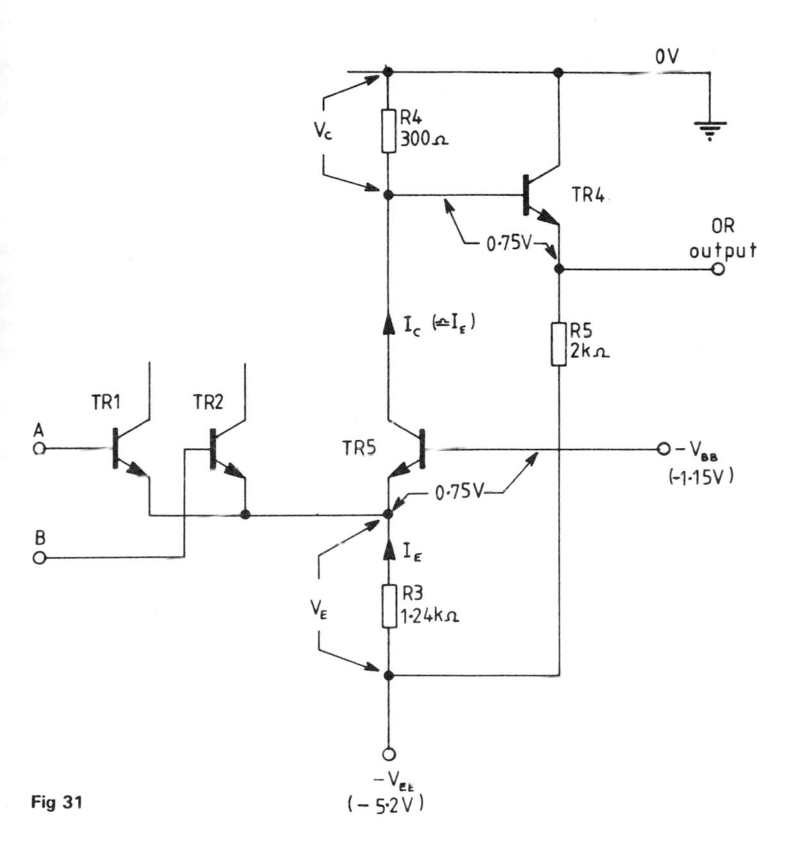

Fig 31

considered:

(a) p.d. across R_4

$$V_E = V_{EE} - (V_{BB} + V_{BE_{SAT}})$$
$$= 5.2 - (1.15 + 0.75) = \mathbf{3.3\ V}$$

(b) Current through R_3

$$I_E = V_E/R_3$$
$$= \frac{3.3\ V}{1.24 \times 10^3}\ \text{A} = \mathbf{2.66\ mA}$$

If the gain of TR5 is high, such that $I_B \ll I_C$, it may be assumed that $I_C = I_E$, therefore.

(c) p.d. across R_4

$$V_C = I_E \times R_4$$
$$= 2.66 \times 10^3 \times 300 = \mathbf{0.798\ V}$$

(d) Output voltage

$$V_{OL} = 0 - V_C - V_{BE_{SAT}}$$
$$= 0 - 0.798 - 0.75 = \mathbf{-1.55\ V}$$

Therefore a logical 0 is represented at this gate output by a potential of approximately -1.55 V.

The logical 1 output condition, when TR5 is cut off may now be considered:

(a) I_E is diverted through one (or more) of the gate input transistors.

(b) V_E rises towards $-V_{BB}$ and causes TR5 to cut off, causing I_C to become zero.

(c) **p.d. across R_4**

$$V_C = I_B \times R_4$$

but if the gain of TR6 is high, I_B is almost zero, therefore $V_C \approx 0$.

(d) **Output voltage**

$$V_{OL} = 0 - V_{BE_{SAT}} = \mathbf{-0.75\ V}$$

Therefore a logical 1 is represented at this gate output by a potential of approximately -0.75 V.

Using the values just calculated, it can be seen that an output voltage swing of 800 mV results when switching from logical 0 to logical 1 or vice versa.

Problem 20 Construct a table to compare TTL, CMOS and ECL gate circuits in terms of: (a) logic swing (5 V supply); (b) fan out; (c) power dissipation; and (d) propagation delay.

Table 3

	Standard TTL	*Schottky TTL*	*LS TTL*	*ECL*	*CMOS*
Logic swing	3.8 V	3.8 V	3.8 V	0.8 V	5 V
Fan out	10	10	10	25	50*
Power dissipation	10 mW	20 mW	2 mW	25 mW	10 nW
Propagation delay	10 ns	3 ns	10 ns	2 ns	50 ns

* Lower for faster switching rates.

C FURTHER PROBLEMS ON LOGIC FAMILIES

(a) SHORT ANSWER PROBLEMS

1 The name given to a type of logic circuit which makes use of multiple-emitter transistors is . . . .

2 Logic circuits which make use of a mixture of *p*-channel and *n*-channel FETs are known as . . . . circuits.

3 A logic gate output stage which has its load resistor replaced by a transistor is said to have . . . .

4 A logic gate output stage which uses a transistor for its load is frequently known as a . . . . output stage.

5 Schottky transistors may be used in a TTL output stage to cause . . . .

6 For low power consumption, ... circuits are most likely to be used.

7 For very high speed switching, ... circuits are preferred.

8 If TTL gate outputs are connected in parallel, ... circuits must be used.

9 The current flowing in a TTL input when held at logical 0 is known as . . . ., and is equal to . . . . mA.

10 When driving a standard TTL input from a CMOS output, a . . . . circuit may be required.

11 When driving a CMOS input from a TTL output, a . . . . may be required.

12 Incorrect handling of CMOS circuits may cause them to be damaged by . . . .

13 TTL circuits may be operated with a supply of . . . .

14 CMOS circuits may be operated with a supply voltage between . . . . and . . . .

15 The power consumption of a CMOS logic circuit . . . . as the switching rate is increased.

16 The increase or decrease in potential at a gate input required to cause a change in logic output is known as the . . . . of the circuit.

17 The number of logic inputs that may be connected to a single output is known as the . . . . of a gate.

18 Unused TTL inputs should not be left floating, otherwise problems may arise due to . . . .

19 The main advantage of buffered, compared with unbuffered CMOS gates is that they have a constant . . . .

20 The power consumption of a standard TTL circuit is typically . . . . per gate.

21 The switching time for a standard TTL circuit is typically . . . .

22 The switching time for a low power Schottky (LS) circuit is typically . . . .

23 The switching time for a CMOS logic circuit is typically . . . .

24 The switching time for a Schottky TTL circuit is typically . . . .

25 The power consumption of an ECL gate circuit is typically . . . .

(b) CONVENTIONAL PROBLEMS

1. Explain two techniques used in TTL circuits to increase their operating speeds.
2. Explain why TTL circuits must operate with a 5 V supply, but CMOS circuits may be used with 3 V–15 V supplies.
3. Describe the factors which may limit the fan out of (a) TTL; (b) CMOS; and (c) ECL circuits.
4. Explain the term 'noise margin' when applied to the logic circuits, and determine the noise margin of (a) TTL; (b) CMOS; and (c) ECL circuits when used with a 5 V supply.
5. State *two* applications where CMOS ICs may be used to advantage, and explain the particular advantages of CMOS over other forms of logic that makes it must appropriate for these applications.
6. Explain why input damping diodes are used in (a) TTL and (b) CMOS circuits.
7. Draw the circuit diagram and explain the function of each component in a practical 'totem pole' output stage.
8. State *three* advantages of using ECL circuits and explain how each of these advantages is obtained.
9. Explain the problems associated with leaving unused inputs floating on (a) TTL and (b) CMOS circuits and show how these problems may be overcome.
10. Explain why, although the noise margin in ECL circuits is smaller than that for TTL circuits, ECL circuits are less likely to suffer from noise problems.

4 Semiconductor memories

A MAIN POINTS CONCERNED WITH SEMICONDUCTOR MEMORIES

1 The basic requirements for all microelectronic stores are:

(a) A number of individual storage elements known as **memory cells**, each capable of temporary or permanent storage of a single binary digit.
(b) A **system of addressing** which provides a means of selecting a specified memory cell (or group of cells) within the memory device.
(c) A **method of storing** data in specified locations.
(d) A **method of reading** data from specified locations.

2 Microelectronic stores may be classified as:

(a) **Volatile**.
(b) **Non-volatile**.

A volatile memory loses its stored data if its power supply is not continously maintained. A non-volatile memory retains its stored data permanently, and removal of its power supply does not result in the loss of its stored data.

3 The following five types of microelectronic stores are commonly found in microcomputer systems:

(a) Random access memory (RAM)
This is a type of memory in which any particular storage location (address) may be directly selected without first having to sequence through other locations, i.e., one storage location may be accessed as easily as any other regardless of its physical position within the particular memory device. Any type of memory may employ random access of data, although the term RAM has become generally accepted (incorrectly) to mean **read-write** memory, i.e., memory in which the user may either be read or write to specified locations whilst in its normal circuit environment.
Two different tupes of semiconductor read-write memory are used in microcomputer systems, and both types are volatile. These are:

(i) **static RAM** in which bistable or flip-flop circuits are used as storage elements. With this type of memory, the stored data is retained permanently provided its power supply is continuously maintained; and

(ii) **dynamic RAM (DRAM)** in which capacitors are used as temporary storage elements. Information is stored as a charge or non-charge of a capacitor, but leakage effects cause loss of charge and hence loss of stored information. Therefore the charges on these capacitors must be restored at frequent intervals (≈ 2 ms) to prevent loss of stored information, and this process is known as **refreshing**.

(b) Read-only memory (ROM)
This is a non-volatile memory, used for the storage of permanent unalterable data. Data are stored in ROM during the final metallizing stages of manufacture, by means of a mask programming process, using a mask constructed according to information supplied by the customer. The construction of a mask is an expensive process, hence ROMs are only used in high volume applications.

(c) Programmable read-only memory (PROM)
A PROM fulfils basically the same function as a ROM except that whereas a ROM must be programmed by its manufacturer, a PROM is programmed by its user, i.e., it is **field programmable**. A PROM device is supplied by its manufacturer in a blank state, with all bits held at logical 0 or logical 1, by means of small fusible links. Programming a PROM involves its user blowing (or fusing) links in selected locations, thus changing the logic level in these locations. Like a ROM, once programmed a PROM cannot be reprogrammed.

(d) Erasable programmable read-only memory (EPROM)
An EPROM also fulfils basically the same functions as a ROM, and like a PROM, it is also field programmable. Its main advantage, however, is that it may be erased and reprogrammed. An EPROM uses floating gate avalanche (FAMOS) technology, and is supplied in its blank state by its manufacturer, with all locations at logical 0 or logical 1. Programming an EPROM involves storing electrical charges on selected FET floating gates to change these particular locations to their opposite logic state.

Erasure of an EPROM (changing all locations to the same logic state) is achieved by exposing the entire memory chip to ultraviolet light. The main disadvantages of an EPROM are:

(i) selected locations cannot be erased, which means that the entire memory must be erased and reprogrammed if changes are required in only one or two locations; and
(ii) the method of erasure prevents this type of memory from being reprogrammed in circuit.

(e) Electrically alterable read-only memory
An EAROM performs the same basic functions as ROM, PROM and EPROM devices, but its stored data may be changed electrically by the application of suitable control signals. It is therefore sometimes called a **read-mostly** memory. Erasure of an EAROM is carried out electrically by manipulation of its control inputs and voltages, but, unlike an EPROM, an EAROM may have selected blocks or individual locations erased without having to erase the entire memory.

Since an EAROM is erased electrically, it is possible to arrange for reprogramming to take place without removal from its normal circuit environment.

B WORKED PROBLEMS ON SEMICONDUCTOR MEMORIES

Problem 1 With reference to microelectronic stores, explain the following terms: (a) **serial access**; and (b) **random access**.

(a) A store with **serial (or sequential) access** is one in which a particular location cannot be accessed directly, but in which data must be brought out in its stored sequence until the required information is reached. Therefore, data in this type of store are clocked out in sequence rather than being directly selected by means of an address. The response time of such a store is slow and variable, and depends upon the position of the selected data within the store.

(b) A store with **random access** is one in which each location is as readily accessible as any other, and may be selected directly without having to sequence through any other locations. This means that any location within the store may be selected, at random, in any order, with a virtually constant response time, by applying an appropriate address to the store.

Problem 2 Draw a block diagram to illustrate the main sections of a typical 1 K bit static RAM (read/write memory).

A typical 1 K bit static RAM is the type 2102 which uses a 32 × 32 bit memory array, arranged as shown in *Fig. 1*.

Problem 3 With the aid of diagrams, explain how memory devices are arranged internally to reduce the number of external address lines.

The number of address lines required for a memory device is reduced by organizing the memory cells in the form of a matrix. A further reduction in the number of address lines is obtained by decoding binary address inputs into individual outputs (see *Fig. 2*).

The address input lines are divided into two groups which are decoded separately within the memory device to form two sets of memory cell select signals. These are known as **row select (X)** and **column select (Y)** signals, and they are arranged in the form of a matrix. A memory cell is connected to the X and Y lines at each intersection of the matrix, and is selected when both its X and Y lines are energized (see *Fig. 3*).

Using this system, it can be seen that 64 different locations may be addressed by the use of only 6 address inputs. This principle may be used for larger memories; for example, a 32 × 32 matrix enables 1 K (1024) memory locations to be addressed by the use of only 10 address input lines.

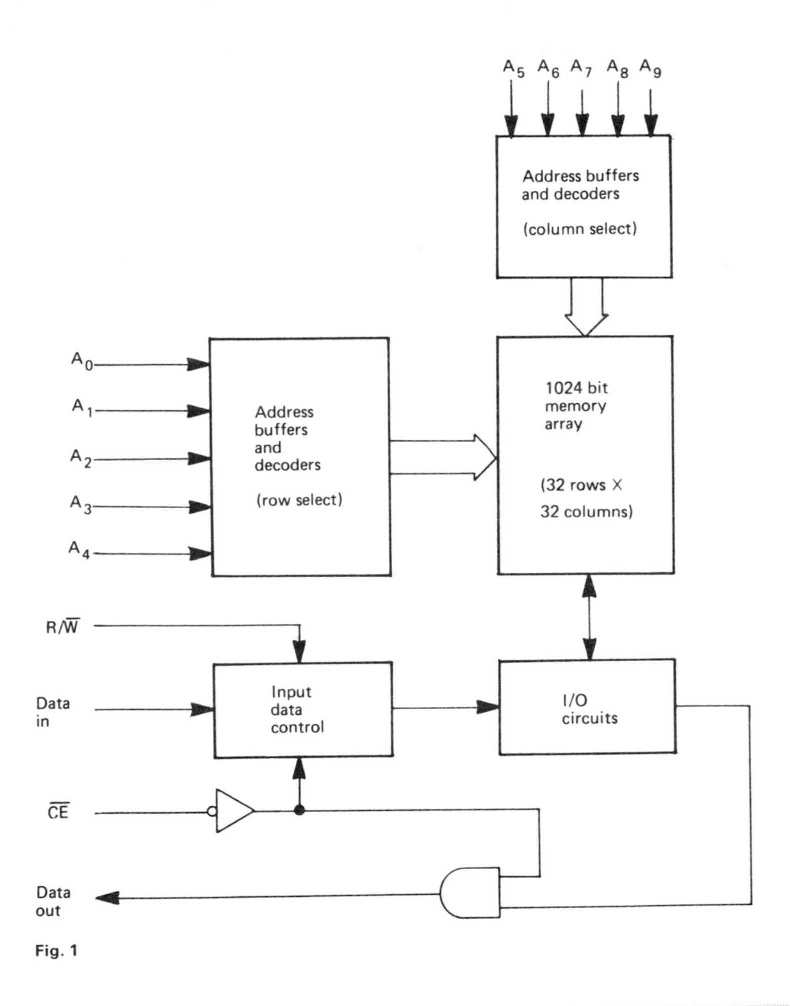

Fig. 1

Problem 4 With the aid of a circuit diagram, explain the operation of a memory cell in a typical static RAM (read-write memory).

The circuit diagram of a 2102 memory cell is illustrated in *Fig. 4*. This device uses NMOS construction and each memory cell consists of two inverter circuits arranged as a flip flop. TR1 and TR2 are the cross-coupled switching transistors, and TR3 and TR4 are used as active pull up resistors. Transistors TR5, 6, 7 and 8 are controlled by row and column select signals and hence determine when this particular cell is selected.

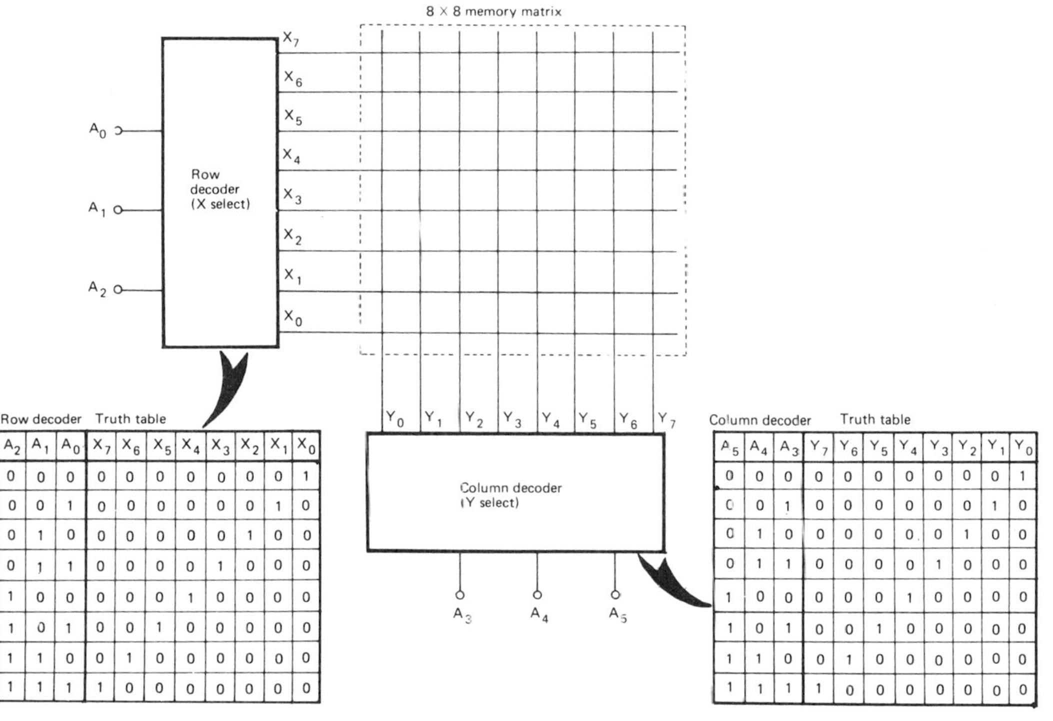

Row decoder Truth table

A_2	A_1	A_0	X_7	X_6	X_5	X_4	X_3	X_2	X_1	X_0
0	0	0	0	0	0	0	0	0	0	1
0	0	1	0	0	0	0	0	0	1	0
0	1	0	0	0	0	0	0	1	0	0
0	1	1	0	0	0	0	1	0	0	0
1	0	0	0	0	0	1	0	0	0	0
1	0	1	0	0	1	0	0	0	0	0
1	1	0	0	1	0	0	0	0	0	0
1	1	1	1	0	0	0	0	0	0	0

Column decoder Truth table

A_5	A_4	A_3	Y_7	Y_6	Y_5	Y_4	Y_3	Y_2	Y_1	Y_0
0	0	0	0	0	0	0	0	0	0	1
0	0	1	0	0	0	0	0	0	1	0
0	1	0	0	0	0	0	0	1	0	0
0	1	1	0	0	0	0	1	0	0	0
1	0	0	0	0	0	1	0	0	0	0
1	0	1	0	0	1	0	0	0	0	0
1	1	0	0	1	0	0	0	0	0	0
1	1	1	1	0	0	0	0	0	0	0

Fig. 2

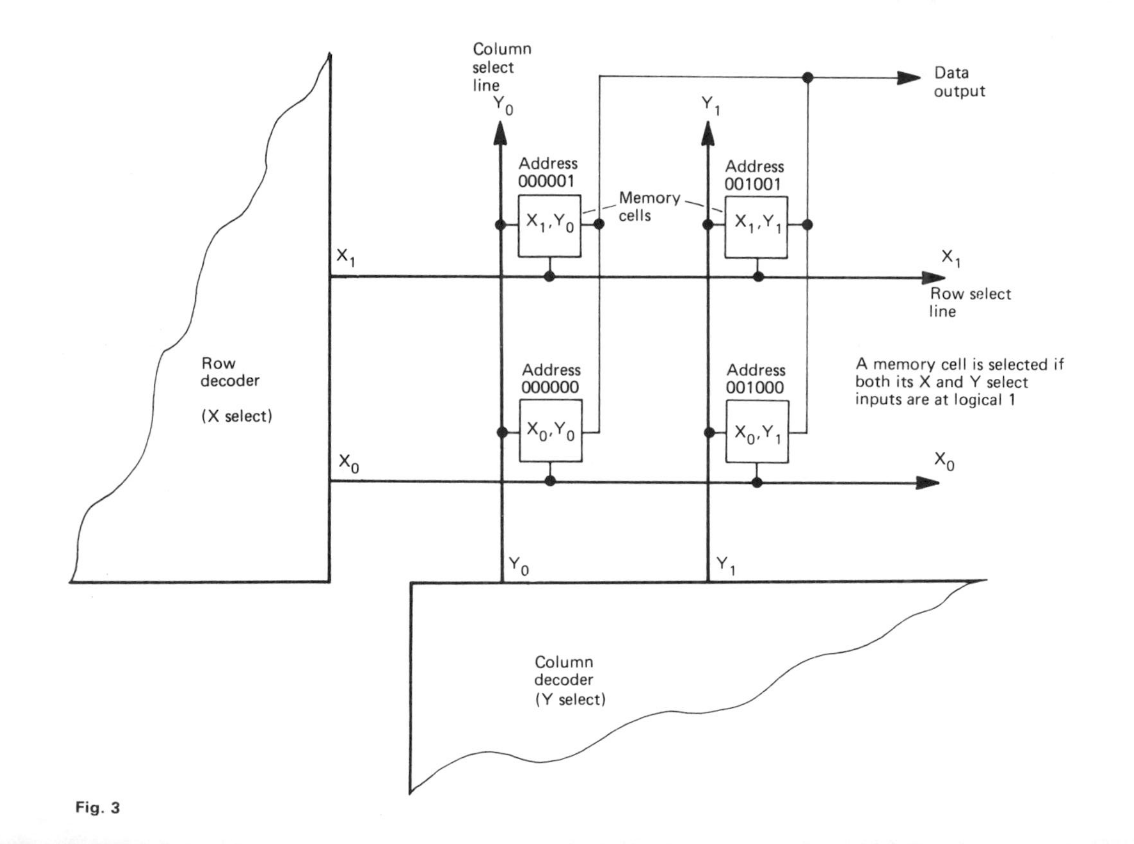

Column select line Y_0
Y_1
Data output
Address 000001
X_1, Y_0
Memory cells
Address 001001
X_1, Y_1
X_1
X_1
Row select line
Row decoder
(X select)
Address 000000
X_0, Y_0
Address 001000
X_0, Y_1
A memory cell is selected if both its X and Y select inputs are at logical 1
X_0
X_0
Y_0
Y_1
Column decoder
(Y select)

Fig. 3

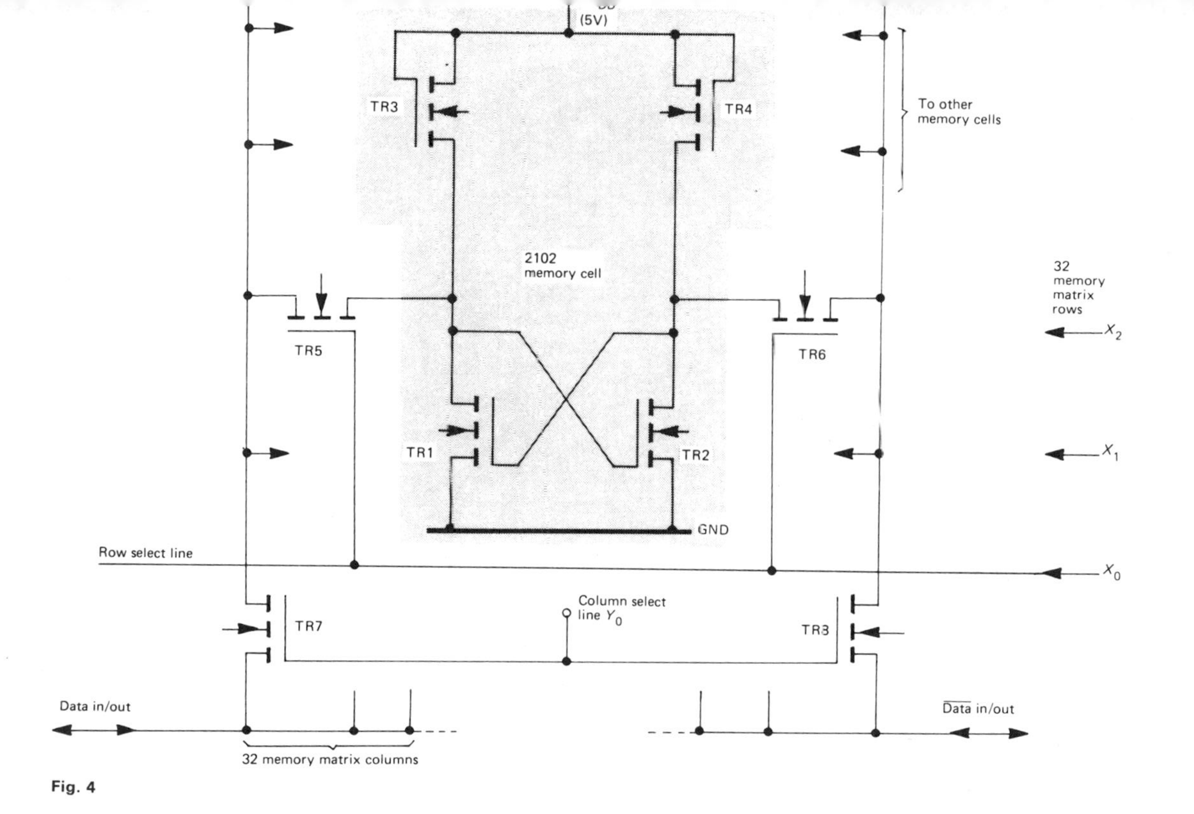
(5V)
TR3
TR4
To other memory cells
2102 memory cell
32 memory matrix rows
TR5
TR6
X_2
TR1
TR2
X_1
GND
Row select line
X_0
Column select line Y_0
TR7
TR3
Data in/out
$\overline{\text{Data}}$ in/out
32 memory matrix columns

Fig. 4

Selection of a particular memory cell takes place in the following manner:

When column select (Y_0) is activated (logical 1), TR7 and TR8 conduct and connect all cells in this particular column of the memory matrix to the data input/output lines.
When row select (X_0) is activated (logical 1), TR5 and TR6 conduct (along with equivalent transistors of all other memory cells in the selected row).

Thus it can be seen that TR5/TR7 and TR6/TR8 effectively behave as AND gates, and only the memory cell which lies at the intersections of the enabled row and column lines is selected. Assuming the memory cell illustrated in *Fig. 4* is selected, and the memory is set in the **write** mode, data may be stored in the following manner:

A logical 1 applied to the DATA IN/OUT terminal causes a positive potential to be applied to TR2 gate via TR5 and TR7. This causes TR2 to conduct and its drain potential falls to approximately 0 V, which, in turn, causes TR1 to become non-conductive (since its gate is connected to TR2 drain). Thus TR1 drain potential remains at $+V_{DD}$ (logical 1) after the input signal is removed, and a logical 1 is stored on TR1 drain.
A logical 0 may be stored by a similar process.

With the RAM switched to the **read** mode, data stored at TR1 drain (or its complement at TR2 drain) may be non-destructively interrogated by enabling TR5/7 and TR6/8, i.e., by selecting this particular memory cell.

Problem 5 Draw a block diagram to illustrate the main sections of a typical 4 K bit dynamic RAM (read/write memory).

A typical 4 K bit dynamic RAM is the Type 2107 which uses a 64 × 64 bit memory array, arranged as shown in *Fig. 5*.

Problem 6 With the aid of a circuit diagram, explain the operation of a typical dynamic RAM memory cell.

With this type of memory a capacitor is used as the basic storage element and a simplified dynamic memory cell is illustrated in *Fig. 6*. The operation of this circuit may be summarized as follows:

(a) **Write data** (S_2 closed, S_1 open)
C charges to a voltage level corresponding to the logic level of the input data.

(b) **Read data** (S_1 open, S_2 closed)
The voltage level on C (which represents the stored logic level) is transferred to the output terminal via S_2.

A typical one-transistor dynamic memory cell is shown in *Fig. 7*.

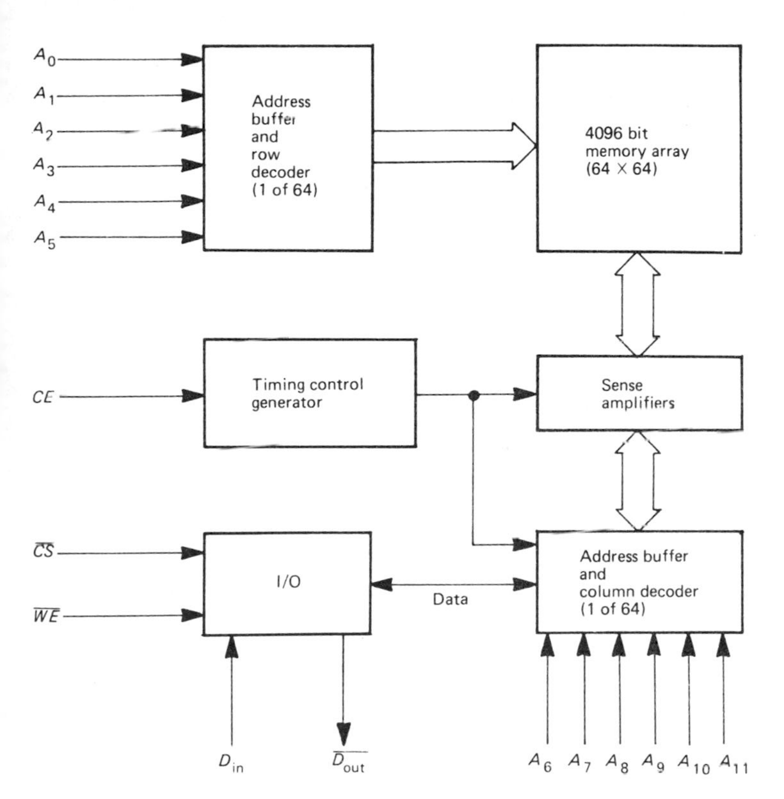

A_0
A_1
A_2
A_3
A_4
A_5
Address buffer and row decoder (1 of 64)
4096 bit memory array (64 × 64)
CE
Timing control generator
Sense amplifiers
$\overline{CS}$
$\overline{WE}$
I/O
Data
Address buffer and column decoder (1 of 64)
D_{in}
$\overline{D_{out}}$
A_6 A_7 A_8 A_9 A_{10} A_{11}

Fig. 5

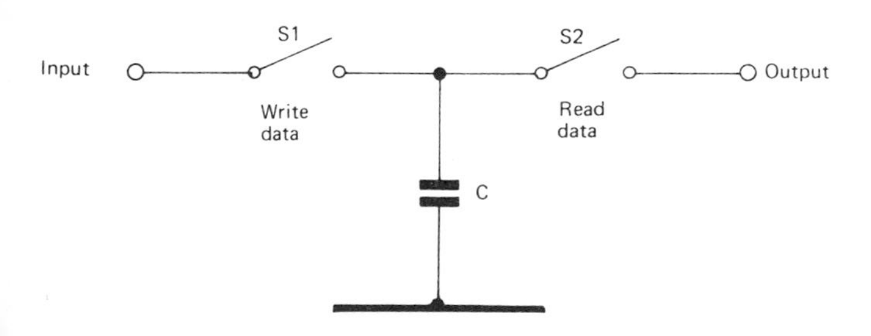

Input
S1
Write data
S2
Read data
Output
C

Fig. 6

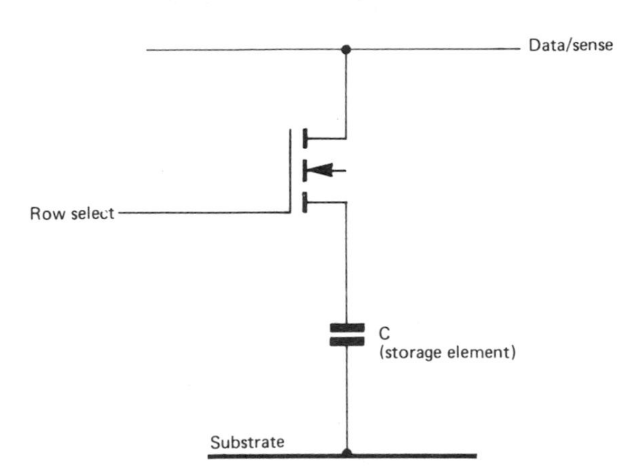

Fig. 7

Problem 7 Explain the main advantages and disadvantages of a dynamic RAM compared with a static RAM.

The main advantages of a dynamic RAM compared with a static RAM are:

(a) A dynamic RAM cell is much smaller physically than that of a static RAM, therefore, during manufacture it is possible to use a much greater packing density, i.e., more memory locations on a given chip size.

(b) A dynamic RAM may consume less power than a static RAM, which is especially important where large amounts of memory are involved. The reason for this is that when a dynamic RAM is not being addressed it is in a virtual 'standby' mode with none of its circuits conducting, whereas a static RAM (consisting of thousands of flip-flop circuits) must have one transistor in each flip flop conducting at all times so that the stored information may be retained.

The main **disadvantage** of a dynamic RAM are:

(a) Due to leakage in its storage capacitors, the information stored in a dynamic RAM is soon lost unless it is repeatedly rewritten into each memory cell at approximately 2 ms intervals of time. This is a process known as **refreshing** which requires additional circuitry both internal and external to the dynamic RAM. For this reason, dynamic RAM is unlikely to be used where less than 16 K locations are involved.

(b) Dynamic RAMs may require multiple supply rails, e.g. +5 V, +12 V and −5 V, whereas most static RAMs use a single (+5 V) rail. The multiple voltages may require a more costly power supply unit if these supplies are not already provided for other sections of the circuit.

Problem 8 Explain why 'refreshing' is required in a dynamic RAM and explain how it takes place.

Logical 1s and 0s are stored in a dynamic RAM memory cell as a charge (or no charge) on a capacitor. Leakage paths across this capacitor (particularly during 'read' operations) cause discharging and consequent loss of data to occur. Therefore, memory cells in a dynamic RAM must be 'topped up' at frequent intervals, and this is a process known as **refreshing**.

Dynamic RAM is organized internally as shown in *Fig. 8*, with one sense amplifier connected to each column of memory cells. Each sense amplifier consists basically of a bistable circuit, connected to the memory cell, and arranged so that it can sense the stored logic level. When a memory cell is read, this bistable latches with its output at logical 0 or logical 1 according to the value it senses, thus connecting the cell storage capacitor to the correct logic level to enable it to be refreshed. Such an arrangement is shown in *Fig. 9*, and the action of this sense amplifier is as follows:

C_2 is charged to approximately 0.5 V by the precharged circuit.
C_1 and C_2 are then simultaneously connected to the sense amplifier which compares the potential across C_1 with that across C_2. If the potential of C_1 is greater than the potential of C_2, Q becomes a logical 1, thus recharging C_1 to V volts. If the potential of C_1 is less than the potential of C_2, Q becomes a logical 0, thus keeping C_1 at 0 V.
Clearly, if C_1 is at logical 1, it must not be allowed to discharge to too low a potential if satisfactory refreshing is to take place.

Refreshing a dynamic RAM, therefore, consists of reading every cell in sequence at time intervals no longer than 2 ms apart. This must be done whether data is required or not (data are usually ignored during refresh cycles anyway). From *Fig. 8* it can be seen that selecting a cell in any particular row results in all cells in that row being connected to their respective column sense amplifiers. Therefore, a dynamic RAM may be refreshed by simply cycling through all row addresses in sequence.

Problem 9 Draw the block diagram of a typical ROM (read only memory).

A typical ROM is the mask programmable type 2316, 16 384 bit (2048 × 8) memory which is illustrated in *Fig. 10*.

Problem 10 With the aid of a diagram, explain the principle of operation of a ROM (read only memory) and state typical methods used to store data in a ROM.

Storage elements for read only memories may be more simple in construction than those necessary for read-write memories. A typical ROM construction is illustrated in *Fig. 11* which uses a memory matrix constructed from diodes.

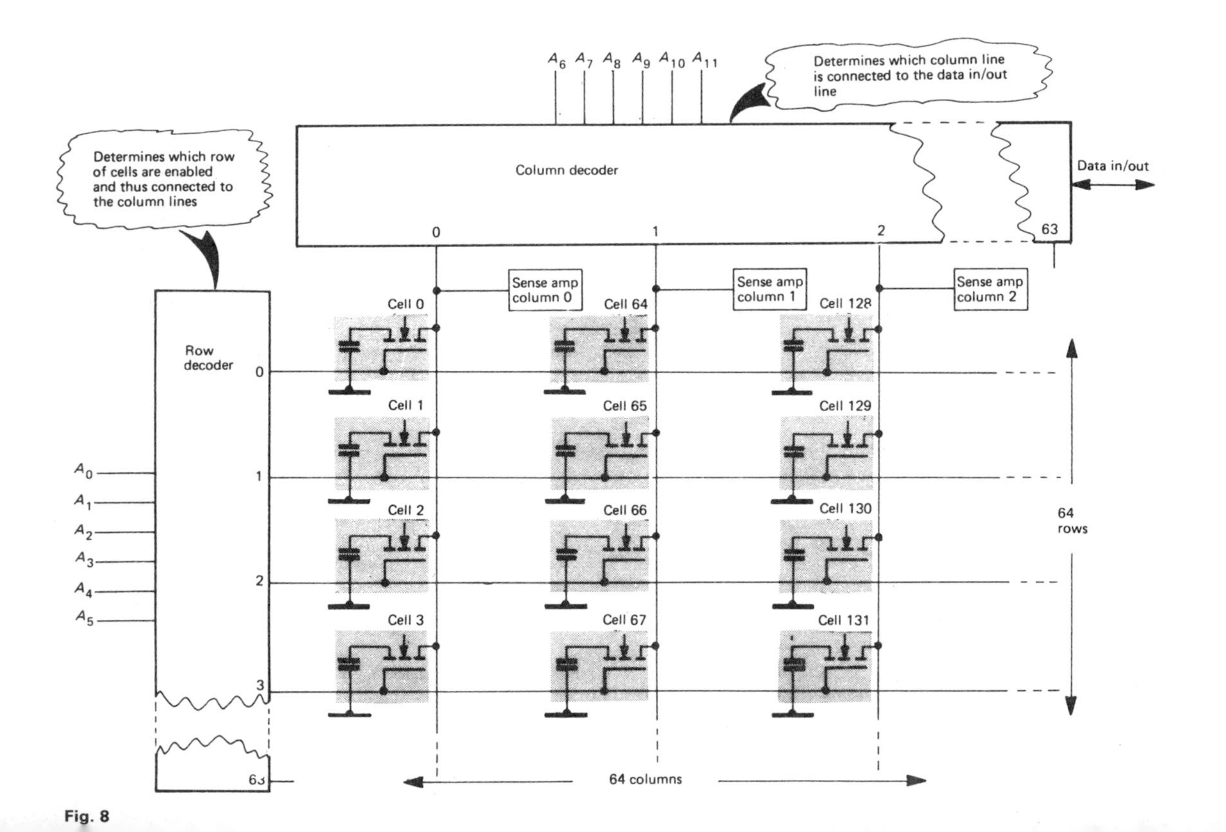

A_6 A_7 A_8 A_9 A_{10} A_{11}
Determines which column line is connected to the data in/out line
Determines which row of cells are enabled and thus connected to the column lines
Column decoder
Data in/out
0
1
2
63
Sense amp column 0
Sense amp column 1
Sense amp column 2
Row decoder
Cell 0
Cell 64
Cell 128
Cell 1
Cell 65
Cell 129
Cell 2
Cell 66
Cell 130
Cell 3
Cell 67
Cell 131
A_0
A_1
A_2
A_3
A_4
A_5
64 rows
64 columns

Fig. 8

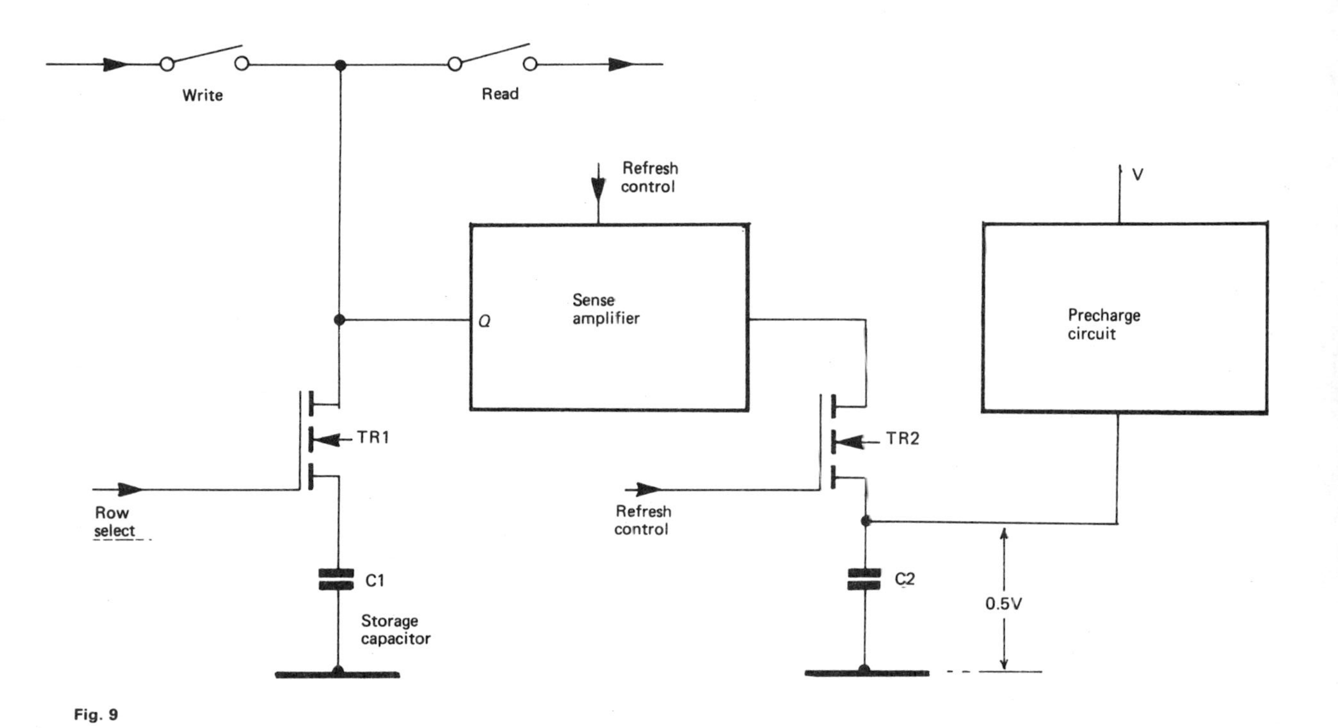
Write
Read
Refresh control
V
Sense amplifier
Q
Precharge circuit
TR1
TR2
Row select
Refresh control
C1
Storage capacitor
C2
0.5V

Fig. 9

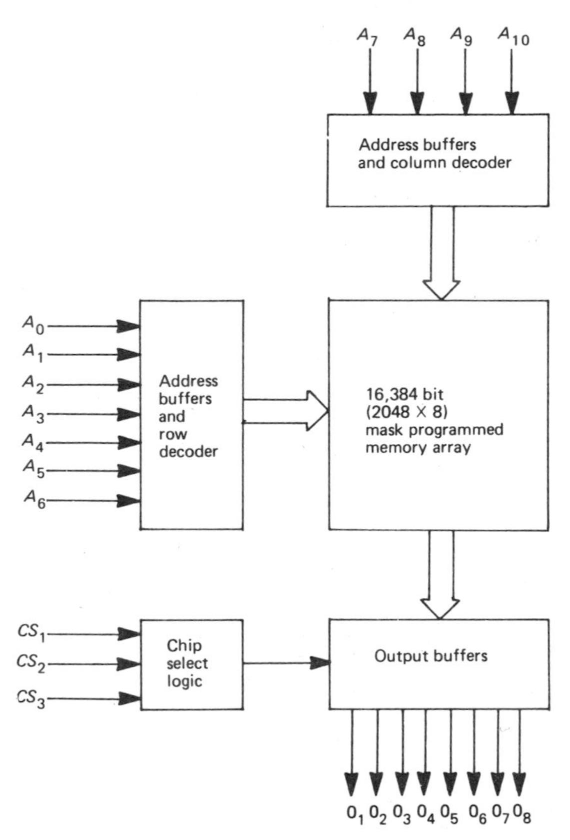

Fig. 10

In this circuit, the lower order address lines (A_0/A_1) are decoded to cause a logical 1 to be applied to one of the row lines (X) of the matrix. This causes each of the column lines (Y) to become logical 1 where a diode is fitted, but to remain at logical 0 where no diode is fitted. Each of the column lines is connected to one input of an AND gate (G_1 to G_4).

The higher order address lines (A_2/A_3) are decoded and this causes a logical 1 to be applied to one input of the selector AND gates (G_1 to G_4). Thus three of these AND gates have a logical 0 on their output, whilst the fourth gate has a logical 0 or a logical 1, determined by the state of its column line.

Each of the diodes shown in the memory matrix in *Fig. 11* represents a unidirectional switch, and the pattern of switches represents the stored information. Various technologies are available for constructing these switches and programming the information into a ROM. These include the following:

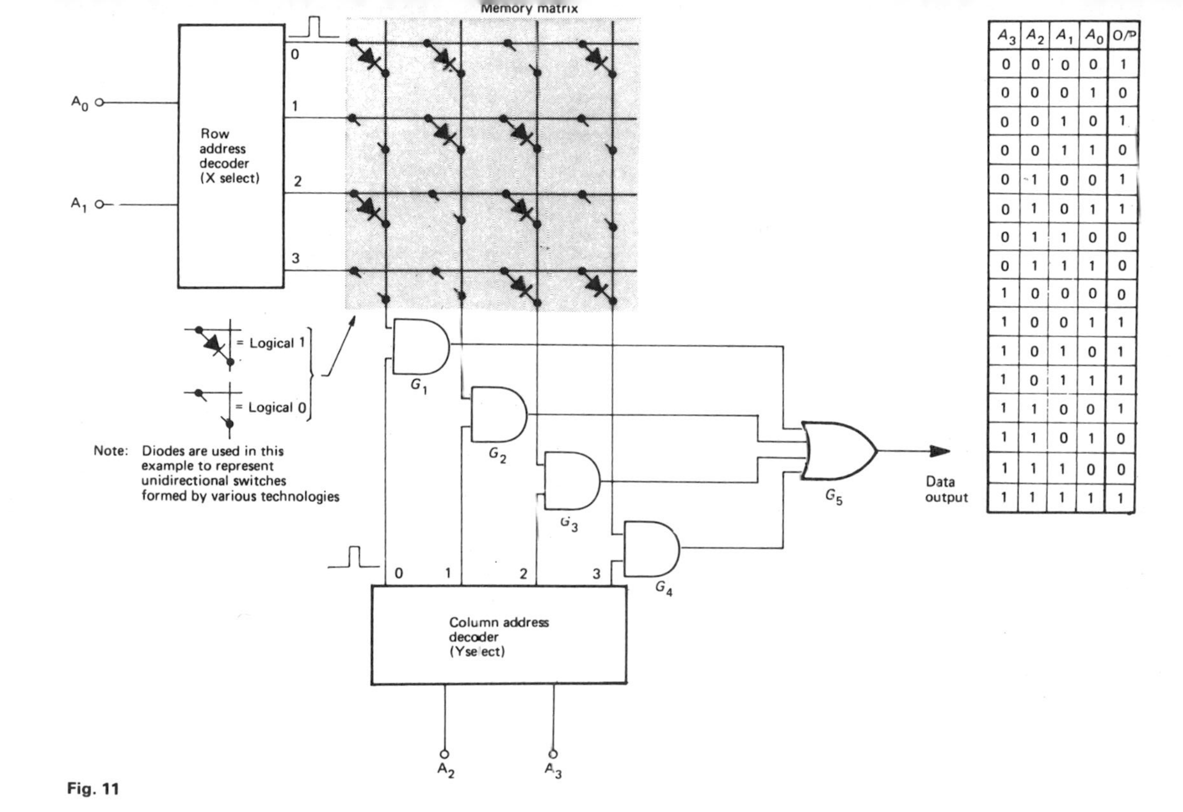

A_3	A_2	A_1	A_0	O/P
0	0	0	0	1
0	0	0	1	0
0	0	1	0	1
0	0	1	1	0
0	1	0	0	1
0	1	0	1	1
0	1	1	0	0
0	1	1	1	0
1	0	0	0	0
1	0	0	1	1
1	0	1	0	1
1	0	1	1	1
1	1	0	0	1
1	1	0	1	0
1	1	1	0	0
1	1	1	1	1

Fig. 11

(a) **Mask programming.** FET or bipolar switches are linked into circuit during the final stages of manufacture of a ROM, using a pattern supplied by the customer. Due to the relatively high cost involved in preparing masks for this process it is only suitable for high volume production applications.
(b) **Fusible link programming.** Fuses in series with closed switches (usually bipolar transistors) are provided in all positions of the matrix. Programming information into this type of memory may be carried out by either the manufacturer or the user by blowing fuses in selected locations.
(c) **Avalanche injection programming.** FET switches in a memory matrix are programmed by selectively turning them on by electrical charges induced in their gates by avalanche injection techniques (see *Problem 13*). This process is carried out by the user, and unlike the previous examples, is reversible for reprogramming purposes.

Problem 11 With reference to a fusible link PROM: (a) describe the construction and operation of a typical memory cell; (b) explain how programming is carried out; and (c) state **three** disadvantages of this type of memory.

(a) Each bit position in this type of memory consists of a transistor switch (bipolar or FET) connected in series with a very small fuse made from nichrome (earlier types), titanium tungsten or polycrystalline silicon. When a PROM is supplied by its manufacturer, all fuses are intact, and all locations are held at the same logic level (0 or 1 according to type). The circuit of a typical bipolar PROM memory cell is shown in *Fig. 12*.
This memory cell is selected by applying a logical 1 to its row address line, and as a result, TR1 is biased on (switch closed). If the fuse is intact, a logical 1 is applied to its column select line (from V_{CC} via TR1) and then on to the data output terminal. If the fuse is blown, however, a logical 0 is applied to its column select line despite TR1 being a conducting state.
(b) A PROM is programmed (usually in special PROM blowing equipment) by carrying out the following sequence of operations:
(i) apply the address of the location to be programmed to the memory address inputs;
(ii) raise V_{CC} to elevated programming value (V_{CCP}) of approximately 14 V;
(iii) using the memory data outputs as inputs, apply V_{CCP} to the desired input;
(iv) apply a pulse to the memory chip enable (*CE*) input to actually blow the fuse;
(v) reduce V_{CC} to its normal value (5 V);
(vi) verify that the fuse is actually blown, and repeat (ii) to (v) as necessary;
(vii) wait a given time interval, and repeat (ii) to (vi) for any further bits which are to be programmed at the same address; and
(viii) change the address inputs to that of the next location to be programmed and repeat the procedure described in (ii) to (vii).

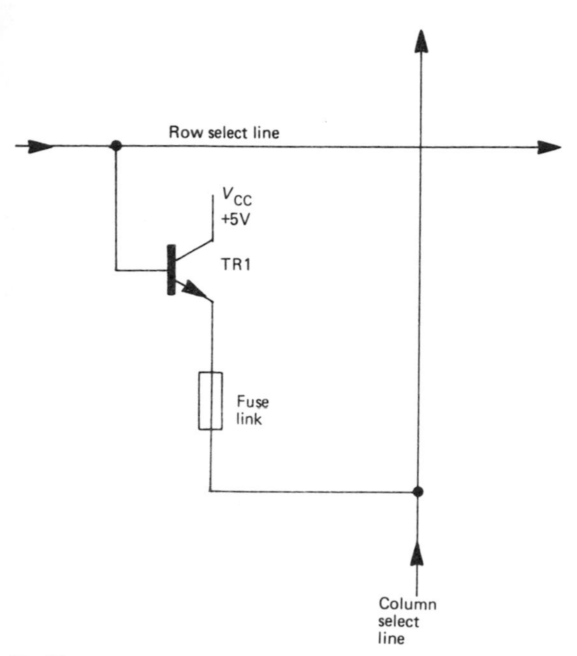

Fig. 12

(c) Three disadvantages of a PROM are:
- (i) they cannot be reprogrammed once used;
- (ii) errors cannot be corrected; and
- (iii) **'grow back'** may occur in which, after a period of time, blown fuses may remake themselves.

Problem 12 Draw the block diagram of a typical EPROM (erasable programmable read-only memory).

A typical EPROM is the type 2716, 16 384 (2048 × 8) memory which is illustrated in *Fig. 13*.

Problem 13 With reference to a typical EPROM: (a) describe the construction and operation of a memory cell; (b) explain how erasure is accomplished; and (c) describe the programming procedure.

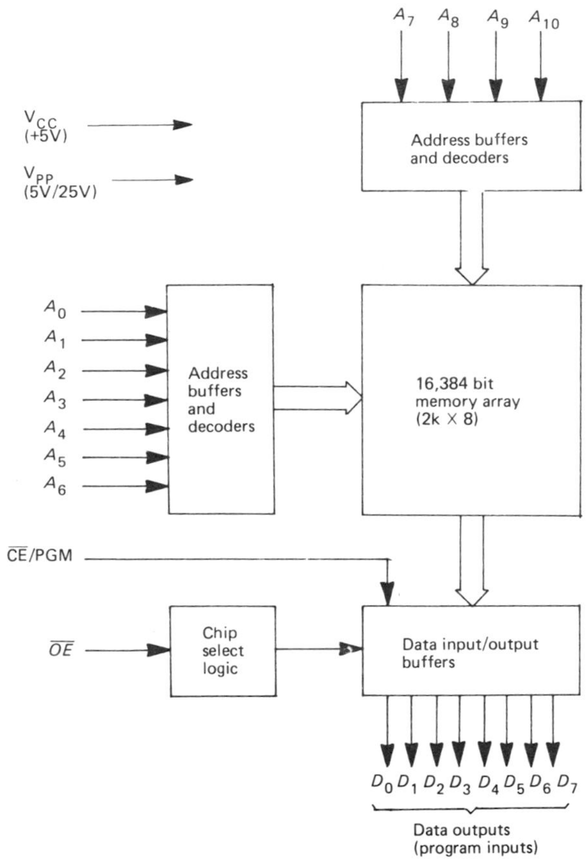

Fig. 13

(a) This type of memory makes use of **floating gate avalanche injection MOS** (FAMOS) technology, and the construction of a typical memory cell is illustrated in *Fig. 14*.

EPROMs are supplied by their manufacturer with all locations at the same logic level (0 or 1 according to type). The logic level at any specified address may be changed by applying a programming potential (V_{PP}) of approximately 12–25 V (depending upon type) to the memory cell via the appropriate read/select lines. This potential is applied between drain and source of an FET in the memory cell, and causes avalanche breakdown of the drain, with consequent injection of high energy current carriers from the surface of the avalanche region into a floating gate.

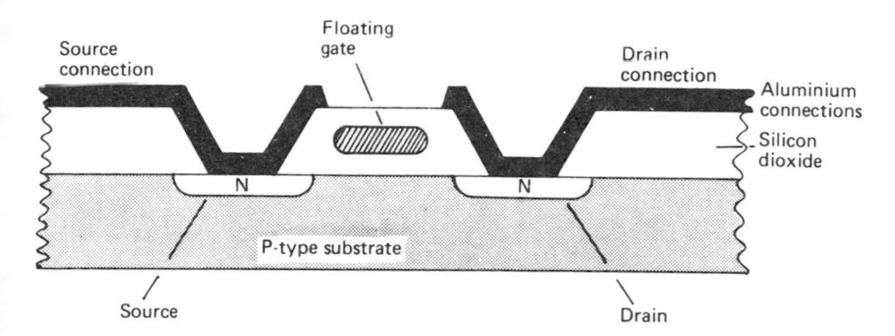

Fig. 14

Since the gate is floating, a charge is accumulated which produces an inversion layer (n-type) at the substrate surface and holds the FET switch in the 'on' state. Once the applied voltage is removed, no path exists for the charge on this gate to leak away and it is retained for a duration of at least 10 years. Thus, once an FET switch is programmed in the 'on' state, it remains in that condition.

(b) One of the advantages of an EPROM is that it may be erased and reprogrammed. Erasure is accomplished by exposing the entire memory chip to ultraviolet light (UV) of 2537 Å wavelength for a period of 10–20 minutes. A window is provided on each EPROM IC to enable its memory chip to be exposed, see *Fig. 15*.

The action of the UV light is to effectively overcome the energy barrier and temporarily short-circuit the gates to the substrate of all FET switches in the memory.

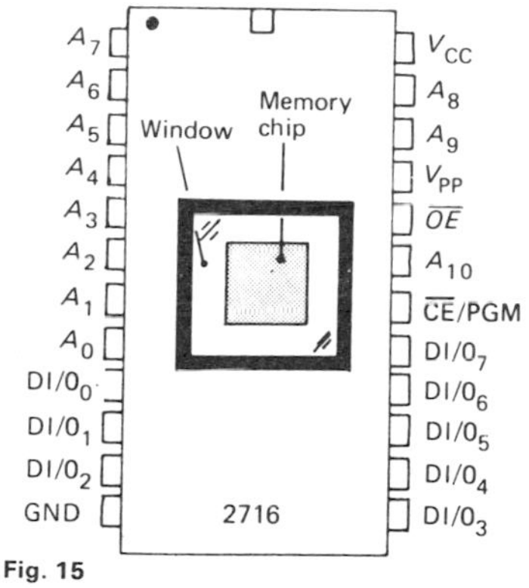

Fig. 15

(c) The methods used for programming EPROMs vary considerably from type to type. As an example, consider the programming sequence used for a 2716 EPROM (see *Fig.* 13):

(i) raise V_{PP} from $+5$ V to $+25$ V;
(ii) connect $\overline{\text{OE}}$ to logical 1 to disable the outputs;
(iii) apply the address of the location to be programmed to the memory address inputs;
(iv) using the data output pins as inputs, apply data to determine which bits are to be changed to logical 0 and which bits are to remain at logical 1;
(v) apply a 50 ms pulse to the $\overline{\text{CE}}$/PGM input;
(vi) repeat (iii) to (v) for all other locations to be programmed.

Problem 14 Draw the block diagram of a typical EAROM (electrically alterable read only memory).

A typical EAROM is the type 3400, 4096 bit (1 K × 4) memory which is illustrated in *Fig. 16*.

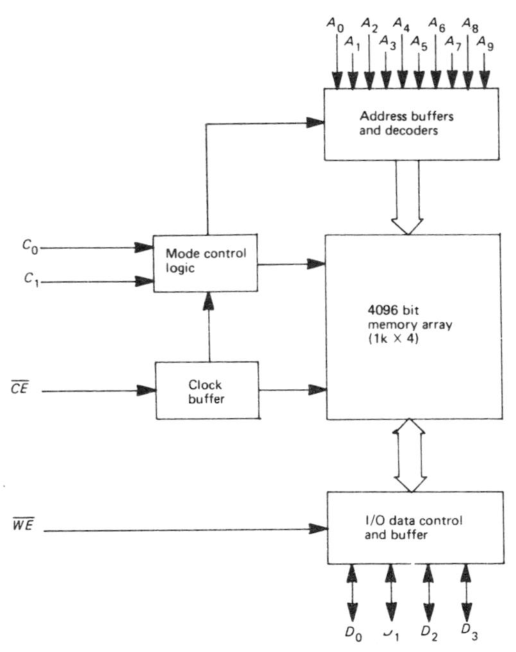

Mode	C_1	C_0
Read	0	0
Write	0	1
Chip erase	1	0
Word erase	1	1

Fig. 16

Problem 15 Describe the construction of an EAROM cell and explain how information is stored.

The gate potential required to switch on a MOSFET (known as its **threshold potential**) is typically 3–4 V, which is too high to allow compatibility with TTL circuits.

The threshold of a MOSFET may be reduced by replacing its silicon dioxide insulating layer with **silicon nitride**, but this results in unstable operation with different off-to-on and on-to-off thresholds, i.e., a **hysteresis effect**. This hysteresis is caused by charges tunnelling from the substrate into the nitride layer and remaining trapped there.

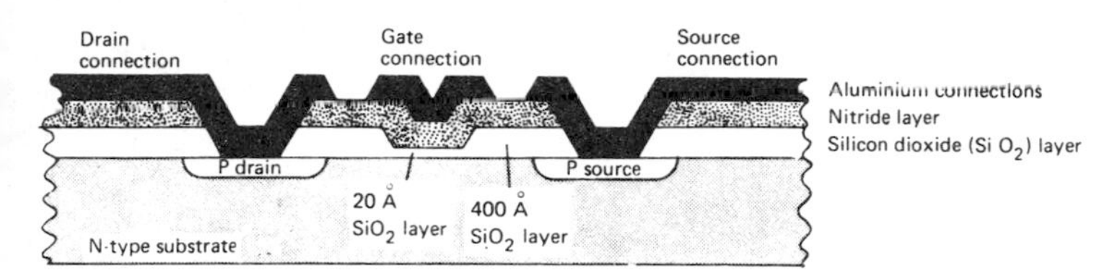

Fig. 17

In most MOSFET memory devices, a thin layer of silicon dioxide is deposited between the substrate and the silicon nitride layer, and this has the effect of stabilizing the threshold potential while still maintaining TTL compatibility.

In an EAROM cell, use is made of the hysteresis effect as a means of storing data in a non-volatile manner, and a typical EAROM cell is illustrated in *Fig. 17*. In this construction, the silicon dioxide layer is made very thin ($\approx$25 Å) so that at normal gate potentials charge tunnelling does not occur and TTL compatibility is maintained. If gate potentials of the order -25 V to -30 V are used, however, electrons are driven out of the silicon dioxide/nitride interface and into the substrate material with the result that positive charges are stored in trap sites at the interface. Since silicon dioxide and silicon nitride are both good insulators, the trapped charges are retained for up to 10 years.

A trapped positive charge has the same effect as a positive gate bias on a MOSFET and holds the transistor in a low conduction (off) state. This positive charge must also be overcome by externally applied negative gate signals with the result that the switching threshold is raised to approximately -12 V.

An EAROM cell may be erased into a low threshold (high conduction) state by applying a positive potential to its gate which has the effect of attracting electrons from the substrate into the silicon dioxide/nitride interface and trapping a negative charge there. A trapped negative charge has the same effect as negative gate bias on a MOSFET and therefore aids the normal negative gate bias with the result that the transistor is held in a high conduction (on) state and has a switching threshold of approximately -2 V.

Therefore information is stored in an EAROM in a similar manner to that used in a EPROM, except that the trapped charges are induced by gate potentials rather than by avalanching. For this reason an EAROM may be selectively erased and reprogrammed since all gates are electrically accessible via a matrix decoding mechanism.

C FURTHER PROBLEMS ON SEMICONDUCTOR MEMORIES

(a) SHORT ANSWER PROBLEMS

1 A microelectronic store which can only deliver data in the fixed order in which it was stored is known as a . . . . memory.

2 A microelectronic store which can deliver data in any specified order is known as a . . . . memory.

3 A microelectronic store which loses its information if its power supply is interrupted is known as a . . . . memory.

4 ROM, PROM and EPROM are all examples of a . . . . memory.

5 The technique used by a manufacturer to store information in a ROM is known as . . . .

6 An EPROM may be erased by the use of . . . .

7 In order to retain the stored data in a dynamic RAM, a system of . . . . is required.

8 The advantages of using dynamic RAMs in a microcomputer are that they . . . . and they have . . . .

9 It is not usual to use a custom built ROM for small volume applications due to the . . . .

10 A typical access time for a static RAM is . . . .

(b) CONVENTIONAL PROBLEMS

1 Explain the uses of the following types of microelectronic stores in a microcomputer:
 (i) static RAM
 (ii) dynamic RAM
 (iii) ROM
 (iv) EPROM

2 Using manufacturers' data sheets, compare a static RAM with a dynamic RAM in terms of:
 (i) type of access
 (ii) access time
 (iii) volatility
 (iv) packing density
 (v) relative cost
 (vi) power consumption

3 Compare the different techniques to program the following types of microelectronic stores:
 (i) ROM
 (ii) PROM
 (iii) EPROM

4 Explain the differences between:
 (i) volatile and non-volatile microelectronic stores; and
 (ii) static and dynamic microelectronic stores.

5 Interfacing devices

A MAIN POINTS CONCERNED WITH INTERFACING DEVICES

1 Circuits which are designed to allow two otherwise incompatible systems to be interconnected are known as **interface circuits**. An interface circuit is used between a microcomputer and each of its peripheral devices, since it is unlikely that any peripheral device could be connected directly to the data bus of a microcomputer. Devices within a microcomputer which enable digital information to be transferred to and from external circuits are known as **I/O devices** (input/output devices). Most devices of this type operate at TTL levels, and their use in relation to the other parts of a microcomputer system is shown in *Fig. 1*.

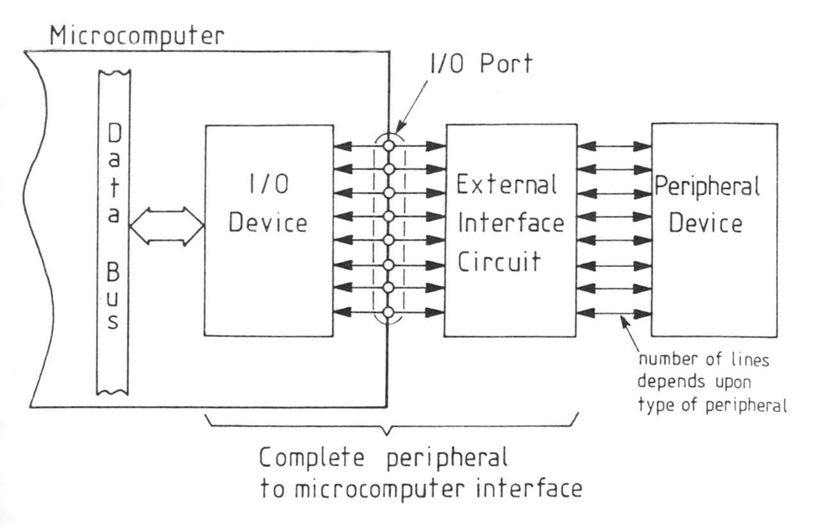

Fig. 1

2 Many methods are used to transfer data between a microcomputer and its peripheral devices, but they all come under one of the following categories:

(a) **Programmed I/O**, in which all data transfers are controlled by the microcomputer program,
(b) **Interrupt I/O**, in which peripheral devices force a microcomputer to leave its primary task to deal with data transfers, or
(c) **Direct memory access (DMA)**, which is a specialized form of I/O in which peripheral devices take over the system busses, and the microprocessor takes no part in the data transfer.

This chapter concentrates on programmed I/O (interrupt I/O is discussed in Chapter 2, and DMA is outside the scope of this syllabus).

3 Special programmable I/O devices are available which permit various different input and output options to be selected by means of appropriate software. These are known as **peripheral interface adapters (PIA)** or **parallel input/output (PIO)** devices. The internal construction of each I/O device varies considerably in complexity according to the particular facilities offered, but devices may include the following components:

(i) a **data input register**,
(ii) a **data output register**,
(iii) a **data direction register**,
(iv) a **control register**, and
(v) a **status flag register**.

A block diagram of an I/O device which uses these components is shown in *Fig. 2*. By assigning a unique address, or by other means, each of these components within an I/O device may be treated as an individual entity.

4 Most microprocessors have an 8-bit or a 16-bit data bus. Peripheral devices may require any number of lines from one upwards, depending upon their type. The data input and data output registers in an I/O device may be used to perform an

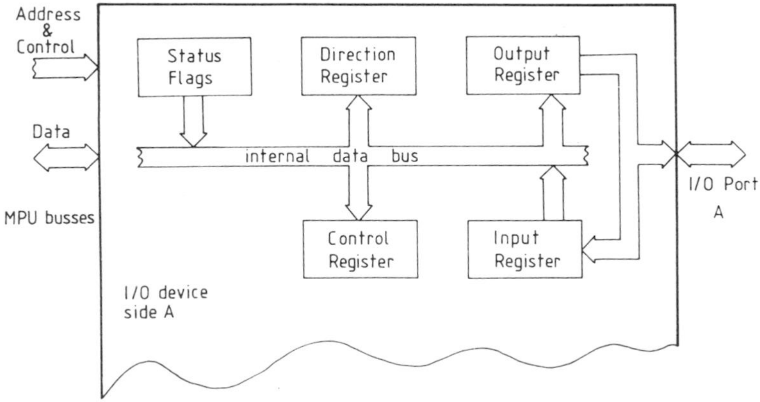

Fig. 2

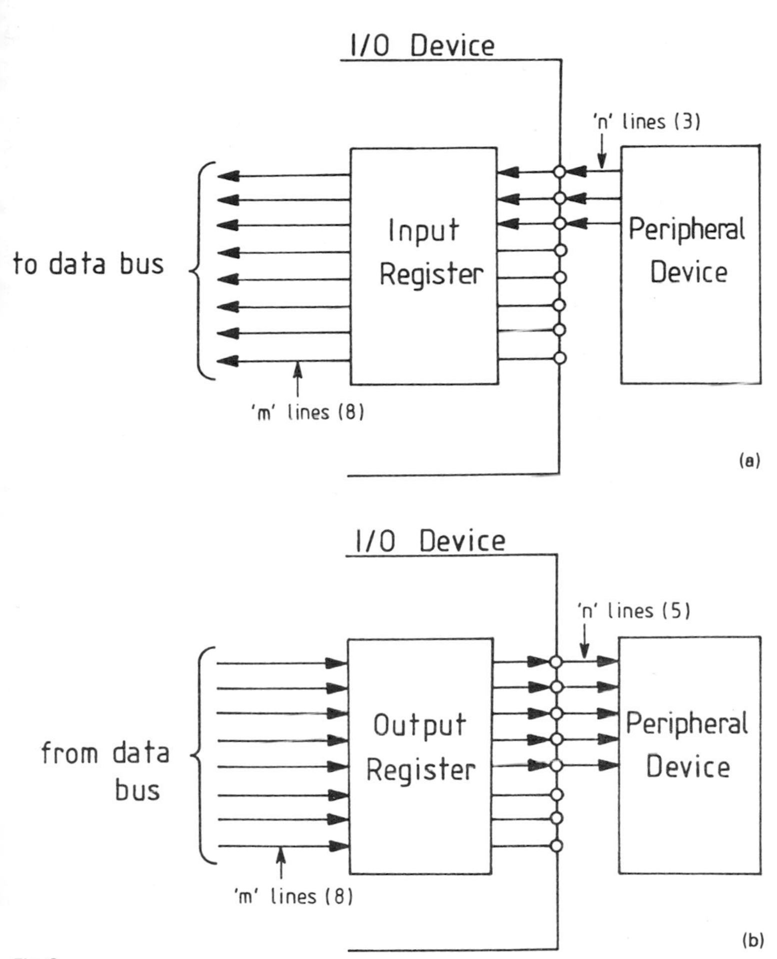

Fig. 3

m line to n line conversion, where 'm' is the number of lines in the data bus, and 'n' is the number of lines to or from the peripheral device (see *Fig. 3*(*a*) and (*b*)).

5 Sometimes communication between a microcomputer and its peripheral devices takes place using **serial data** transfers. This form of communication involves sending each bit of a data word, one after another, along a single data line, as shown in *Fig. 4*.

For converting parallel data into serial data, or serial data into parallel data, a **shift register** may be used, and such a device may be included in certain I/O

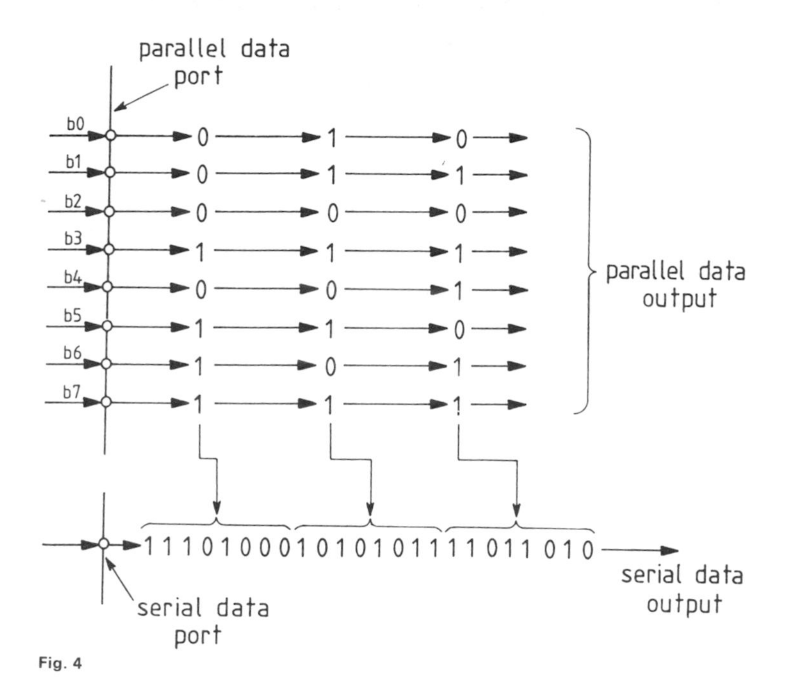

Fig. 4

devices, e.g. 6522 PIA. Special purpose I/O devices are available which provides all of the logic required to implement serial data communications. These are called **UARTs** (universal asynchronous receiver/transmitter) or **ACIAs** (asynchronous communications interface adapter).

6 An I/O device cannot usually handle all aspects of data transfer between a microcomputer and its peripheral devices, since its data inputs and outputs normally operate at TTL levels. Additional circuits may therefore be necessary between I/O and peripheral devices to deal with the following aspects of interfacing:

(a) **changes in voltage levels**,
(b) **changes in current levels**,
(c) **electrical isolation**,
(d) **timing of data transfers**,
(e) **digital to analogue conversion (D to A)**, and
(f) **analogue to digital conversion (A to D)**.

7 Each group of conductors from an I/O device are terminated with a connector to which peripheral devices may be attached. This is known as an **I/O port**, and an I/O device may have more than one port (typically two). A number of two-state switching devices may be connected to an I/O port as shown in *Fig. 5*.

If the PIA is **memory mapped** (i.e., is assigned an address within the memory address space), the two-state devices may be interrogated by using an

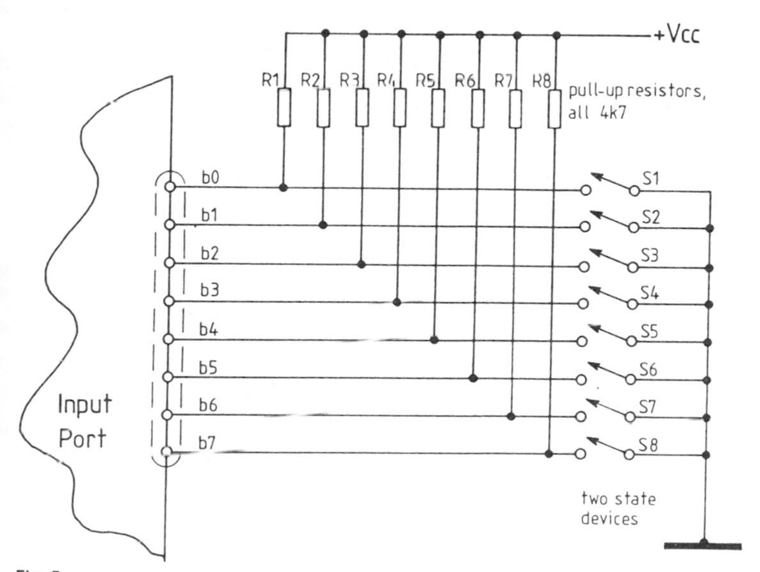

Fig. 5

appropriate LOAD instruction. If the PIA is not memory mapped, but is assigned to separate I/O address space, as is possible with the Z80 microprocessor, the two-state devices may be interrogated by means of an appropriate IN (input) instruction. A number of two-state indicator devices may be connected to an I/O port as shown in *Fig. 6*.

If the PIA is memory mapped, the indicators may be controlled by means of an appropriate STORE instruction. If the PIA is not memory mapped, an OUT (output) instruction may be used to control the indicators. The circuits shown in *Figs. 5* and *6* may be combined, as shown in *Fig. 7*, and programs which enable the condition of the two-state devices to be displayed on the indicators are as follows:

(a) 6502 microprocessor

```
;Memory mapped 6530 PIA in the address range
;$1700 to $1702
;
PAD      = $1700      ;port A data I/O register
PADD     = $1701      ;port A data direction register
PBD      = $1702      ;port B data I/O register
PBDD     = $1703      ;port B data direction register
       * = $0200      ;program starting address
;
```

```
0200  A9 FF                 LDA #$FF
0202  8D 01 17              STA PADD      ;port A all outputs
0205  A9 00                 LDA #$00
0207  8D 03 17              STA PBDD      ;port B all inputs
020A  AD 02 17    LOOP      LDA PBD       ;read the two-state devices
020D  8D 00 17              STA PAD       ;display on the indicators
0210  4C 0A 02              JMP LOOP      ;repeat if required
```

(b) Z80 microprocessor

```
                  ;Non-memory mapped Z80 PIO with I/O addresses
                  ;in the range 04 to 07
                  ;
                    PAD     EQU 4          ;port A data I/O register
                    PBD     EQU 5          ;port B data I/O register
                    PAC     EQU 6          ;port A control section
                    PBC     EQU 7          ;port B control section
                            ORG 0C90H      ;program starting address
                  ;
0C90  3E 0F                 LD A,0FH
0C92  D3 06                 OUT (PAC),A    ;port A all outputs
0C94  3E 4F                 LD A,4FH
0C96  D3 07                 OUT (PBC),A    ;port B all inputs
0C98  DB 05       LOOP      IN A,(PBD)     ;read the two-state devices
0C9A  D3 04                 OUT (PAD),A    ;display on the indicators
0C9C  18 FA                 JR LOOP        ;repeat if required
```

Note, for this program to operate correctly, the PIO strobe line for port B ($\overline{\text{BSTB}}$) must be at a logical 0 level.

(c) 6800 microprocessor

```
                  ;Memory mapped 6820 PIA in the address range
                  :$8004 to $8007
                  ;
                    DRA     EQU $8004     ;port A I/O or direction register
                    CRA     EQU $8005     ;port A control register
                    DRB     EQU $8006     ;port B I/O or direction register
                    CRB     EQU $8007     ;port B control register
                            ORG $0100     ;program starting address
                  ;
0100  4F                    CLR A
0101  B7 80 05              STA A CRA     ;select direction register A
0104  B7 80 07              STA A CRB     ;select direction register B
0107  B7 80 06              STA A DRB     ;port B all inputs
010A  43                    COM A
010B  B7 80 04              STA A DRA     ;port A all outputs
010E  86 04                 LDA A #$04    ;set bit 2 of accumulator A
```

```
0110  B7 80 05          STA A CRA   ;select I/O register A
0113  B7 80 07          STA A CRB   ;select I/O register B
0116  B6 80 06   LOOP   LDA A DRB   ;read the two-state devices
0119  B7 80 04          STA A DRA   ;display on the indicators
011C  20 F8             BRA LOOP    ;repeat if required
```

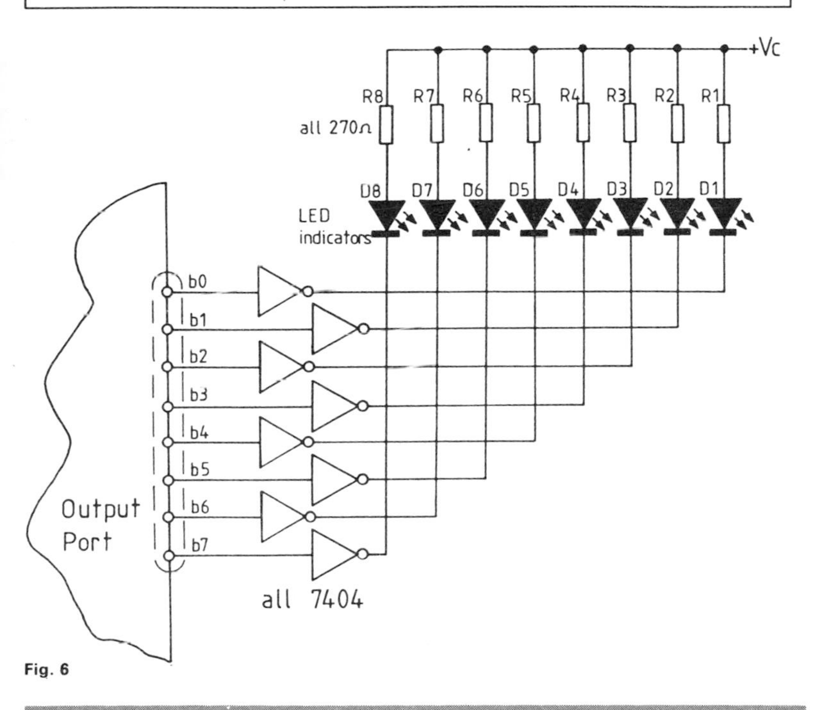

Fig. 6

B WORKED PROBLEMS ON INTERFACING DEVICES

Problem 1 With the aid of a diagram, explain how **four** SPST switches may be directly interfaced to bits 0, 1, 2 and 3 of the data bus of a microcomputer.

An 81LS95 tri-state buffer device may be used to transfer the switch data to the microcomputer data bus during a memory read cycle (or Z80 input cycle), as shown in *Fig. 8*. The address decoder output and the inverted $R/\overline{W}$ signal (or Z80 $\overline{RD}$ signal) are used as enabling inputs to the tri-state buffer. Only when both inputs are at logical 0 will the tri-state buffer be enabled. Resistors R_1 to

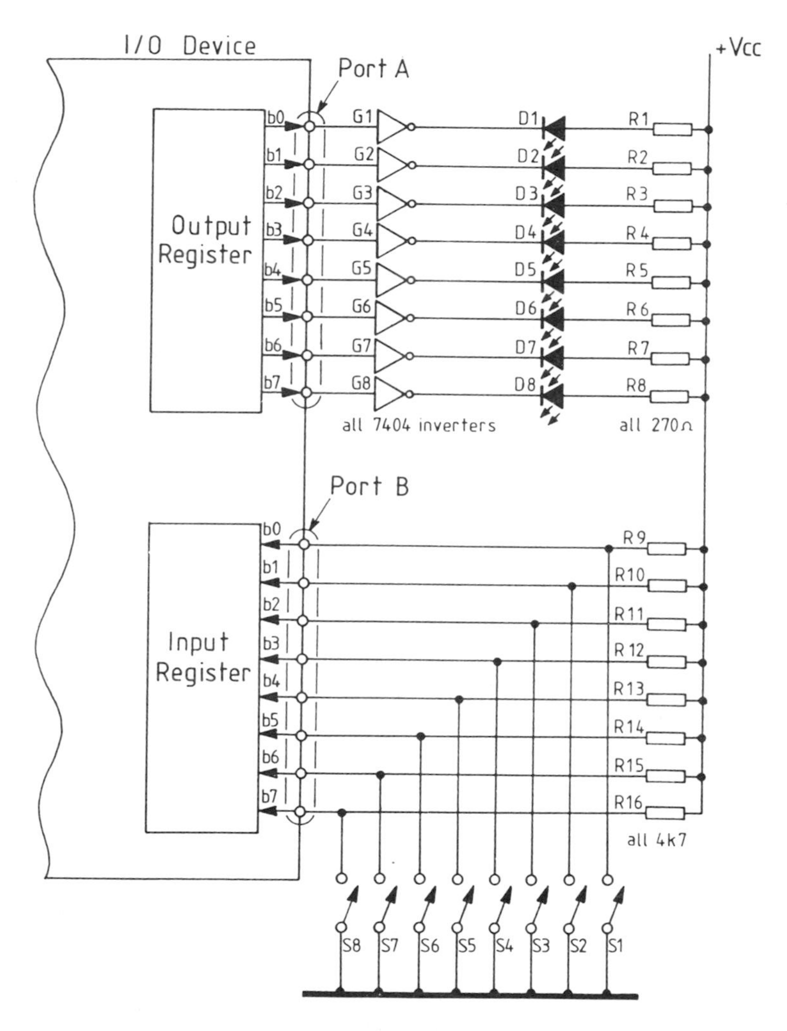

Fig. 7

R_4 are used as pull-ups to ensure that the inputs to the tri-state buffer are at logical 1 when their input switches are in the open condition. When the switches are closed, a logical 0 is applied to each input of the tri-state buffer. The data from these switches is therefore applied directly to bits 0, 1, 2 and 3 of the data bus each time that the tri-state buffer is enabled.

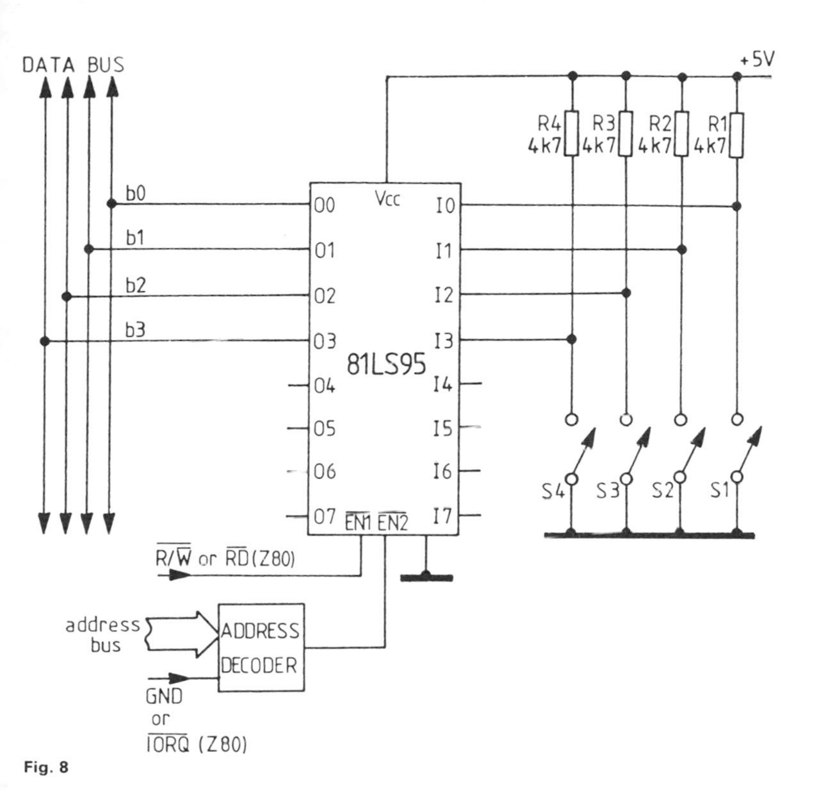

Fig. 8

Problem 2 With the aid of a diagram, explain how **four** LEDs may be directly interfaced to bits 4, 5, 6 and 7 of the data bus of a microcomputer.

A 7475 quadruple TTL latch may be used to capture information from the microcomputer data bus during a memory write cycle (or Z80 output cycle), as shown in *Fig. 9*. The address decoder output is gated with the $R/\overline{W}$ signal (or Z80 $\overline{WR}$ signal), and used to generate a positive pulse which enables information from the data bus to be transferred into the 7475 latches. Data is latched in the 7475 on the falling edge of this pulse. A 7404 quadruple TTL inverter is used to buffer and invert the Q outputs of the latches and sink current through the selected LEDs.

Problem 3 The logic block diagram of an 8212 I/O port is shown in *Fig. 10*. With the aid of diagrams, show how this device may be used as: (a) **a system input port**, (b) **a system output port**.

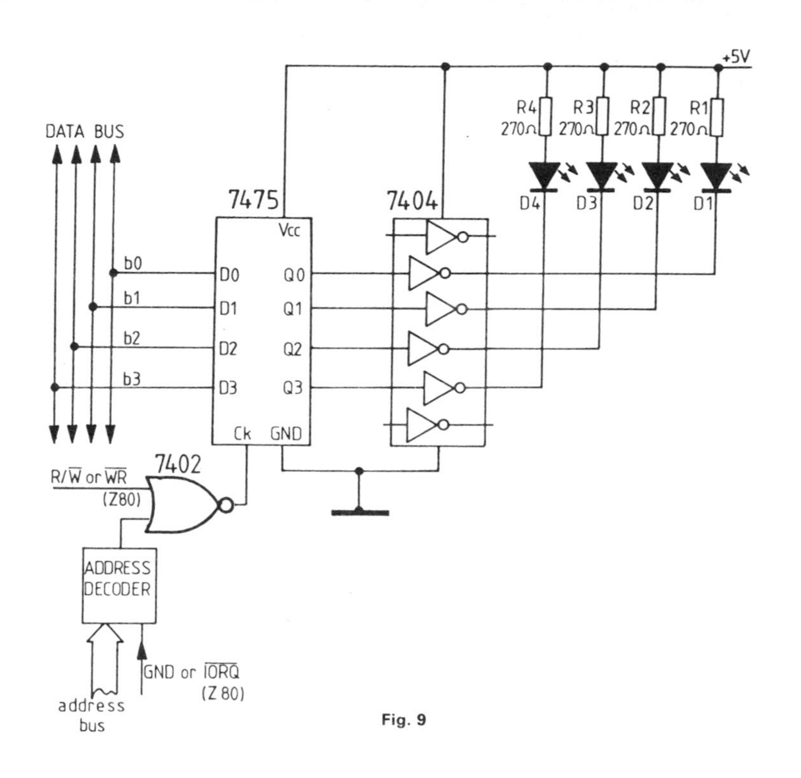

Fig. 9

(a) The 8212 may be used as a system input port when connected as shown in *Fig. 11(a)*.

When used as an input port, the MD (mode) input of the 8212 must be connected to a logical 0 level. Data is read from the peripheral device, and transferred into the data latches on the falling edge of a positive STB (strobe) pulse which is usually generated by the peripheral itself. When the MPU wishes to read the peripheral data, it enables the output buffers by generating the necessary device select signals and applying these to DS1 and DS2. The 8212 responds by transferring the contents of its data latches to the data bus. The $\overline{\text{INT}}$ output of an 8212 may be used to interrupt an MPU (see Chapter 2), and this output becomes active low in response to a STB pulse.

(b) The 8212 may be used as a system output port when connected as shown in *Fig. 11(b)*.

When used as an output port, the MD (mode) input of the 8212 must be connected to a logical 1 level. The output buffers are thus permanently enabled, and may transfer data to the peripheral device. When the MPU wishes to send new data to the peripheral, it transfers information from the data bus into the data latches by generating appropriate signals and

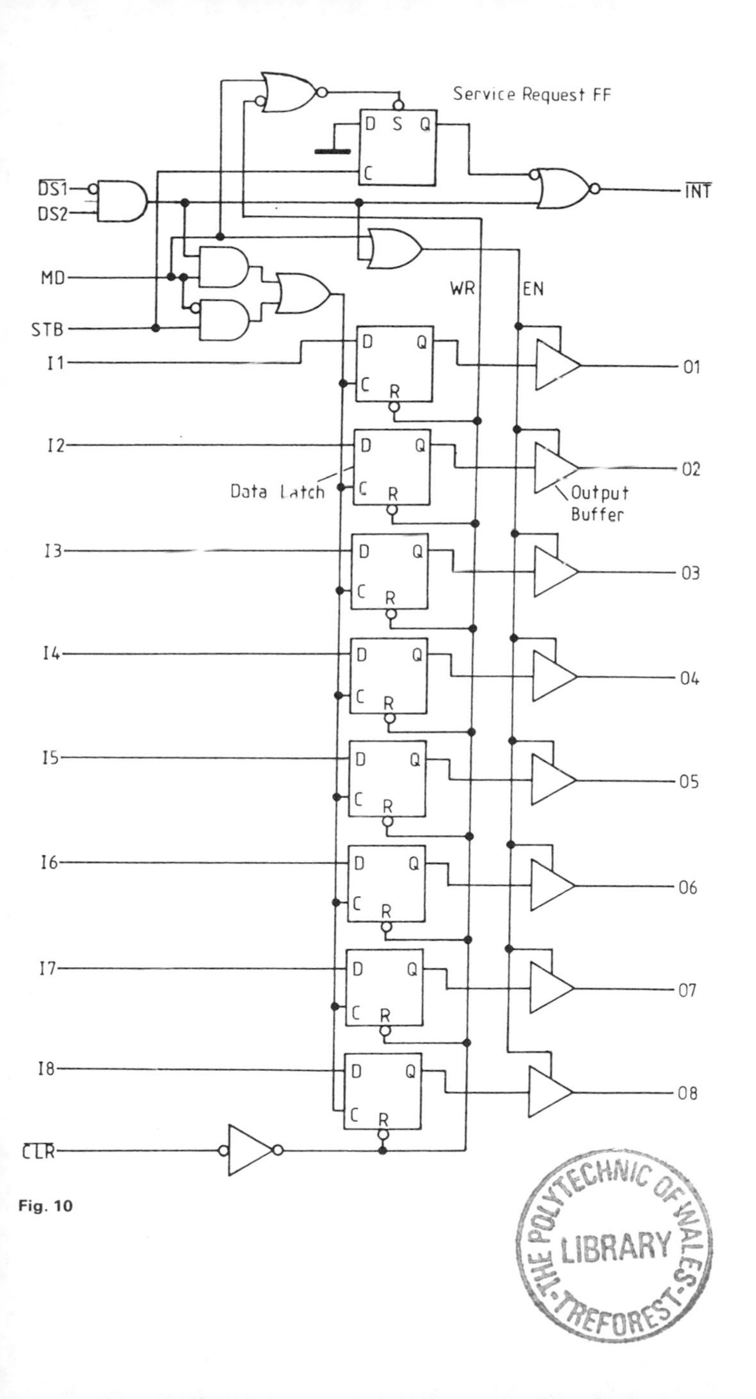
Service Request FF
D S Q
C
DS1
DS2
INT
MD
STB
WR
EN
I1
I2
I3
I4
I5
I6
I7
I8
D Q
C R
Data Latch
Output
Buffer
O1
O2
O3
O4
O5
O6
O7
O8
CLR

Fig. 10

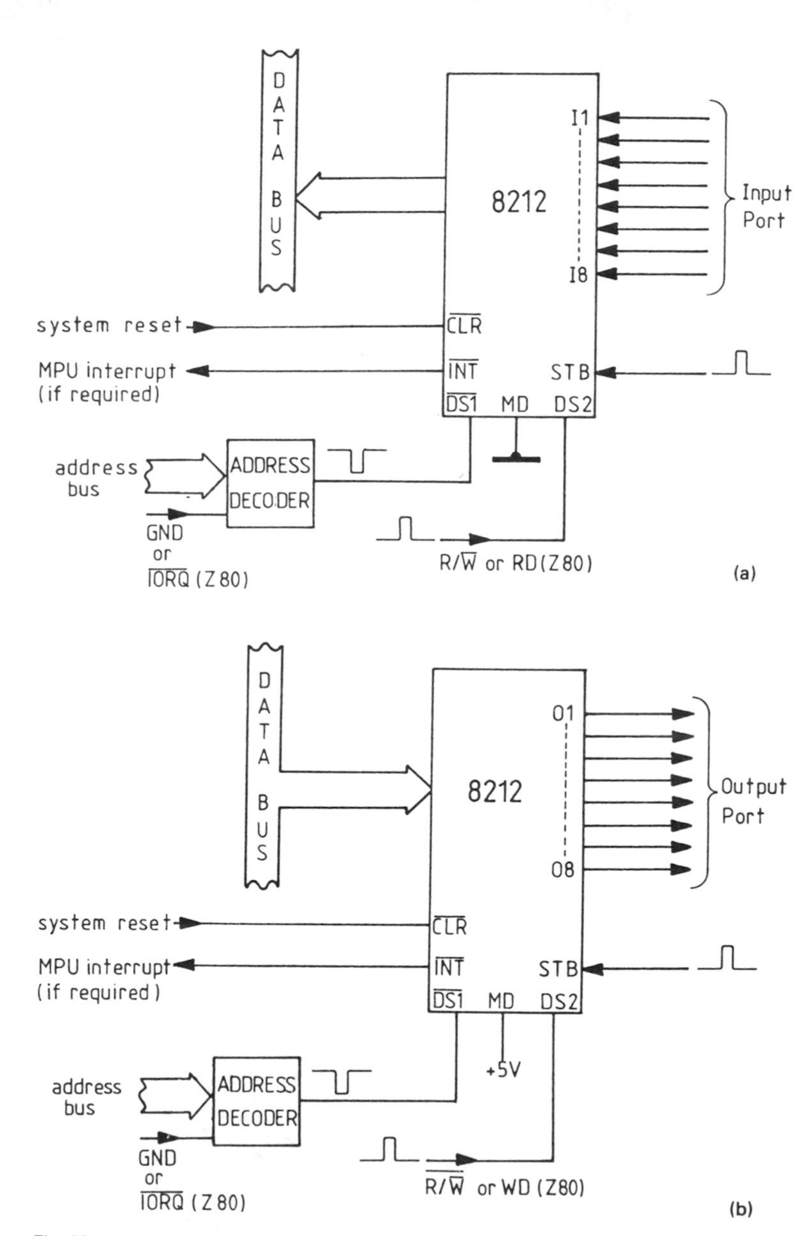
DATA BUS
8212
I1
I8
Input Port
system reset
CLR
MPU interrupt (if required)
INT
STB
DS1
MD
DS2
address bus
ADDRESS DECODER
GND or IORQ (Z80)
R/W or RD(Z80)
(a)
DATA BUS
8212
O1
O8
Output Port
system reset
CLR
MPU interrupt (if required)
INT
STB
DS1
MD
DS2
+5V
address bus
ADDRESS DECODER
GND or IORQ (Z80)
R/W or WD (Z80)
(b)

Fig. 11

applying these to DS1 and DS2. These input signals act as a clock signal for the data latches. If desired, the STB (strobe) input may be used so that the peripheral device can indicate to the MPU that it has received the data. It does this by generating an active $\overline{INT}$ (interrupt) output, which causes the MPU to be interrupted, and thus initiate the transfer of new data to the output port.

Problem 4 Describe the function of the following components in a programmable I/O device: (a) **data input register**; (b) **data output register**; (c) **data direction register**; (d) **control register**; (e) **status flags.**

(a) A data input register is used to retain **peripheral generated data**, thus allowing time for the MPU to read the data. This arrangement allows the peripheral data to be applied to the input port asynchronously, i.e., the peripheral device does not need to be controlled by the MPU clock. The peripheral device may determine when the data is transferred to the input register if the device is equipped with a strobe (STB) input. This arrangement is shown in *Fig. 12(a)*.

(b) A data output register is used to retain **MPU generated output data**, thus allowing time for a peripheral device to read the data. This arrangement allows the MPU to continue with other tasks without having to wait for a peripheral device to accept the data. A ready (RDY) signal may be available to indicate to a peripheral device that the output register is loaded with new data. This arrangement is shown in *Fig. 12(b)*.

(c) A data direction register contains information which allows individual I/O lines to be assigned as either input lines or output lines. Each bit in a data direction register determines whether its corresponding bit in the I/O port functions as an output or as an input. This register is accessible to the programmer, and storing the required information in it forms part of the PIA (or PIO) configuring procedure. The logic levels required to define input bits or output bits may differ between one device and another, and reference to the appropriate data sheet should be made for information regarding this. The use of a data direction register is shown in *Fig. 12(c)*.

(d) A control register contains information which specifies the programmable options within an I/O device. For example, a microprocessor may store information in this register to specify whether an I/O port functions as an input port, as an output port or as a bidirectional port. In some cases, bits in a control register are used to assign an address to one of several different registers, i.e., all registers have the same address, but bits in the control register determine which register responds when this address is accessed.

(e) Status flags are bistables with an I/O device which provide information concerning data transfers (or transfer requests) between a microcomputer and its peripheral device. For example, when a microprocessor writes new data to an output port, a flag associated with that port may be set and used to prevent the microprocessor from sending further data until it is cleared. When a peripheral device reads data from the output port, the status flag is cleared by this action, thus allowing the microprocessor to send new data. Other flags may be included to deal with interrupts, e.g. interrupt enable or interrupt pending flags.

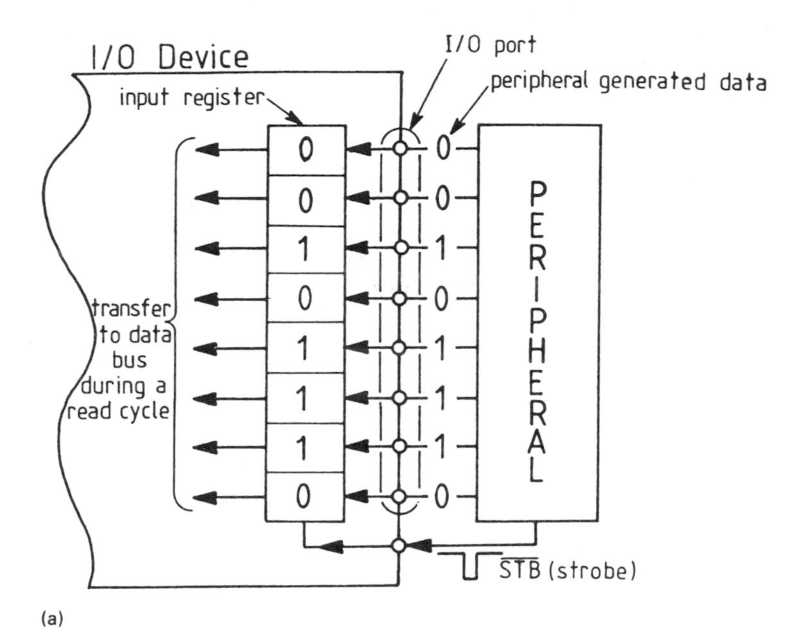
I/O Device
I/O port
peripheral generated data
input register
0
0
1
0
1
1
1
0
transfer to data bus during a read cycle
PERIPHERAL
STB (strobe)

(a)

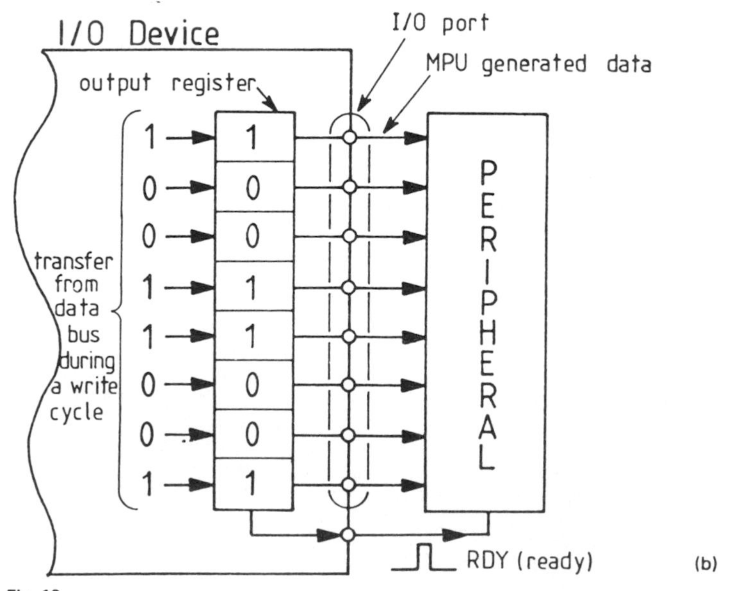
I/O Device
I/O port
MPU generated data
output register
1
0
0
1
1
0
0
1
transfer from data bus during a write cycle
PERIPHERAL
RDY (ready)

(b)

Fig. 12

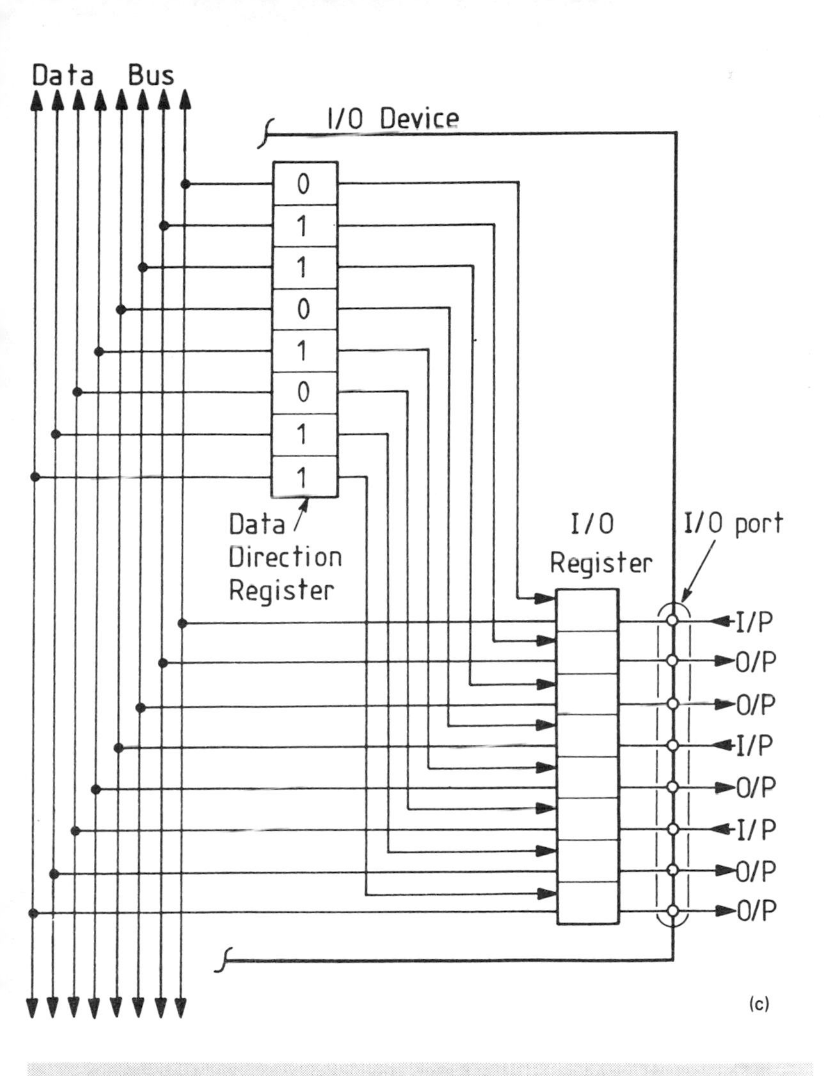

Problem 5 The block diagram of a **6530 I/O device** is shown in *Fig. 13*. (a) State the advantages of using an I/O device of this type; (b) State one disadvantage associated with the use of this device, and explain how this may be overcome; (c) Describe the I/O facilities provided by this device.

(a) The **6530** device contains 1 K bytes of ROM, 64 bytes of RAM, two 8-bit I/O ports and an interval timer. It is therefore sometimes known as an

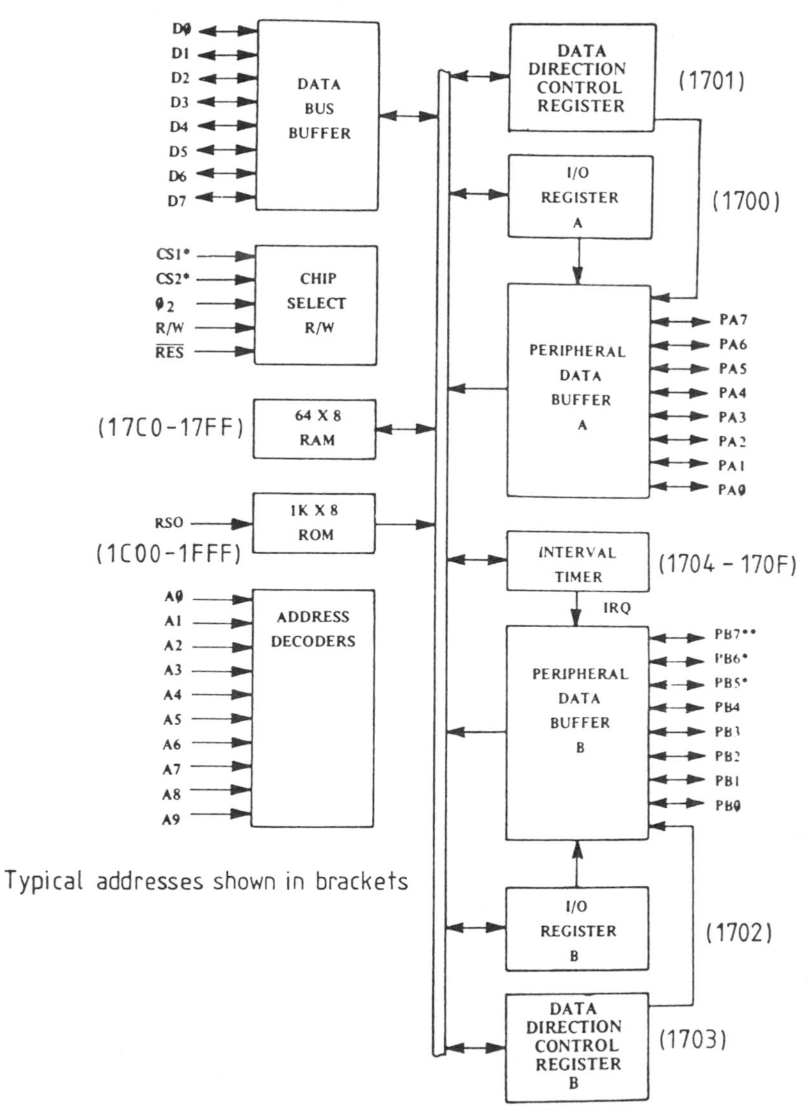
D0
D1
D2
D3
D4
D5
D6
D7
DATA
BUS
BUFFER
CS1*
CS2*
Ø2
R/W
RES
CHIP
SELECT
R/W
(17C0-17FF)
64 X 8
RAM
RSO
1K X 8
ROM
(1C00-1FFF)
A0
A1
A2
A3
A4
A5
A6
A7
A8
A9
ADDRESS
DECODERS
Typical addresses shown in brackets
DATA
DIRECTION
CONTROL
REGISTER
(1701)
I/O
REGISTER
A
(1700)
PERIPHERAL
DATA
BUFFER
A
PA7
PA6
PA5
PA4
PA3
PA2
PA1
PA0
INTERVAL
TIMER
(1704 - 170F)
IRQ
PERIPHERAL
DATA
BUFFER
B
PB7**
PB6*
PB5*
PB4
PB3
PB2
PB1
PB0
I/O
REGISTER
B
(1702)
DATA
DIRECTION
CONTROL
REGISTER
B
(1703)
*CS1/CS2 ARE MASK OPTIONS IN PLACE OF PB6/PB5
**PB6 MAY BE USED AT IRQ

Fig. 13

RRIOT, and contains all of the essential elements which, when combined with a microprocessor, form a complete microcomputer. The main advantage of a using a 6530, therefore, is that a complete microcomputer may be constructed with only two main devices (6502 MPU + 6530 RRIOT).

(b) The main disadvantage of using a **6530** device is that it uses a mask programmed ROM whose contents are fixed during manufacture and is therefore only suitable for use in high volume applications where the expense of creating special masks can be justified.

One way of overcoming this problem is to use a **6532** device which is logically similar to the **6530**, but which does not contain any ROM (this device is therefore sometimes known as a **RIOT**, and actually contains 128 bytes of RAM). The read-only memory requirements of a microcomputer may be provided by an EPROM, thus increasing the chip count to three for a complete microcomputer.

(c) The **6530** device has 16 pins available for peripheral I/O operations. Each pin may be programmed to act as either an input or an output. The 16 pins are divided into two 8 bit ports, PAO to PA7 and PBO to PB7. When an I/O line is defined as an input, its corresponding buffer in the I/O register is disabled and the microprocessor reads the peripheral input pin directly. When an I/O line is defined as an output, the microprocessor stores data in its corresponding bit in the I/O register.

The I/O pins are all TTL compatible, and, in addition, PAO and PBO are capable of sourcing 3 mA at 1.5 V, thus making them suitable for Darlington drive. PB7 may be used as an interrupt pin from the internal timer, and this line has no internal pull-up resistor, thus permitting several 6530 devices to be wire-ORed to the 6502 $\overline{\text{IRQ}}$ pin. In addition, PB5 and PB6 may be used as chip select inputs and may therefore be unavailable for peripheral I/O applications.

Problem 6 (a) Describe the function of the data direction control registers on a **6530** device, and (b) using the addresses indicated in *Fig. 13*, show how the I/O section of a **6530** device may be configured according to *Table 1*.

(a) Each of the data I/O lines of port A or port B of a **6530** device may be defined as an **output line** or an **input line**. The contents of data direction registers A and B determine how each of the I/O lines of their respective I/O ports are defined. A logical 0 in a particular bit position in a data direction register defines its corresponding port bit as an input line, whereas a logical 1 defines its corresponding port bit as an output line.

(b) In order to configure a **6530** device as indicated in *Table 1*, the contents of the data direction registers must be as shown in *Fig. 14*.

The **6502** machine code required to configure the ports in this manner is as follows:

```
A9 0F      LDA #$0F
8D 01 17   STA $1701     ;configure port A
A9 2D      LA #$2D
8D 03 17   STA $1703     ;configure port B
```

Table 1

Port	Bit no.	Direction	
A	0	output	
	1	output	
	2	output	
	3	output	
	4	input	
	5	input	
	6	input	
	7	input	
B	0	output	
	1	input	
	2	output	
	3	output	
	4	input	
	5	output	
	6	X	← Not available (used as a chip select)
	7	input	

Problem 7 (a) Explain the principle of operation of a programmable timer. (b) State the advantages of using a programmable timer device.

(a) A programmable timer is a device which may be used for the accurate measurement of time intervals by counting the system clock time periods. The timer usually contains an 8-bit register which may be loaded, under program control, with any value in the range 0 to FF_{16}. Once loaded, the contents of this register are automatically decremented at a rate determined by the system clock. An output from the timer indicates when the register contents reach zero, and this output may be polled by the microprocessor, or used to initiate an interrupt. Thus time intervals, which are a programmable multiple of the block period, are available for real time control of devices.

(b) The main advantages of using a programmable timer are:

(i) Since the system clock is usually crystal controlled, it provides an accurate time reference, therefore a programmable timer is capable of generating accurate time intervals.

(ii) Once loaded, a timer counts down automatically without further attention from the microprocessor. Therefore microprocessor time is not wasted processing long delay loops.

Problem 8 Describe the operation of the interval timer in a **6530** device.

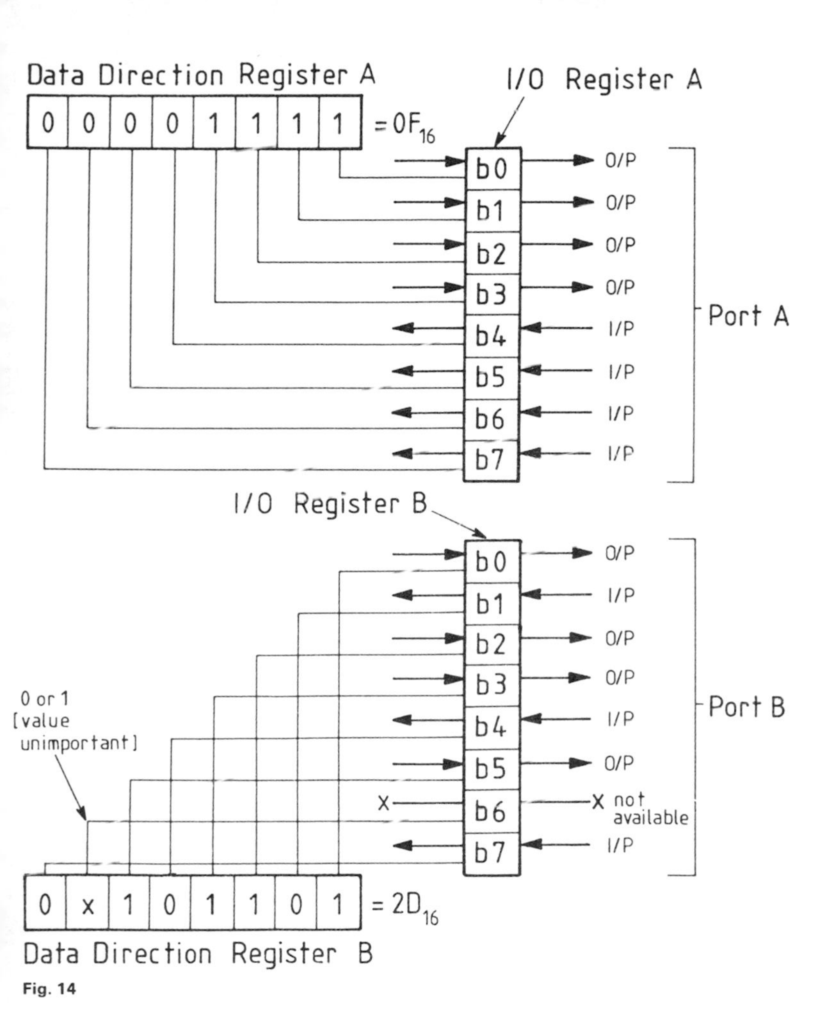

Fig. 14

The interval timer section of a 6530 device contains the following three basic parts:

(a) a preliminary divide down register;
(b) a programmable 8-bit register; and
(c) interrupt logic.

These parts are organized as shown in *Fig. 15*.

The interval timer allows the user to specify a preset count of up to 256_{10} and a **clock divide rate** of 1, 8, 64, 1024 by writing to a particular memory location.

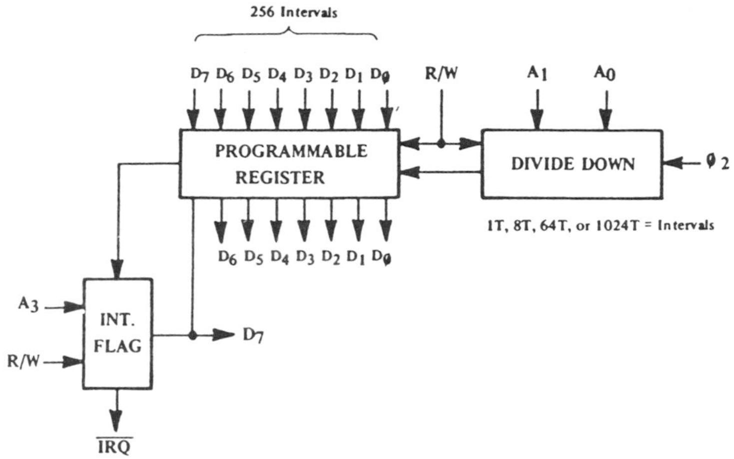

256 Intervals

D_7 D_6 D_5 D_4 D_3 D_2 D_1 D_0 R/W A_1 A_0

PROGRAMMABLE REGISTER

DIVIDE DOWN

$\phi 2$

1T, 8T, 64T, or 1024T = Intervals

D_6 D_5 D_4 D_3 D_2 D_1 D_0

A_3

R/W

INT. FLAG

D_7

$\overline{IRQ}$

Fig. 15

As soon as the write occurs, counting at the specified rate begins. The timer counts down by one for every 1, 8, 64 or 1024 clock cycles, according to the divide ratio selected. The current timer count may be read at any time, and, at the user's option, the timer may be programmed to generate an interrupt when a count of zero is passed. Once the timer has passed zero, the divide rate is automatically set to 1, and the counter continues to count down at the clock rate, starting at a count of FF_{16}. This allows the user to determine how many clock cycles have passed since a count of zero.

Problem 9 With reference to the interval timer in a **6530** device, explain how: (a) the timer may be loaded; (b) the timer status may be determined; and (c) the count in the timer may be read.

Table 2

Address	*Divide ratio*	*Interrupt capability*	*Maximum time interval*
1704	1	Disabled	255 μs
1705	8	Disabled	2 ms
1706	64	Disabled	16 ms
1707	1024	Disabled	260 ms
170C	1	Enabled	255 μs
170D	8	Enabled	2 ms
170E	64	Enabled	16 ms
170F	1024	Enabled	260 ms

Note: All time intervals assume to a 1 MHz clock frequency.

(a) The divide rate and the interrupt enable/disable option are programmed by decoding the least significant address bits A_0, A_1 and A_3. The starting count for the timer is determined by the value written to a specified address, and the options available are shown in *Table 2*.

(b) After timing has started, the **timer status** may be determined by reading address 1707_{16}. If the counter has passed the count of zero, bit 7 of this location is set to a logical 1, otherwise bit 7 (and all other bits in 1707_{16}) is at a logical 0. This allows a program to monitor location 1707_{16} and determine when the timer has timed out.

(c) If the timer has not counted past zero, reading location 1706_{16} provides the current timer count and **disables** the interrupt option. Reading location $170E_{16}$ provides the current timer count and **enables** the interrupt option, thus the interrupt option may be changed whilst the timer is counting down. Once the timer has passed zero, reading locations 1706_{16} or $170E_{16}$ restores the divide ratio to its previously programmed value, disables or enables the interrupt option and leaves the timer with its current count. The timer **never** stops counting, and continues to count down at the clock frequency unless new information is written to it.

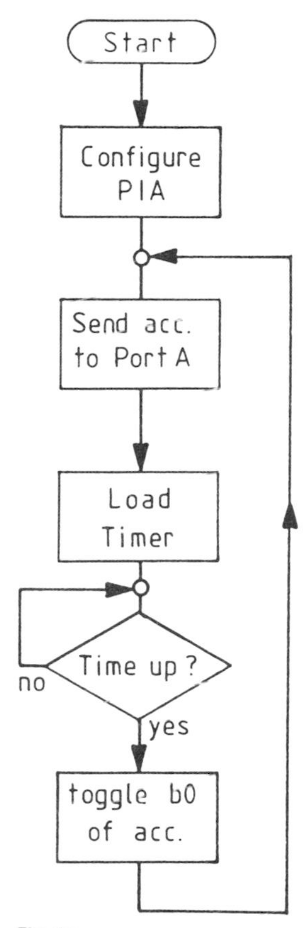

Fig. 16

Problem 10 (a) Write a **6502** machine code routine to show how the internal timer of a **6530** device may be used to generate a 5 kHz square wave signal on bit 0 of port A, using polling techniques to determine when the timer has timed out. Assume a 1 MHz system clock is used.
(b) Repeat (a), but arrange for the timer to generate an interrupt when it has timed out (see Chapter 2 for information on polling and interrupts).

(a) The timer in a **6530** device may be used to determine the frequency of a square wave generated at the output of bit 0 of port A by means of the following **6502** machine code program (see *Fig. 16* for the flowchart):

```
                    ;5 kHz square wave generator
                    ;6530 timer with polled status
                    ;
                    PAD         = $1700
                    PADD        = $1701
                    TIME        = $1704
                              * = $0200
                    ;
0200   A9 01                    LDA #$01       ;configure PIA
0202   8D 01 17                 STA PADD       ;b0 of port A output
0205   49 01        SQWV        EOR #$01       ;toggle b0
0207   8D 00 17                 STA PAD        ;and send to port A
020A   A2 64                    LDX #$64       ;X = clock periods
020C   8E 04 17                 STX TIME       ;load timer
020F   2C 07 17     POLL        BIT TIME + 3   ;test timer status
0212   10 FB                    BPL POLL       ;time up?
0214   30 EF                    BMI SQWV       ;next half cycle
```

(b) If the timer interrupt is selected, bit 7 of port B goes to a logical 0 each time that the counter passes zero, therefore this line should be connected to $\overline{\textbf{IRQ}}$ or $\overline{\textbf{NMI}}$ on the 6502 microprocessor in order to create an interrupt driven timing system. This arrangement is shown in *Fig. 17*.

An interrupt service routine is used in which a routine to toggle bit 0 of port A and reload the timer is included. This means that once the timing process is initiated in the main program, the microprocessor is free to run any desired program, since the timer automatically requests the appropriate I/O operation when required. A **6502** machine code program to implement this system is as follows:

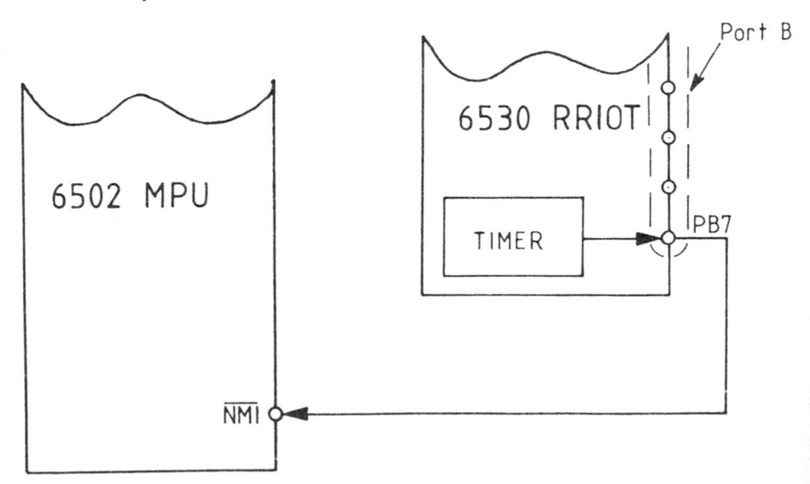

Fig. 17

```
                 ;5 kHz square wave generator
                 ;6530 timer with interrupt option
                 ;
                 PAD     =$1700
                 PADD    =$1701
                 TIME    =$170C
                 NMIV    =$17FA
                        *=$0200
                 ;
0200  A9 50              LDA #$50       ;low byte of interrupt
0202  8D FA 17           STA NMIV       ;routine address
0205  A9 02              LDA #$02       ;high byte of interrupt
0207  8D FB 17           STA NMIV+1     ;routine address
020A  A9 01              LDA #$01       ;configure b0 of port A
020C  8D 01 17           STA PADD       ;as an output
020F  49 01              EOR #$01       ;toggle b0
0211  8D 00 17           STA PAD        ;and send to port A
0214  A2 64              LDX #$64       ;X=clock periods
0216  8E 0C 17           STX TIME       ;load timer
      ..........
      ..........
      ..........                        ;go to any user's program
      ..........
      ..........
                 ;interrupt service routine
                 ;toggle b0 of port A when
                 ;timer counter reaches zero
                 ;
                        *=$0250
0250  49 01              EOR #$01       ;toggle b0
0252  8D 00 17           STA PAD        ;and send to port A
0255  8E 0C 17           STX TIME       ;reload the timer
0258  40                 RTI            ;and return to user's program
```

Note: bit 7 of port B must be configured as an input. This will normally be the case after a system reset, but if in doubt, port B should be configured during the initial program sequence.

It may be seen from this program, that once started, the square wave output may be maintained using only four instructions, i.e., those instructions forming the interrupt service routine. These instructions are executed at regular intervals during a user's program, at times dictated by the 6530 interval timer.

Problem 11 The block diagram of a **6522 versatile interface adapter** (VIA) is shown in *Fig. 18*. Describe the basic facilities provided by this device.

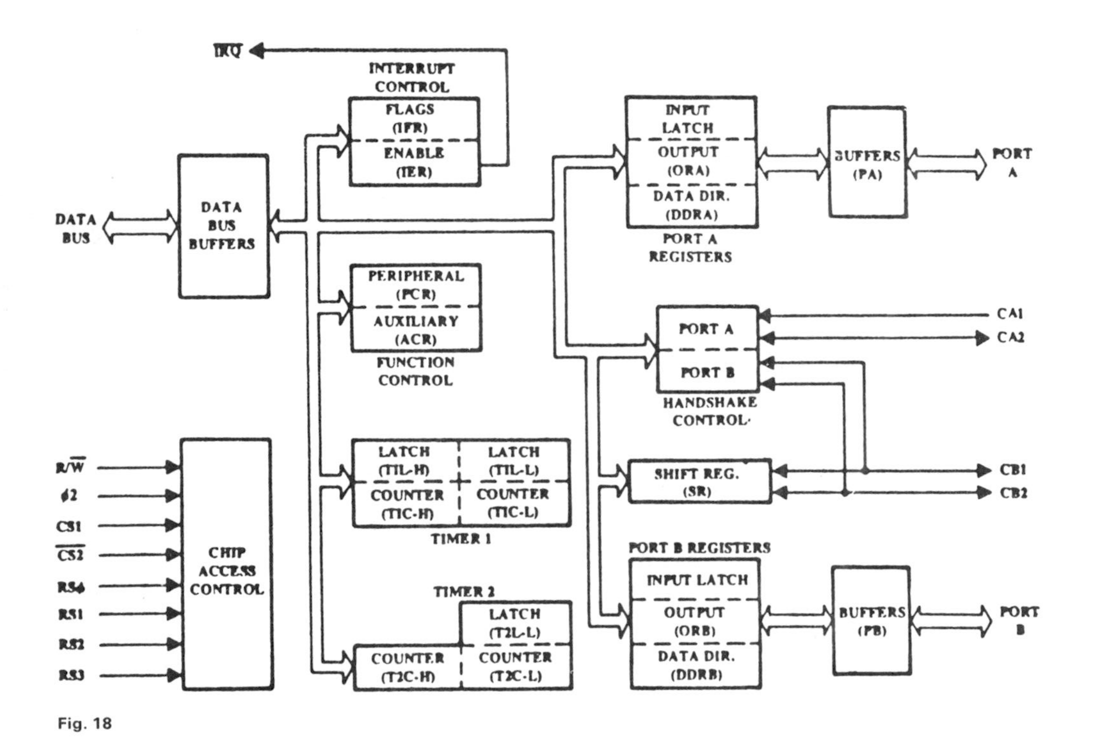
IRQ
INTERRUPT CONTROL
FLAGS (IFR)
ENABLE (IER)
DATA BUS
DATA BUS BUFFERS
PERIPHERAL (PCR)
AUXILIARY (ACR)
FUNCTION CONTROL
INPUT LATCH
OUTPUT (ORA)
DATA DIR. (DDRA)
PORT A REGISTERS
BUFFERS (PA)
PORT A
PORT A
PORT B
HANDSHAKE CONTROL
CA1
CA2
SHIFT REG. (SR)
CB1
CB2
LATCH (T1L-H)
LATCH (T1L-L)
COUNTER (T1C-H)
COUNTER (T1C-L)
TIMER 1
TIMER 2
LATCH (T2L-L)
COUNTER (T2C-H)
COUNTER (T2C-L)
PORT B REGISTERS
INPUT LATCH
OUTPUT (ORB)
DATA DIR. (DDRB)
BUFFERS (PB)
PORT B
R/W
φ2
CS1
CS2
RS4
RS1
RS2
RS3
CHIP ACCESS CONTROL
Fig. 18

The **6522 VIA** offers the following facilities:

(a) An **I/O section** which consists of two 8-bit bidirectional ports, A and B, each with handshaking facilities. Associated with each port is an input register, an output register and a data direction register. In addition, port A has an output register without handshake facilities. Each port is configured by loading its data direction register in a similar manner to that already described for the **6530** device (see *Problem 6*).

(b) Two **interval timers** which may be used as either outputs or inputs. When used as an input, a timer may be used to measure the duration or count the number of pulses applied to port B bit 6 (timer 2 only). When used as an output, each timer may generate a **single pulse** of programmable duration (**one-shot mode**), and timer 1 may also generate a **continuous** train of pulses of programmable duration (**free-running mode**).

(c) A **shift register** which may be used to perform serial data transfers into or out of the CB2 pin under the control of an internal modulo-8 counter. Shift pulses may be generated by either timer 2, the system clock ($\Phi 2$), or by externally generated pulses which are applied to CB1. The various shift register operating modes are determined by the contents of an **auxiliary control register** (ACR).

(d) A set of **control registers** which enable the **6522** device to operate under a wide variety of conditions, all under software control, to provide the flexibility required when interfacing different peripherals to a microcomputer system.

The memory assignment of the programmable elements of a **6522** device are shown in *Table 3*.

Problem 12 Describe the use of the **peripheral control register** (PCR) of a **6522 VIA**.

The contents of the PCR determine the operation of the control lines CA1/CA2 and CB1/CB2 associated with I/O ports A and B. Each bit in the PCR relates to one of the control lines, as indicated in *Fig. 19*.

The control lines CA1/CA2 and CB1/CB2 enable the timing of the data transfers between peripheral devices and a microcomputer to be controlled by means of interrupts or by full handshaking. The contents of the PCR determine whether the control lines act simply as interrupt lines or as handshake lines, and also allow the active edge of the control signals to be defined. Full details of the use of the PCR are shown in *Table 4*.

Problem 13 **Write 6502 machine code routines to show how port A of a 6522 device may be used to act as an input port under the following conditions: (a) simple input port with no control options; (b) an input port with an active low-to-high 'data ready' strobe applied to CA1, and (c) an input port with full handshaking, in which CA2 generates a positive going pulse to indicate to the peripheral device that the microprocessor is ready to accept new data, and in which the peripheral device indicates that new data are available by applying an active low-to-high 'data ready' strobe to CA1.**

Table 3

Location	Function		
A000	Output data register B (ORB)		
A001	Output data register A (ORA): controls handshake		
A002	Data direction register B (DDRB)		0 = input
A003	Data direction register A (DDRA)}		1 = output
	Timer	R/W = 0	R/W = 1
A004	T1	Write to T1 latch low	Read T1 counter low Reset T1 interrupt flag
A005	T1	Write to T1 latch high Write to T1 counter high Latch low → counter low Reset T1 interrupt flag	Read T1 counter high
A006	T1	Write to T1 latch low	Read T1 latch low
A007	T1	Write to T1 latch high Reset T1 interrupt flag	Read T1 latch high
A008	T2	Write to T2 latch low	Read T2 counter low Reset T2 interrupt flag
A009	T2	Write to T2 counter high Latch low → counter low Reset T2 interrupt flag	Read T2 counter high
A00A	Shift register (SR)		
A00B	Auxilliary control register (ACR)		
A00C	Peripheral control register (PCR)		
A00D	Interrupt flag register (IFR)		
A00E	Interrupt enable register (IER)		
A00F	Output data register A (ORA): no effect on handshake		

Note: high order addresses may be different according to the memory mapping of the system used, but all registers still have the same relative position in the memory map.

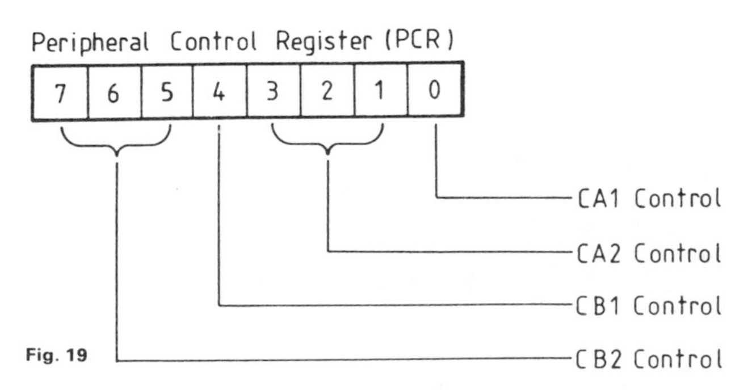

Fig. 19

Assuming that the **6522 VIA** is assigned to memory locations $A000_{16}$ to $A00F_{16}$, the machine code routines are as follows:

(a)

```
                    ;simple input port – no protocol
                    ;
                    ORA     = $A001
                    DDRA    = $A003
                    BUFF    = $0080
                          * = $0200
                    ;
0200  A9 00                     LDA #0        ;configure VIA
0202  8D 03 A0                  STA DDRA      ;port A all inputs
0205  AD 01 A0                  LDA ORA       ;read port A
0208  85 80                     STA BUFF      ;save data in memory
      ...............
      ...............
```

(b)

```
                    ;input port with data ready strobe
                    ;polls interrupt flag register (IFR)
                    ;for strobe
                    ;
                    ORA     = $A001
                    DDRA    = $A003
                    PCR     = $A00C
                    IFR     = $A00D
                    BUFF    = $0080
                          * = $0200
                    ;
0200  A9 00                     LDA #0        ;configure VIA
0202  8D 03 A0                  STA DDRA      ;port A all inputs
0205  A9 01                     LDA #1        ;define CA1 active edge
0207  8D 0C A0                  STA PCR       ;as low to high
020A  AD 0D A0      FLAG        LDA IFR       ;read CA1 interrupt flag
020D  29 02                     AND #2        ;mask off unwanted bits
020F  F0 F9                     BEQ FLAG      ;keep looking if not ready
0211  AD 01 A0                  LDA ORA       ;read port A
0214  85 80                     STA BUFF      ;save data in memory
      ...............
      ...............
```

In this example, CA1 is connected to the peripheral device. CA1 is programmed to accept an active low-to-high transition by storing a logical 1 in the bit 0 position of the peripheral control register (PCR) at address $A00C_{16}$. When the peripheral device is ready to send new data to port A, it activates CA1 which causes the CA1 interrupt flag in the **interrupt flag register** (IFR) to be set. The IFR is located at address $A00D_{16}$, and the CA1 interrupt flag is bit 1 of this register. The program polls this flag, and waits until a logical 1 is detected before reading port A and transferring data into memory at address 0080_{16}.

(c)

```
                          ;input port with handshake control
                          ;of data transfer; data inputs not latched
                          ;
                          ORA      = $A001
                          DDRA     = $A003
                          PCR      = $A00C
                          IFR      = $A00D
                          BUFF     = $0080
                                 * = $0200
                          ;
0200  A9 00                        LDA #0      ;configure VIA
0202  8D 03 A0                     STA DDRA    ;port A all inputs
0205  A9 0C                        LDA #$0C    ;select CA2 low output
0207  8D 0C A0                     STA PCR     ;mode
020A  49 01                        EOR #1      ;set bit 0 of A to generate
020C  8D 0C A0                     STA PCR     ;MPU ready pulse
020F  49 01                        EOR #1      ;reset bit 0 of A to terminate
0211  8D 0C A0                     STA PCR     ;MPU ready pulse (CA2 high)
0214  AD 0D A0  FLAG               LDA IFR     ;read CA1 interrupt flag
0217  29 02                        AND #2      ;mask off unwanted bits
0219  F0 F9                        BEQ FLAG    ;keep looking for data strobe
021B  AD 01 A0                     LDA ORA     ;read port A
021E  85 80                        STA BUFF    ;save data in memory
      ............
      ............
```

Note, in examples (b) and (c), the interrupt flag register is polled in the programs in order to determine when the peripheral has sent a data ready strobe. By appropriate changes to these programs, an interrupt driven input may be implemented. This involves removal of the polling sequence, inclusion of an appropriate interrupt service routine, and loading the **6522 interrupt enable register** (IER) at address $A00E_{16}$ with 02 to enable CA1 to generate an interrupt. Also, in examples (b) and (c), data may be latched onto the 6522 inputs by setting bit 0 of the auxiliary control register (ACR) at address $A00B_{16}$.

Problem 14 (a) Explain the operation of the **6522 VIA** shift register,
(b) Write a **6502** machine code routine to show how the contents of a specified memory location may be shifted out of CB2 as **serial data** at a rate determined by the microprocessor clock Φ2, and
(c) Write a **6502** machine code program to show how serial data applied to CB2 may be shifted into a specified memory location using the microprocessor clock Φ2.

CB1 control		
PCR4	= 0	The CB1 interrupt flag (IFR4) will be set by a negative transition (high to low) on the CB1 pin.
	= 1	The CB1 interrupt flag (IRF4) will be set by a positive transition (low to high) on the CB1 pin.

CB2 control

PCR7	PCR6	PCR5	*Mode*
0	0	0	CB2 negative edge interrupt (IFR3/ORB clear) mode – set CB2 interrupt flag (IFR3) on a negative transition of the CB2 input signal. Clear IFR3 on a read or write of the ORB or by writing logic 1 into IFR3.
0	0	1	CB2 negative edge interrupt (IFR3 clear) mode – set IFR3 on a negative transition of the CB2 input signal. Clear IFR3 by writing logic 1 into IFR3.
0	1	0	CB2 positive edge interrupt (IFR3/ORB clear) mode – set CB2 interrupt flag (IFR3) on a positive transition of the CB2 input signal. Clear IFR3 on a read or write of the ORB or by writing logic 1 into IFR3.
0	1	1	CB2 positive edge interrupt (IFR3 clear) mode – set IFR3 on a positive transition of the CB2 input signal. Clear IFR3 by writing logic 1 into IFR3.
1	0	0	CB2 handshake output mode – set CB2 output low on a write of the peripheral B output register. Reset CB2 high with an active transition on CB1.
1	0	1	CB2 pulse output mode – CB2 goes low for one cycle following a read or write of the peripheral B output register.
1	1	0	CB2 low output mode – CB2 output is held low in this mode.
1	1	1	CB2 high output mode – CB2 output is held high in this mode.

CA1 control		
PCR0	= 0	The CA1 interrupt flag (IFR1) will be set by a negative transition (high to low) on the CA1 pin.
	= 1	The CA1 interrupt flag (IFR1) will be set by a positive transition (low to high) on the CA1 pin.

CA2 control

PCR3	PCR2	PCR1	*Mode*
0	0	0	CA2 negative edge interrupt (IFR0/ORA clear) mode – set CA2 interrupt flag (IFR0) on a negative transition of the CA2 input signal. Clear IFR0 on a read or write of the ORA or by writing logic 1 into IFR0.
0	0	1	CA2 negative edge interrupt (IFR0 clear) mode – set IFR0 on a negative transition of the CA2 input signal. Clear IFR0 by writing logic 1 into IFR0.
0	1	0	CA2 positive edge interrupt (IFR0/ORA clear) mode – set CA2 interrupt flag (IFR0) on a positive transition of the CA2 input signal. Clear IFR0 on a read or write of the ORA or by writing logic 1 into IFR0.
0	1	1	CA2 positive edge interrupt (IFR0 clear) mode – set IFR0 on a positive transition of the CA2 input signal. Clear IFR0 by writing logic 1 into IFR0.
1	0	0	CA2 handshake output mode – set CA2 output low on a write of the peripheral A output register. Reset CA2 high with an active transition on CA1.
1	0	1	CA2 pulse output mode – CA2 goes low for one cycle following a read or write of the peripheral A output register.
1	1	0	CA2 low output mode – CA2 output is held low in this mode.
1	1	1	CA2 high output mode – CA2 output is held high in this mode.

(a) The **6522 VIA** contains an 8-bit shift register which may be used to shift serial data into or out of the CB2 pin under the control of an internal modulo-8 counter. The rate at which data is transferred may be controlled by either the microprocessor clock Φ2, interval timer 2 (one-shot or free-run modes), or by an external clock connected to CB1. The source of shift pulses and the direction of the serial data is determined by the contents of the auxiliary control register (ACR), as defined in *Fig. 20* and *Table 5*.

Table 5

Port A latch enable			
ACR0	= 1		Port A latch is enabled to latch input data when CA1 interrupt flag (IFR1) is set.
	= 0		Port A latch is disabled, reflects current data on PA pins.
Port B latch enable			
ACR1	= 1		Port B latch is enabled to latch the voltage on the pins for the interrupt lines or the ORB contents for the output lines when CB1 interrupt flag (IFR4) is set.
	= 0		Port B latch is disabled, reflects current data on PB pins.
Shift register control			
ACR4	ACR3	ACR2	*Mode*
0	0	0	Shift register disabled.
0	0	1	Shift is under control of timer 2.
0	1	0	Shift is under control of Φ2
0	1	1	Shift is under control of external clock.
1	0	0	Free running output at rate determined by timer 2.
1	0	1	Shift out under control of timer 2.
1	1	0	Shift out under control of Φ2.
1	1	1	Shift out under control of external clock.
Timer 2 control			
ACR5	= 0		T2 acts as an interval timer in the one-shot mode.
	= 1		T2 counts a predetermined number of pulses on PB6.
Timer 1 control			
ACR7	ACR6		*Mode*
0	0		T1 one-shot mode – generate a single time-out interrupt each time T1 is loaded. Output to PB7 disabled.
0	1		T1 free-running mode – generate continuous interrupts. Output to PB7 disabled.
1	0		T1 one-shot mode – generate a single time-out interrupt and an output pulse on PB7 each time T1 is loaded.
1	1		T1 free-running mode – generate continuous interrupts and a square wave output on PB7

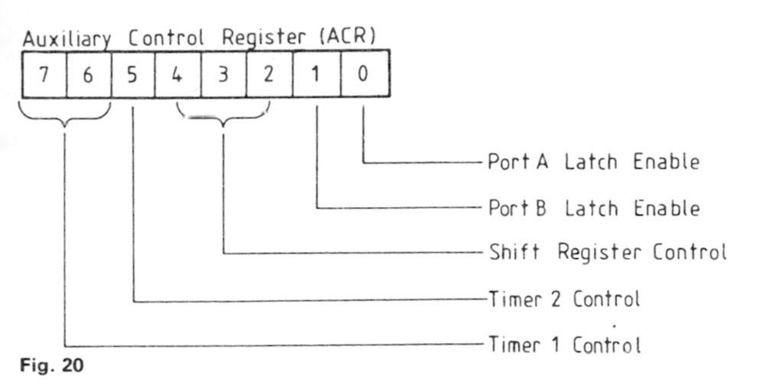

Fig. 20

(b)

```
                          ;serial data output from CB2 under
                          ;control of MPU clock Φ2
                          ;
                          ACR        = $A00B
                          SR         = $A00A
                          BUFF       = $0080
                                   * = $0200
                          ;
0200   A9 00                         LDA #0
0202   8D 0B A0                      STA ACR      ;
0205   A9 18                         LDA #$18     ;select shift out under
0207   8D 0B A0                      STA ACR      ;control of Φ2
020A   A5 80                         LDA BUFF     ;get parallel data
020C   8D 0A A0                      STA SR       ;shift out in serial form
       ...............
       ...............
```

The $\Phi 2$ shift out mode is selected by loading the ACR with 18_{16} which puts the binary combination 110_2 into bits 4, 3 and 2. The shift register is loaded with data from memory location 0080_{16}, and this action automatically starts the shifting operation at a rate determined by $\Phi 2$. The program should wait until all 8 bits have been shifted out before loading the shift register with new data. The **6522 VIA** automatically sets bits 2 of the **interrupt flag register** (IFR) once all eight bits have been shifted out, and the state of this flag should be tested to determine when the shift register may be reloaded. Bit 2 of the IFR is cleared by writing new data in the shift register.

(c)

```
                        ;serial data input from CB2 under
                        ;control of MPU clock Φ2
                        ;
                        ACR      = $A00B
                        SR       = $A00A
                        IFR      = $A00D
                        BUFF     = $0080
                               * = $0200
                        ;
0200  A9 00                      LDA #0
0202  8D 0B A0                   STA ACR       ;clear SR
0205  A9 0C                      LDA #$0C      ;select shift in under
0207  8D 0B A0                   STA ACR       ;control of Φ2
020A  AD 0D A0          FLAG     LDA IFR       ;read 8-shifts flag
020D  29 04                      AND #$04      ;mask off unwanted flags
020F  F0 F9                      BEQ FLAG      ;all 8 shifts done?
0211  AD 0A A0                   LDA SR        ;read shift register
0214  85 80                      STA BUFF      ;save data in memory
      ...............
      ...............
```

Reading the parallel data from the shift register automatically clears bit 2 of the IFR and initiates another shift-in sequence.

In both examples (b) and (c), shift pulses generated internally are available as an output from CB1 for controlling external circuits.

Problem 15 The block diagram of an 8255 **programmable peripheral interface (PPI)** device is shown in *Fig. 21*.

Describe the three modes of operation available with this device.

The **8255 PPI** has **three I/O ports** (port A, port B and port C), providing a total of twenty-four I/O pins which may be programmed in two groups of twelve, and which may be used in one of the following three modes:

(a) **Mode 0:** basic I/O mode;
(b) **Mode 1:** strobed I/O mode; and
(c) **Mode 2:** bidirectional I/O mode.

The characteristics of each of these operating modes may be summarized as follows:

Mode 0

This mode provides **simple I/O** operations without handshake facilities, data being simply written to or read from each of the three I/O ports as required. Ports A and B operate as two 8-bit ports, whilst port C is operated as two 4-bit ports, and any port may be configured as either an input port or an output port.

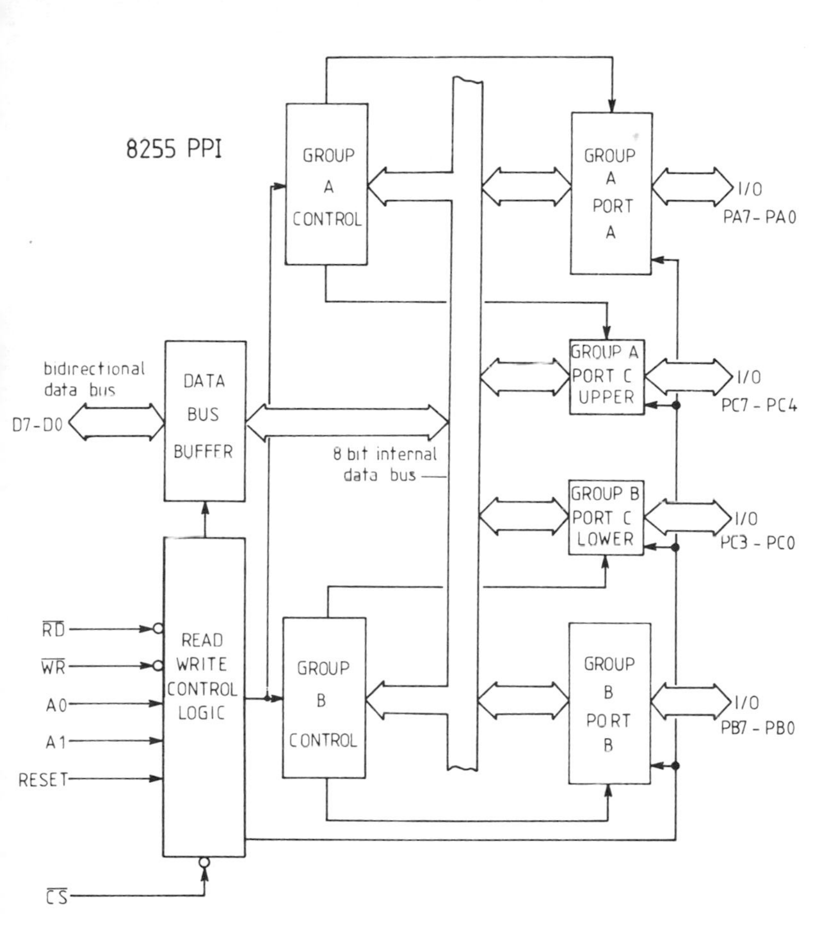

Fig. 21

Mode 1
This mode provides I/O operations with **handshake** facilities. Ports A and B may be defined as either input ports or output ports, and handshaking is provided by port C. Six bits of port C are used for handshaking and interrupt control, three bits for port A, and three bits for port B.

Mode 2
This mode provides **bidirectional I/O** operations on port A only. Handshaking is provided by five bits of port C.

These modes of operations are summarized in *Table 6*.

Table 6

	Mode 0		*Mode 1*		*Mode 2*	
	In	*Out*	*In*	*Out*	*Group A only*	
PA0	in	out	in	out	↔	
PA1	in	out	in	out	↔	
PA2	in	out	in	out	↔	
PA3	in	out	in	out	↔	
PA4	in	out	in	out	↔	
PA5	in	out	in	out	↔	
PA6	in	out	in	out	↔	
PA7	in	out	in	out	↔	
PB0	in	out	in	out	—	
PB1	in	out	in	out	—	
PB2	in	out	in	out	—	Mode 0
PB3	in	out	in	out	—	or mode 1
PB4	in	out	in	out	—	only
PB5	in	out	in	out	—	
PB6	in	out	in	out	—	
PB7	in	out	in	out	—	
PC0	in	out	INTRA B	INTRA B	I/O	
PC1	in	out	IBF B	$\overline{\text{OBF B}}$	I/O	
PC2	in	out	$\overline{\text{STB B}}$	$\overline{\text{ACK B}}$	I/O	
PC3	in	out	INTRA A	INTRA A	INTRA A	
PC4	in	out	$\overline{\text{STB A}}$	I/O	$\overline{\text{STB A}}$	
PC5	in	out	IBF A	I/O	IBF A	
PC6	in	out	I/O	$\overline{\text{ACK A}}$	$\overline{\text{ACK A}}$	
PC7	in	out	I/O	$\overline{\text{OBF A}}$	$\overline{\text{OBF A}}$	

INTR = Interrupt
IBF = Input buffer full
$\overline{\text{OBF}}$ = Output buffer full
$\overline{\text{STB}}$ = Strobe
$\overline{\text{ACK}}$ = Acknowledge

Problem 16 (a) Explain how an **8255 PPI** is configured to obtain specified operating conditions, and (b) write a **Z80** machine code routine to show how an 8255 may be configured to operate in mode 0 as follows:
Port A input mode;
Port B output mode;
Port C (low) output mode, and
Port C (high) input mode.

(a) The **8255 PPI** has two port select signals (A_0 and A_1) which, in conjunction with the $\overline{\text{RD}}$ and $\overline{\text{WR}}$ inputs control selection of one of the three I/O ports

Table 7

A_1	A_0	$\overline{RD}$	$\overline{WR}$	$\overline{CS}$	*Input operation (read)*
0	0	0	1	0	Port A → data bus
0	1	0	1	0	Port B → data bus
1	0	0	1	0	Port C → data bus
					Output operation (write)
0	0	1	0	0	Data bus → Port A
0	1	1	0	0	Data bus → Port B
1	0	1	0	0	Data bus → Port C
1	1	1	0	0	Data bus → Control
					Disable function
X	X	X	X	1	Data bus → 3-state
1	1	0	1	0	Illegal condition

or the control word register. Selection of an I/O port or the control word register is achieved by using an appropriate address in conjunction with an IN or OUT instruction (or LD in memory mapped I/O systems), and the least significant two bits of the address identify the port or control register as shown in *Table 7*.

The contents of the control word register determine the mode selection and the port configuration, as shown in *Fig. 22*. Therefore, configuring an **8255 PPI** involves storing an appropriate control word in this register.

(b) Configuring an **8255 PPI** for the conditions stated involves the following two instructions:

```
3E 98    LDA,98H      ;control word
D3 03    OUT (3),A    ;send to control register
```

The derivation of the control word used in this example is shown in *Fig. 23*.

Problem 17 Explain the operation of the **single bit set/reset** features of an **8255 PPI**.

For normal mode selection and configuring operations on an **8255 PPI**, bit 7 of the control word register is set to a logical 1. This bit behaves as a **mode set** flag and ensures that the data sent to the control address is stored in the **mode definition register** of the PPI. If a control word with bit 7 reset to logical 0 is used, a different function is provided at the control address. This function enables any of the eight bits of port C to be set or reset using a single OUT instruction, as shown in *Fig. 24*.

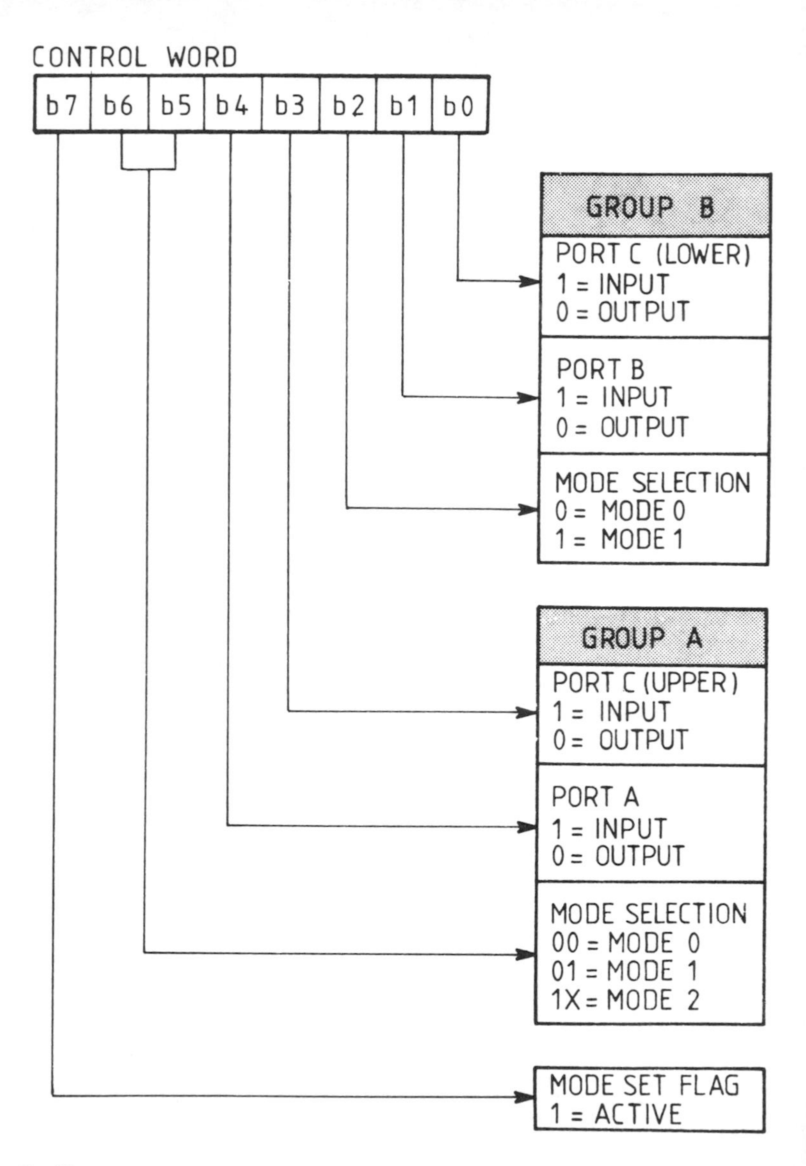
CONTROL WORD
b7
b6
b5
b4
b3
b2
b1
b0
GROUP B
PORT C (LOWER)
1 = INPUT
0 = OUTPUT
PORT B
1 = INPUT
0 = OUTPUT
MODE SELECTION
0 = MODE 0
1 = MODE 1
GROUP A
PORT C (UPPER)
1 = INPUT
0 = OUTPUT
PORT A
1 = INPUT
0 = OUTPUT
MODE SELECTION
00 = MODE 0
01 = MODE 1
1X = MODE 2
MODE SET FLAG
1 = ACTIVE

Fig. 22

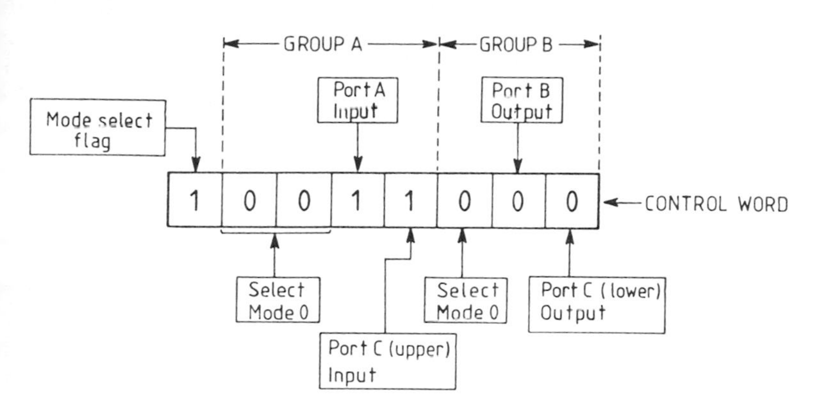

Fig. 23

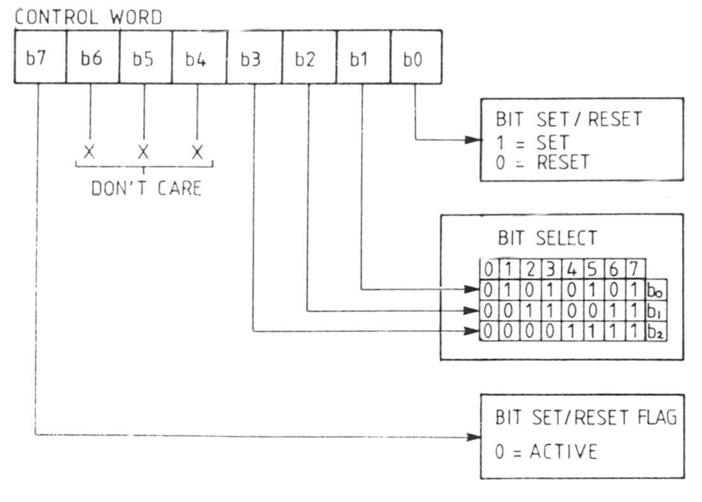

Fig. 24

When the **8255 PPI** is used in mode 1 or mode 2, control signals are provided which may be used as interrupt requests to the microprocessor. The interrupt request signals generated by port C may be enabled or disabled by setting or resetting the appropriate **INTE flip-flop** using the bit set/reset facility of port C.

Problem 18 The block diagram of a **Z80 PIO** is shown in *Fig. 25*.
Explain the function of the following registers:
(a) **mode control register**;
(b) **input/output select register**;
(c) **mask control register**;
(d) **mask register**.

(a) The **mode control register** is a 2-bit register whose contents are used to select one of the four operating modes of the **Z80 PIO**. The four modes are:
(i) **Output mode** (mode 0)
(ii) **Input mode** (mode 1)
(iii) **Bidirectional mode** (mode 2)
(iv) **Bit mode** (mode 3)
The mode numbers have been chosen to have mnemonic significance, i.e., 0 = out, 1 = in, and 2 = bidirectional. Port A may be operated in any of these four modes; port B may operate in all modes except mode 2.
(b) The **input/output select register** is an 8-bit register which is used in mode 3 only to specify which I/O lines of a port act as inputs, and which lines act as outputs. A logical 0 in a particular bit position in this register defines the corresponding I/O line as an output line, and a logical 1 defines the corresponding I/O line as an input line. Note the mnemonic significance of this, i.e., 0 = output, 1 = input.
(c) The **mask control register** is a 2-bit register which is used in mode 3 only. In mode 3 (bit mode), the I/O lines may function as interrupt inputs if programmed to do so. One bit in the mask control register determines the active state of the inputs for initiating an interrupt (i.e., active high or active low). The other bit determines whether the PIO requires all specified inputs to be active to initiate an interrupt (AND condition), or if any single line active may initiate an interrupt (OR condition).
(d) The **mask register** is an 8-bit register whose contents determine which I/O lines in mode 3 must not be used to initiate interrupts, i.e., which bits are to be masked. Only those port bits whose corresponding bits in this register are at logical 0 will be monitored for generating an interrupt.

Problem 19 Explain how the following operations are carried out when a **Z80 PIO** is configured: (a) **load interrupt vector**; (b) **set mode**; and (c) **set interrupt control**.

A **Z80 PIO** is arranged such that all control words are sent to the same I/O address. It is therefore necessary for the PIO to distinguish between the different control words so that they may be directed to the correct register within the control section. This is achieved by considering the format of a control word, or the order in which it is sent to the PIO. The operations listed may be carried out in the following manner:

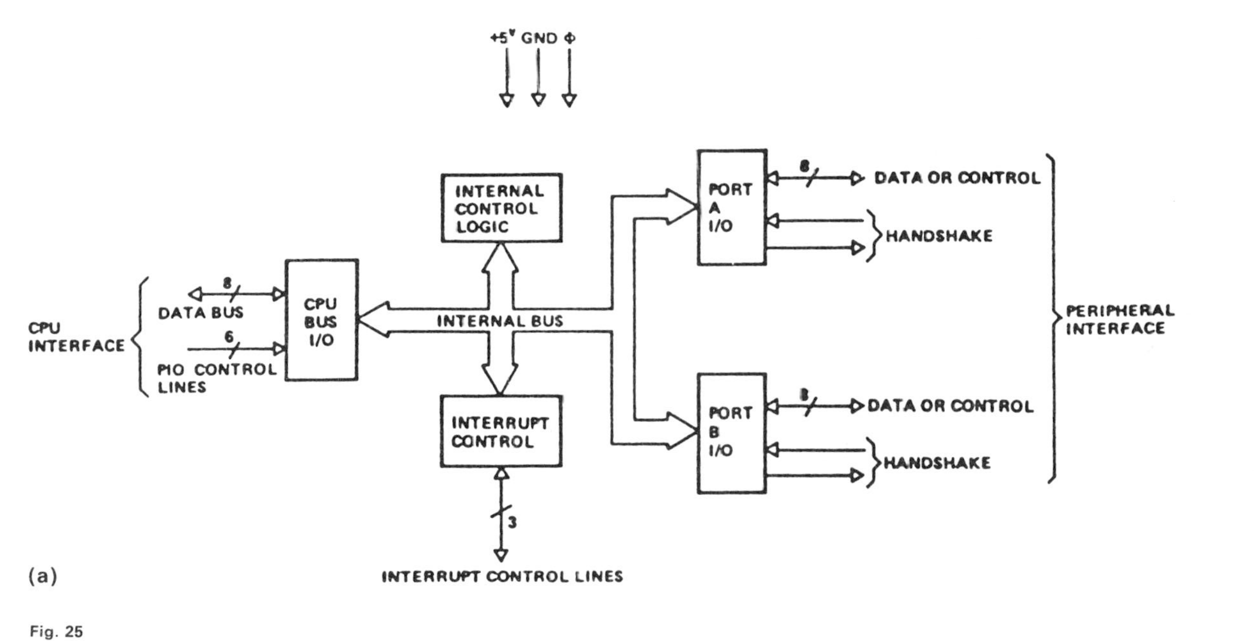
+5V GND Φ
INTERNAL
CONTROL
LOGIC
CPU
INTERFACE
8
DATA BUS
6
PIO CONTROL
LINES
CPU
BUS
I/O
INTERNAL BUS
INTERRUPT
CONTROL
3
INTERRUPT CONTROL LINES
PORT
A
I/O
8
DATA OR CONTROL
HANDSHAKE
PORT
B
I/O
8
DATA OR CONTROL
HANDSHAKE
PERIPHERAL
INTERFACE

(a)

Fig. 25

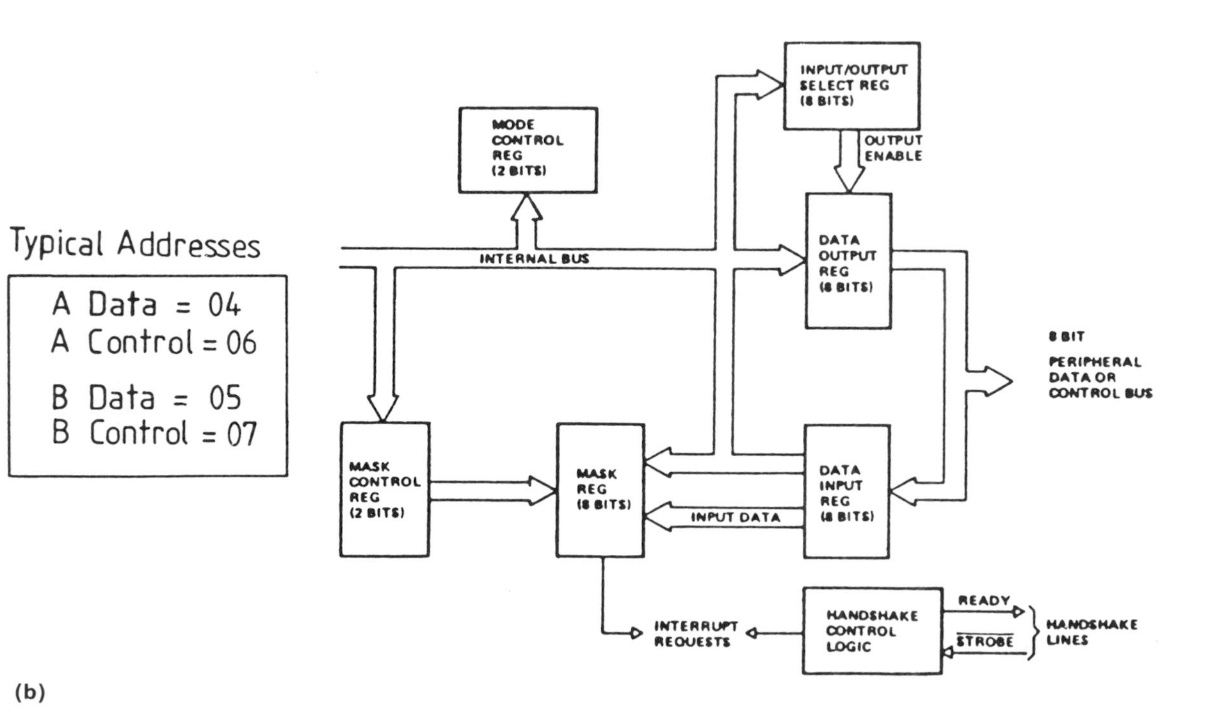

(b)

Fig. 25 – continued

(a) An **interrupt vector** may be loaded into the PIO by writing a control word to the desired port of the PIO, using the format shown in *Fig. 26(a)*.

It may be seen from this diagram that any **even number ($D_0 = 0$)** which is sent to the PIO control section is treated as an interrupt vector.

(b) The **Z80 PIO** is capable of operating in any of the four distinct modes. These are:

(i) **byte output mode** (mode 0);
(ii) **byte input mode** (mode 1);
(iii) **directional mode** (mode 2) – port A only, and
(iv) **bit mode** (mode 3).

The mode of operation may be selected by writing a control word to the PIO with the format shown in *Fig. 26(b)*. It may be seen from this diagram that bits D_3 and D_0 must be set to **1111** to indicate to the PIO control section that this is a **set mode** word.

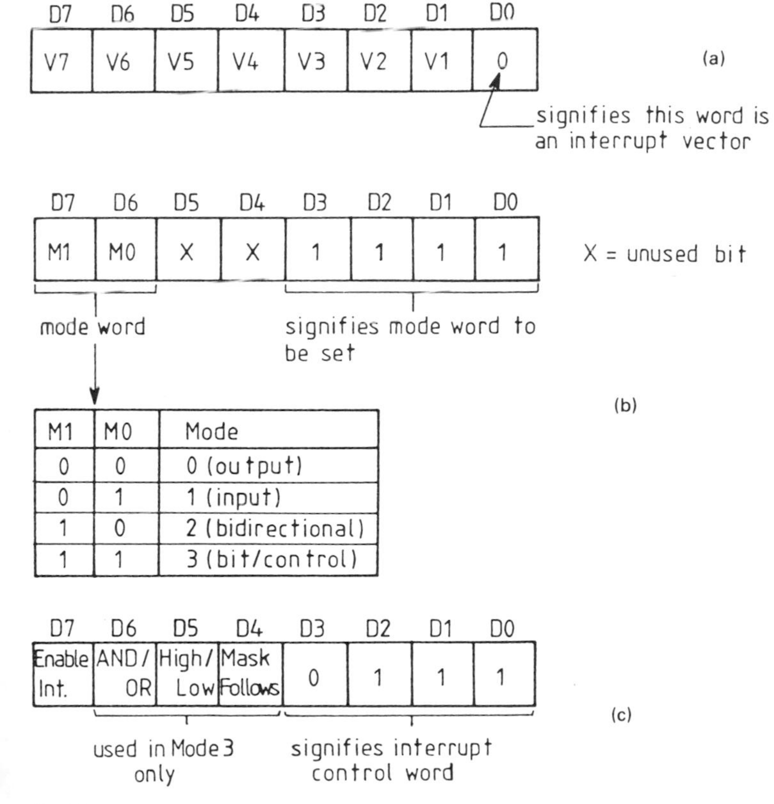

Fig. 26

Bits D_5 and D_4 are ignored and may therefore be any value chosen (typically both 0). If mode 3 (bit mode) is selected, the next word written to the PIO control section **must** define which port data lines are input lines and which are output lines. Each bit in this word defines its corresponding port bits as an output if it is a logical 0 or an input if it is a logical 1.

(c) The **interrupt control word** for each port has the format shown in *Fig. 26(c)*. It may be seen from this diagram that bits D_3 to D_0 must be **0111** to indicate to the PIO control section that this is a **set interrupt control** word. Bits D_6 and D_4 are only used in mode 3 and are ignored by other modes. In mode 3, bits D_6 to D_4 are used to select different forms of port monitoring for interrupts, and these are indicated in *Table 8*.

Table 8

Bit		*0*	*1*
Mask follows	D_4	No mask to follow	Next word sent to control port must be an interrupt mask
High/low	D_5	Interrupt active low	Interrupt active high
And/or	D_6	Any unmasked input active generates an interrupt	All unmasked inputs must be active to generate an interrupt

If the control word sent to the PIO has its **mask follows** bit (bit D_4) set, the next control word sent to the port is interpreted as a mask. Port lines whose corresponding mask bit is 0 are monitored for generating an interrupt. The remaining lines are not monitored for generating an interrupt.

Problem 20 Write **Z80** machine code routines to show how a **Z80 PIO** may be configured to enable the following operations to take place:

(a) Read eight bits of data in from port A, invert the data and send it back out through port B.

(b) Read in four bits of data from port A bits 0 to 3, invert the data and send it back out through port A bits 4 to 7.

(The PIO is located at I/O addresses 04 to 07.)

(a)

```
                ;Non-memory mapped Z80 PIO with I/O addresses
                ;in the range 04 to 07
                ;STB must be connected to logical 0 to transfer
                ;data to port A input register
                ;
                 PAD    EQU 4         ;port A data I/O register
                 PBD    EQU 5         ;port B data I/O register
                 PAC    EQU 6         ;port A control section
                 PBC    EQU 7         ;port B control section
                        ORG 0C90H     ;
                ;
0C90   3E 4F            LD A,4FH
0C92   D3 06            OUT (PAC),A   ;port A byte input (mode 1)
0C94   3E 0F            LD A,0FH
0C96   D3 07            OUT (PBC),A   ;port B byte output (mode 0)
0C98   DB 04    READ    IN A,(PAD)    ;read port A
0C9A   2F               CPL           ;invert data
0C9B   D3 05            OUT (PBD),A   ;and send to port B
0C9D   18 F9            JR READ       ;repeat if required
```

(b)

```
                ;Non-memory mapped Z80 PIO with I/O addresses
                ;4 and 6
                ;Uses bit mode (Mode 3), therefore STB is
                ;inoperative
                ;
                 PAD    EQU 4         ;port A data I/O register
                 PAC    EQU 6         ;port A control section
                        ORG 0C90H     ;
                ;
0C90   3E CF            LD A,0CFH
0C92   D3 06            OUT (PAC),A   ;port A bit mode (mode 3)
0C94   3E 0F            LD A,0FH      ;I/O definition word must
0C96   D3 06            OUT (PAC),A   ;follow selection of mode 3
0C98   DB 04    READ    IN A,(PAD)    ;read data b0-b3
0C9A   87               ADD A,A       ;equivalent to shift left
OC9B   87               ADD A,A       ;instruction
0C9C   87               ADD A,A       ;do four times
0C9D   87               ADD A,A
0C9E   EE F0            XOR 0F0H      ;invert data
0CA0   D3 04            OUT (PAD),A   ;send shifted data to b4-b7
0CA2   18 F4            JR READ       ;repeat if required
```

Problem 21 A **Z80-based microcomputer** with a **Z80 PIO** is used to monitor an industrial process, as shown in *Fig. 27*.

If an abnormal operating condition is detected, the **PWR. FAIL, TEMP** or **PRESS** alarm inputs to the PIO become active (high) and an interrupt is generated. With the aid of a Z80 machine code routine, explain how a **Z80 PIO** may be configured for this application.

D7	D6	D5	D4	D3	D2	D1	D0
Special Test	Turn On Power	Power Failure Alarm	Halt Processing	Temp. Alarm	Turn Heaters On	Pressurize System	Pressure Alarm

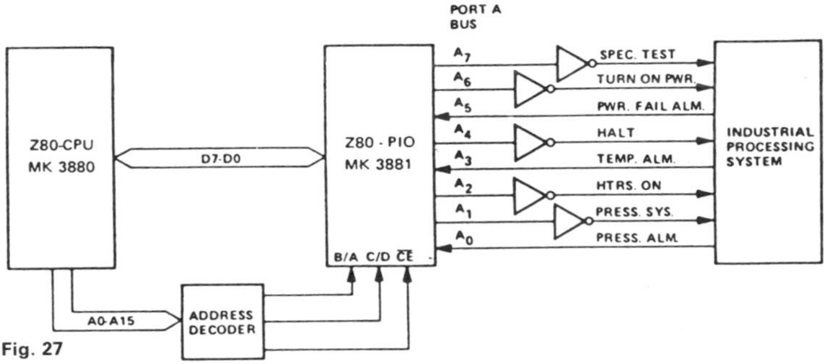

Fig. 27

The sequence of operations required to configure a **Z80 PIO** for this application is shown in *Fig. 28*. From this it may be determined that a suitable configuring program for this system may be as follows:

```
                ;Z80 control application
                ;Non-memory mapped I/O with port A addresses 4 and 6
                ;Mode 2 interrupt with ISR vector at 0D20H
                ;Mode 3 PIO operation, STB ignored
                ;
                  PAD   EQU 4          ;port A data I/O register
                  PAC   EQU 6          ;port A control section
                        ORG 0C90H
                ;
0C90  3E CF             LD A,0CFH
0C92  D3 06             OUT (PAC),A    ;port A bit mode (mode 3)
0C94  3E 29             LD A,29H
0C96  D3 06             OUT (PAC),A    ;define I/O lines
0C98  3E 20             LD A,20H
0C9A  D3 06             OUT (PAC),A    ;load peripheral vector
0C9C  3E B7             LD A,0B7H
0C9E  D3 06             OUT (PAC),A    ;define interrupt control word
0CA0  3E D6             LD A,0D6H
0CA2  DE 06             OUT (PAC),A    ;define interrupt mask
      ..........
      ..........
```

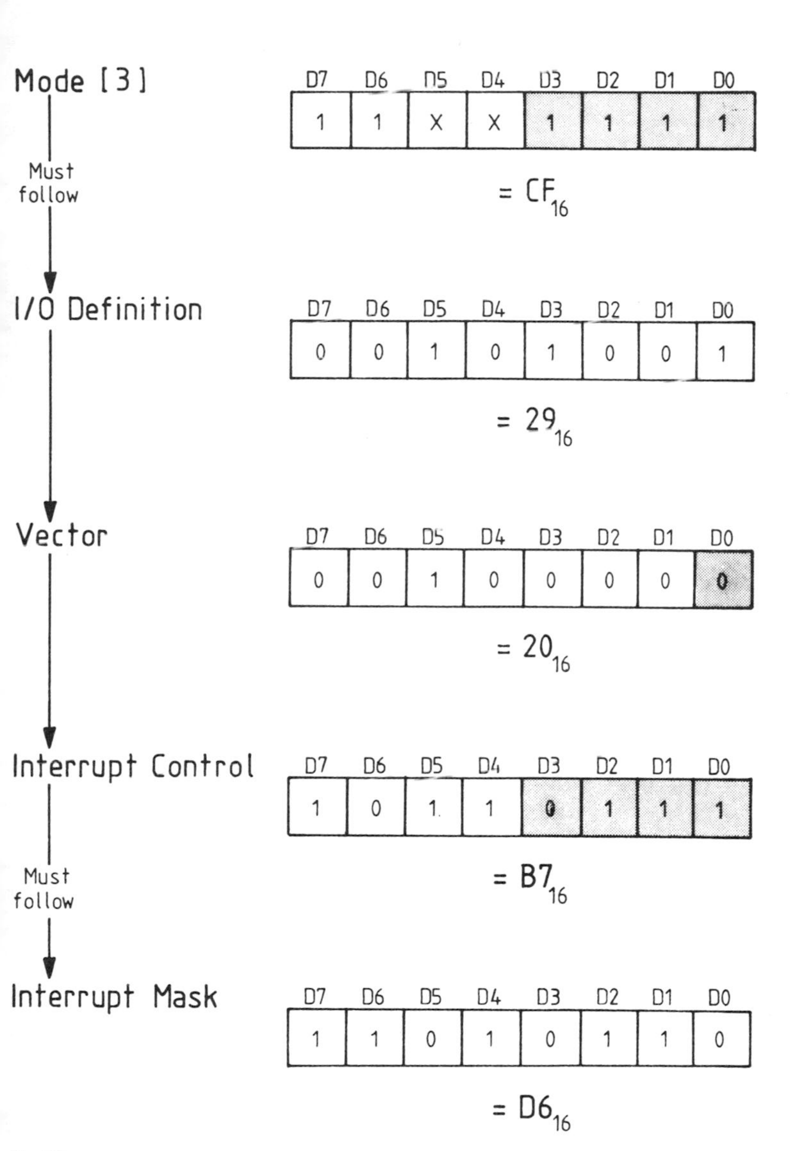

Mode [3]
Must follow
D7 D6 D5 D4 D3 D2 D1 D0
1 1 X X 1 1 1 1
= CF_{16}
I/O Definition
D7 D6 D5 D4 D3 D2 D1 D0
0 0 1 0 1 0 0 1
= 29_{16}
Vector
D7 D6 D5 D4 D3 D2 D1 D0
0 0 1 0 0 0 0 0
= 20_{16}
Interrupt Control
Must follow
D7 D6 D5 D4 D3 D2 D1 D0
1 0 1 1 0 1 1 1
= $B7_{16}$
Interrupt Mask
D7 D6 D5 D4 D3 D2 D1 D0
1 1 0 1 0 1 1 0
= $D6_{16}$

Fig. 28

Since this configuring routine consists of sending a stream of data to the same location, a program loop may be used to reduce its length. This is particularly so if the data to be sent to the control section of the PIO can be located in such a position as to simplify detection of the last data word. A routine of this type is as follows:

```
                      ;Z80 control application
                      ;Non-memory mapped I/O with port A addresses 4 and 6
                      ;Mode 2 interrupt with ISR vector at 0D20H
                      ;Mode 3 PIO operations, STB ignored
                      ;Data table located at address 0D01H
                      ;
                        PAD     EQU 4            ;port A data I/O register
                        PAC     EQU 6            ;port A control section
                        TABLE   EQU 0D05         ;top address in table
                                ORG 0C90H
                      ;
0C90   01 05 0D                 LD BC,TABLE      ;use BC as table pointer
0C93   02             CONFIG    LD A,(BC)        ;get data from table
0C94   D3 06                    OUT (PAC),A      ;send to port A control
0C96   0D                       DEC C            ;decrement pointer and check
0C97   20FA                     JR NZ,CONFIG     ;far end
       ..........
       ..........
       ..........                                ;control program
       ..........
                      ;data table containing configuring information
                      ;
                                ORG 0D01H
                      ;
0D01   D6                       DEFB 0D6H,0B7H,20H   ;interrupt mask word
0D02   B7                                            ;interrupt control word
0D03   20                                            ;peripheral vector
0D04   29                       DEFB 29H,0CFH        ;I/O direction word
0D05   CF                                            ;mode 3 select word
```

Problem 22 The block diagram of a **Z80 CTC** (counter/timer circuit) is shown in *Fig. 29*.

Describe the facilities provided by a CTC of this type.

A **Z80 CTC** (counter-timer circuit) is a programmable four channel device that provides counting and timing functions for a **Z80** microprocessor. Each of the four channels is independent, and consists of two registers, two counters and control logic, as shown in *Fig. 30*.

Each channel has an interrupt vector for automatic selection of an appropriate interrupt service routine (MPU mode 2 interrupt). Channel 0 has the highest

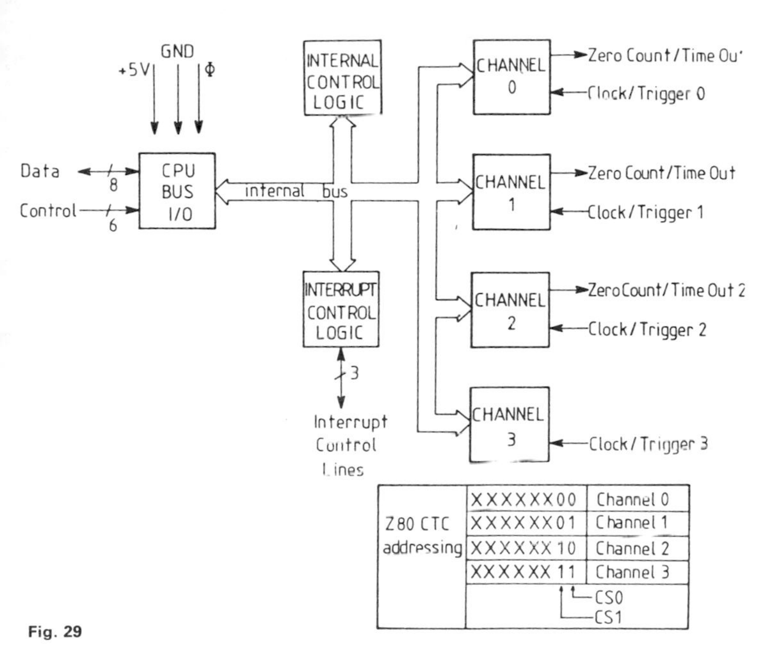

Fig. 29

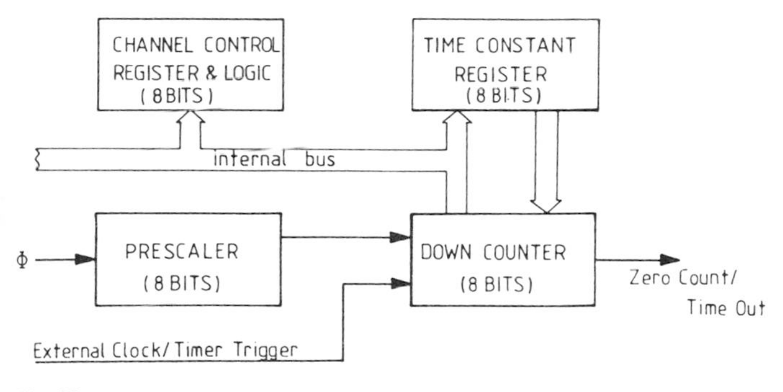

Fig. 30

priority, and each channel is connected in a 'daisy chain' configuration. Each channel of a CTC may be programmed to act as either a counter or as a timer, which operate as follows:

Counter mode

The CTC counts pulses which it receives from an external source. It may be programmed to interrupt its Z80 MPU when a predetermined number of pulses have been counted. The number of pulses counted prior to initiating an interrupt is software selectable.

Timer mode
The CTC counts system clock pulses (Φ). An interrupt may be generated when a predetermined number of clock pulses have been counted. The number of clock pulses counted prior to initiating an interrupt is software selectable, but since the clock pulses are of a precisely known time period, the CTC may be programmed to cause an interrupt after a known time period.

Problem 23 Explain the function of the following components of a **Z80 CTC**:
(a) **channel control register and logic**;
(b) **time constant register**;
(c) **down counter**;
(d) **prescaler**.

(a) The function of each bit of a channel control register is shown in *Fig. 31*. The contents of this register define the operating mode of a particular channel. In order to write data to this register (rather than loading an interrupt vector which shares the same address), bit 0 of the control word must be a logical 1. Basically, the contents of this register determine the following:

(i) selection of counter or timer mode of operation;
(ii) enabling or disabling of CTC generated interrupts;
(iii) prescaling the system clock by 16 or by 256;
(iv) selection of positive or negative edge to trigger the timer;
(v) starting and stopping counting operations.

The selection of each of these facilities requires a particular control word to be sent to the CTC channel in use, and this control word may be determined by referring to *Table 9*.

(b) The time constant register is an 8-bit register which is loaded by the **Z80 MPU** following a channel control word with bit 2 set. The contents of this register are used to initially load the down counter, and then to automatically reload it each time that the down counter reaches a count of zero. The contents of this register therefore determine the number of pulses required to reach a count of zero in the down counter, and hence may be used to determine the time period between interrupts.

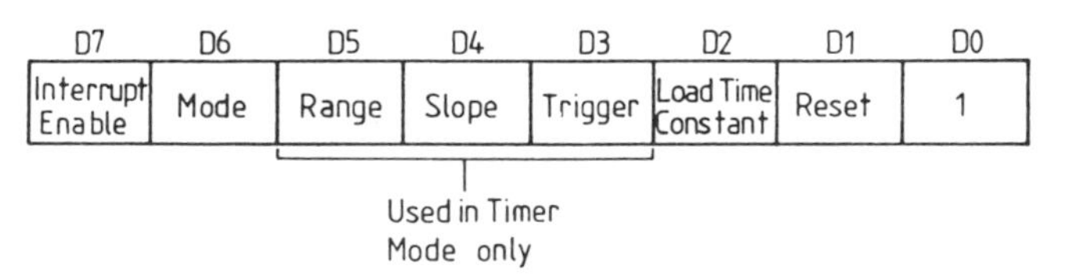

Fig. 31

		1	*0*
Interrupt enable	Bit 7	Channel interrupts enabled to occur every time down counter reaches a count of zero. Setting bit 7 does not let a preceding count of zero cause an interrupt.	Channel interrupts disabled.
Mode	Bit 6	Counter mode – down counter is clocked by external clock. The prescaler is not used.	Timer mode – down counter is clocked by the prescaler. The period of the counter is: $t_c \times P \times TC$ t_c = system clock period P = prescale of 16 or 256 TC = 8-bit binary programmable time constant (256 max)
Range	Bit 5	System clock Φ is divided by 256 in prescaler (timer mode only).	System clock Φ is divded by 16 in prescaler (timer mode only).
Slope	Bit 4	A positive slope trigger starts timer operation (timer mode only).	A negative slope trigger starts timer operation (timer mode only).
Trigger	Bit 3	If bit 2 = 0 the timer begins operation after the current machine cycle when the trigger input makes the specified transition. If bit 2 = 1 the timer begins operation after the load time constant machine cycle when when the trigger makes the specified transition (timer mode only).	If bit 2 = 0 the timer begins operation at the start of the next machine cycle. If bit 2 = 1 the timer begins operation at the start of the machine cycle following the one that loads the time constant (timer mode only).
Load time constant	Bit 2	The time constant for the down counter will be the next word written to the selected channel. If a time constant is loaded while a channel is counting, the present count will be completed before the new time constant is loaded into the down counter.	No time constant will follow the channel control word. One time constant must be written to the channel to initiate counting.
Reset	Bit 1	Channel stops counting until a time constant is loaded. ZC/TO is made inactive. The channel cannot interrupt.	Channel continues counting.
Operating mode/ vector select	Bit 0	Select channel control register.	Select interrupt vector register.

(c) The down counter is loaded from the time constant register, and is decremented towards zero by either:
   (i) external pulses when the CTC is in the counter mode, or
   (ii) pulses from the system clock (via the prescaler) when in the timer mode.

   In either case, when the down counter reaches zero, a positive pulse is generated on the appropriate zero count or time out pin (ZC/TO) of the CTC. Optionally, the CTC may be programmed to initiate a mode 2 interrupt in the **Z80 MPU** whenever a zero count/time occurs. The down counter is automatically reloaded from the time constant register each time that it reaches a zero count.

(d) The prescaler is a binary divider may which be programmed to divide the system clock frequency by either 16 or 256. Therefore, the down counter counts down at either 1/16 or 1/256 the rate of the system clock. Note that the prescaler only operates in this timer mode.

Problem 24 A **Z80 CTC** is required to interrupt a Z80 MPU (mode 2 interrupt) every 200 μs, and execute an interrupt service routine which has a starting address defined by a vector stored at location $0D20_{16}$. Write a **Z80** machine code routine to show how the CTC must be configured, using channel 0, and a 2 MHz system clock (Φ).

The CTC registers are loaded with data as defined in *Fig. 32*. A machine code routine to configure channel 0 of the CTC to interrupt the **Z80 MPU** every 200 μs is as follows:

```
                ;Z80 CTC 200 μs interrupts
                ;assumes CTC channel 0 is at address 00
                ;Mode 2 interrupt with ISR vector at 0D20H
                ;2 MHz system clock, using divide by 16 range
                ;trigger on positive slope
                ;after TC is loaded
                  CTC0  EQU 0
                        ORG 0C90H
                ;
0C90   3E 0D            LD A 0DH
0C92   ED 47            LD I,A          ;load vector high 0DH
0C94   3E 20            LD A,20H
0C96   D3 00            OUT (CTC0),A    ;load vector low into CTC
0C98   3E 95            LD A,95H
0C9A   D3 00            OUT (CTC0),A    ;load control word
0C9C   3E 19            LD A,19H
0C9E   D3 00            OUT (CTC0),A    ;load time constant
0CA0   ED 5E            IM 2            ;mode 2 Z80 INT
0CA2   FB               EI              ;enable interrupt
       ..........
       ..........
       ..........
```

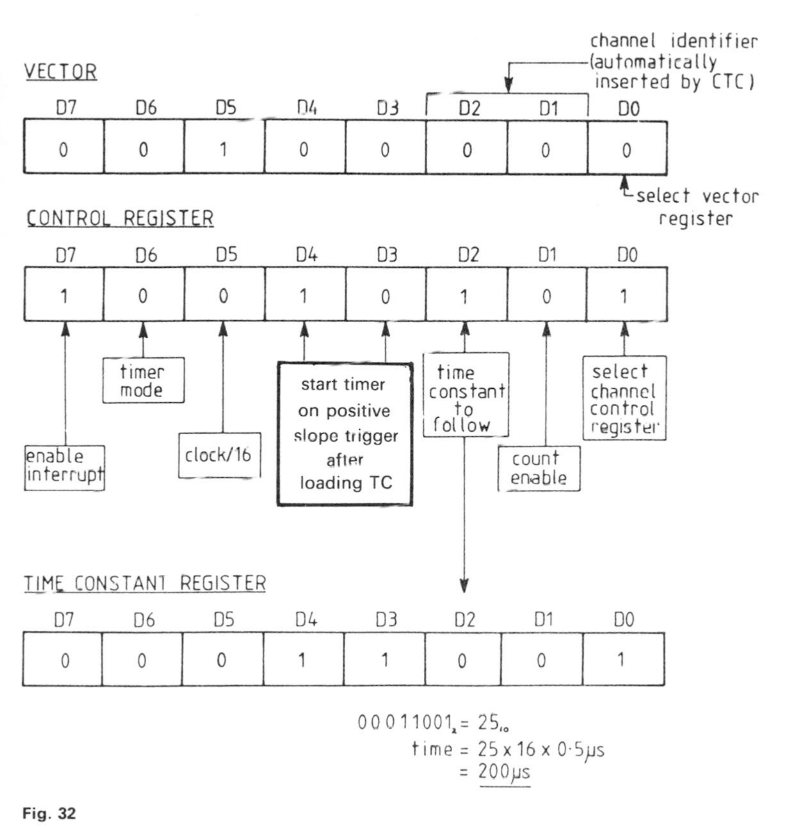

Fig. 32

Problem 25 The block diagram of a **6820 PIA** is shown in *Fig. 33*. Describe the functions of the following registers: (a) **output register**; (b) **data direction register**; (c) **controller register**.

(a) Each of the peripheral data lines PA_0–PA_7 may be programmed to act as either an input or an output by loading **data direction register A** with appropriate bits. The data stored in output register A will appear on the peripheral data lines that are programmed to be outputs. A logical 1 written into output register A causes a logical 1 to appear on the corresponding peripheral data line, and a logical 0 written into output register A causes a logical 0 to appear on the corresponding peripheral data line. During an MPU read peripheral data operation, the data on peripheral lines programmed as inputs appears directly on the

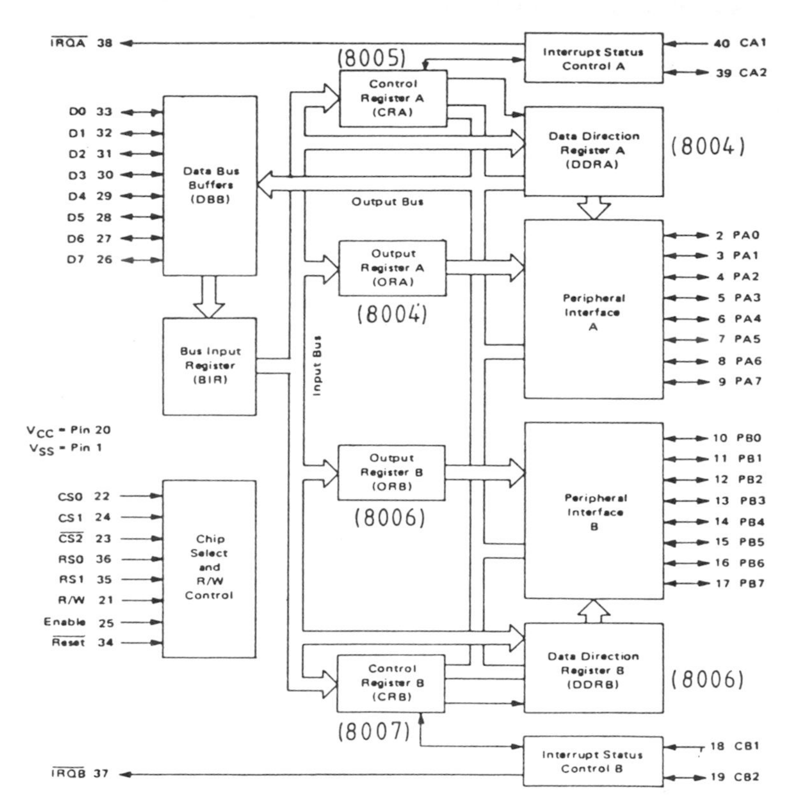

Fig. 33

corresponding MPU data bus lines (D_0–D_7), whilst the contents of the output register are read for those peripheral lines programmed as outputs.

The peripheral data lines of port B (PB_0–PB_7) may be programmed in a similar manner to those of port A. The output buffer driving these lines, however, are different and enter a high impedance state when the peripheral data line is used as an input.

(b) The two **data direction registers** allow the MPU to control the direction of data transfers through each corresponding peripheral data line. A data direction register bit at logical 0 defines the corresponding peripheral data line as an input, and a data direction register bit at logical 1 defines the corresponding peripheral data line as an output.

(c) The two **control registers** (CRA and CRB) allow the MPU to control the operation of the four peripheral control lines CA1, CA2, CB1 and CB2. In addition, the control registers allow the MPU to enable the interrupt lines, monitor the status of the interrupt flags, and allow selection of either a peripheral interface register or a data direction register (these two registers are located at the same address for a specified port).

Problem 26 Explain the operation of the **control registers** in a **6820 PIA**.

The **control registers** (CRA and CRB) of a **6820 PIA** allow the operation of the four peripheral control lines (CA1, CA2, CB1 and CB2) to be controlled by an MPU. In addition, a control register allows interrupts to be enabled or disabled, the status of interrupt lines to be checked, and selection of the **data direction** or **peripheral interface registers** to take place. The format of the control words is shown in *Table 10*. The function of each bit in the control registers may be summarized as follows:

CA1/CB1 control (bits 0 and 1)
These two bits in the **control register** determine the operation of interrupt inputs CA1 and CB1. The interrupt inputs may be programmed to have either positive or negative active edges which cause bit 7 (IRQA1 or IRQB1) of the control register to be set (logical 1). IRQA1 and IRQB1 are **interrupt status flags** which may be polled, or optionally made to initiate an MPU interrupt request. The function of these two bits is summarized in *Table 11*.

DDRA/DDRB access (bit 2)
The **6820 PIA** assigns the **data direction registers** and the **peripheral interface registers** to the **same** address. Bit 2 of the control registers determines which register is selected, such that a logical 0 in this position causes the **data direction register** to be selected, and a logical 1 causes the **peripheral interface register** to be selected. At system reset time the control registers are cleared (loaded with 00), therefore initially the data direction registers are selected.

CA2/CB2 control (bits 3, 4 and 5)
Bits 3, 4 and 5 of the **control registers** are used to control the CA2 and CB2 peripheral control lines. These lines may function as interrupt inputs (similar to CA1 and CB1), or as a peripheral control outputs. When bit 5 of the control register is a logical 0, CA2/CB2 act as interrupt inputs. When bit 5 of the

Table 10 Control word format

	7	6	5 4 3	2	1 0
CRA	IRQA1	IRQA2	CA2 control	DDRA access	CA1 control
	7	6	5 4 3	2	1 0
CRB	IRQB1	IRQB2	CB2 control	DDRB access	CB1 control

control register is at logical 1, CA2/CB2 act as peripheral control outputs. The characteristics of CA2 and CB2 are slightly different to one another when they are used as outputs, and a summary of the operating modes available is shown in *Table 12*(*a*), (*b*) and (*c*).

Table 11 Control of interrupt inputs CA1 and CB1

CRA-1 (CRB-1)	*CRA-0 (CRB-0)*	*Interrupt input CA1 (CB1)*	*Interrupt flag CRA-7 (CRB-7)*	*MPU interrupt request* $\overline{IRQA}$ ($\overline{IRQB}$)
0	0	↓ Active	Set high on ↓ of CA1 (CB1)	Disabled – $\overline{IRQ}$ remains high
0	1	↓ Active	Set high on ↓ of CA1 (CB1)	Goes low when the interrupt flag bit CRA-7 (CRB-7) goes high
1	0	↑ Active	Set high on ↑ of CA1 (CB1)	Disabled – $\overline{IRQ}$ remains high
1	1	↑ Active	Set high on ↑ of CA1 (CB1)	Goes low when the interrupt flag bit CRA-7 (CRB-7) goes high

Notes: 1 ↑ Indicates positive transition (low to high).
2 ↓ Indicates negative transition (high to low).
3 The interrupt flag bit CRA-7 is cleared by an MPU read of the A data register and CRB-7 is cleared by an MPU read of the B data register.
4 If CRA-0 (CRB-0) is low when an interrupt occurs (interrupt disabled) and is later brought high, $\overline{IRQA}$ ($\overline{IRQB}$) occurs after CRA-0 (CRB-0) is written to a 'one'.

Table 12 (a) Control of CA2 and CB2 as interrupt inputs: CRA5 (CRB5) is low

CRA-5 (CRB-5)	*CRA-4 (CRB-4)*	*CRA-3 (CRB-3)*	*Interrupt input CA2 (CB2)*	*Interrupt flag CRA-6 (CRB-6)*	*MPU interrupt request* $\overline{IRQA}$ ($\overline{IRQB}$)
0	0	0	↓ Active	Set high on ↓ of CA2 (CB2)	Disabled – $\overline{IRQ}$ remains high
0	0	1	↓ Active	Set high on ↓ of CA2 (CB2)	Goes low when the interrupt flat bit CRA-6 (CRB-6) goes high
0	1	0	↑ Active	Set high on ↑ of CA2 (CB2)	Disabled – $\overline{IRQ}$ remains high
0	1	1	↑ Active	Set high on ↑ of CA2 (CB2)	Goes low when the interrupt flag bit CRA-6 (CRB-6) goes high

Notes: 1 ↑ Indicates positive transition (low to high).
2 ↓ Indicates negative transition (high to low).
3 The interrupt flag bit CRA-6 is cleared by an MPU read of the A data register and CRB-6 is cleared by an MPU read of the B data register.
4 If CRA-3 (CRB-3) is low when an interrupt occurs (interrupt disabled) and is later brought high, $\overline{IRQA}$ ($\overline{IRQB}$) occurs after CRA-3 (CRB-3) is written to a 'one'.

(b) Control of CB2 as an output: CRB5 is high

CRB-5	*CRB-4*	*CRB-3*	*CB2* *Cleared*	*Set*
1	0	0	Low on the positive transition of the first E pulse following an MPU write 'B' data register operation	High when the interrupt flag bit CRB-7 is set by an active transition of the CB1 signal.
1	0	1	Low on the positive transition of the first E pulse after an MPU write 'B' data register operation.	High on the positive edge of the first 'E' pulse following an 'E' pulse which occured while the port was deselected.
1	1	0	Low when CRB-3 goes low as a result of an MPU write in control register 'B'.	Always low as long as CRB-3 is low. Will go high on an MPU write in control register 'B' that changes CRB-3 to 'one'.
1	1	1	Always high as long as CRB-3 is high. Will be cleared when an MPU write control register 'B' results in clearing CRB-3 to 'zero'.	High when CRB-3 goes high as a result of an MPU write into control register 'B'.

(c) Control of CA2 as an output; CRA5 is high

CRA-5	*CRA-4*	*CRA-3*	*CA2* *Cleared*	*Set*
1	0	0	Low on negative transition of E after an MPU read 'A' data operation.	High when the interrupt flag bit CRA-7 is set by an active transition of the CA1 signal.
1	0	1	Low on negative transition of E after an MPU read 'A' data operation.	High on the negative edge of the first 'E' pulse which occurs during a deselect.
1	1	0	Low when CRA-3 goes low as a result of an MPU write to control register 'A'.	Always low as long as CRA-3 is low. Will go high on an MPU write to control register 'A' that changes CRA-3 to 'one'.
1	1	1	Always high as long as CRA-3 is high. Will be cleared on an MPU write to control register 'A' that clears CRA-3 to a 'zero'.	High when CRA-3 goes high as a result of an MPU write to control register 'A'.

IRQA1/2 and IRQB1/2 (bits 6 and 7)

These are **interrupt flag** bits which are set by active transitions on the interrupt and peripheral control lines (CA1, CB1, CA2 and CB2) when programmed as inputs. These bits cannot be written to by the MPU but may be reset indirectly by reading data from the appropriate port.

Problem 27 Write **6800** machine code routines to show how a **6820 PIA** may be configured to enable the following operations to take place:

(a) Read eight bits of data in from port A, invert the data and send it back out through port B.

(b) Read in four bits of data from port A bits 0 to 3, invert the data and send it back out through port A bits 4 to 7.

The PIA may be assumed to be located in memory at addresses 8004_{16} to 8007_{16}, as indicated on *Fig. 33*.

(a)

```
                         ;Memory mapped 6820 PIA with addresses in
                         ;the range $8004 to $8007
                         ;
                          DRA  EQU $8004    ;data direction/output register A
                          CRA  EQU $8005    ;control register A
                          DRB  EQU $8006    ;data direction/output register B
                          CRB  EQU $8007    ;control register B
                               ORG $0100
                         ;
0100  4F                       CLR A        ;select data direction registers
0101  B7 80 05                 DTA A CRA    ;(normally selected after
0104  B7 80 07                 SRA A CRB    ;reset)
0107  B7 80 04                 STA A DRA    ;port A all inputs
010A  43                       COM A
010B  B7 80 06                 DTA A DRB    ;port B all outputs
010E  86 04                    LDA A #$04
0110  B7 80 05                 STA A CRA    ;select I/O register A
0113  B7 80 07                 STA A CRB    ;select I/O register B
0116  B6 80 04  READ           LDA A DRA    ;read port A
0119  43                       COM A        ;invert the data
011A  B7 80 06                 STA A DRB    ;and send to port B
011D  20 F7                    BRA READ     ;repeat if required
```

(b)

```
                         ;Memory mapped 6820 PIA with addresses in
                         ;the range $8004 to $8007
                         ;
                          DRA  EQU $8004    ;data direction/output register A
                          CRA  EQU $8005    ;control register A
                               ORG $0100
                         ;
0100  4F                       CLR A        ;select data direction
0101  B7 80 05                 STA CRA      ;register A
0104  86 F0                    LDA A #$F0   ;I/O definition word
0106  B7 80 04                 STA A DRA    ;b0-b3 inputs, b4-b7 outputs
0109  86 04                    LDA A #$04
010B  B7 80 05                 STA A CRA    ;select I/O register A
010E  B6 80 04  READ           LDA A DRA    ;read port A
0111  48                       ASL A
0112  48                       ASL A
0113  48                       ASL A
0114  48                       ASL A
0115  43                       COM A        ;invert the data
0116  B7 80 04                 STA A DRA    ;and send back to port A
0119  20 F3                    BRA READ     ;repeat if required
```

Problem 28 Write **6800** machine code routines to show how a **6820 PIA** may be configured to enable the following operations to take place:

(a) Increment the binary number displayed on eight LEDs connected to port A each time that a negative going transition is applied to control input CA1.

(b) Cause an LED connected to control line CA2 to flash on and off at a 1 Hz rate. The system clock is 614.4 kHz.

The PIA may be assumed to be located in memory at addresses 8004_{16} to 8007_{17}, as indicated on *Fig. 33*.

(a) A circuit which is suitable for implementing the binary counter is shown in *Fig. 34*.

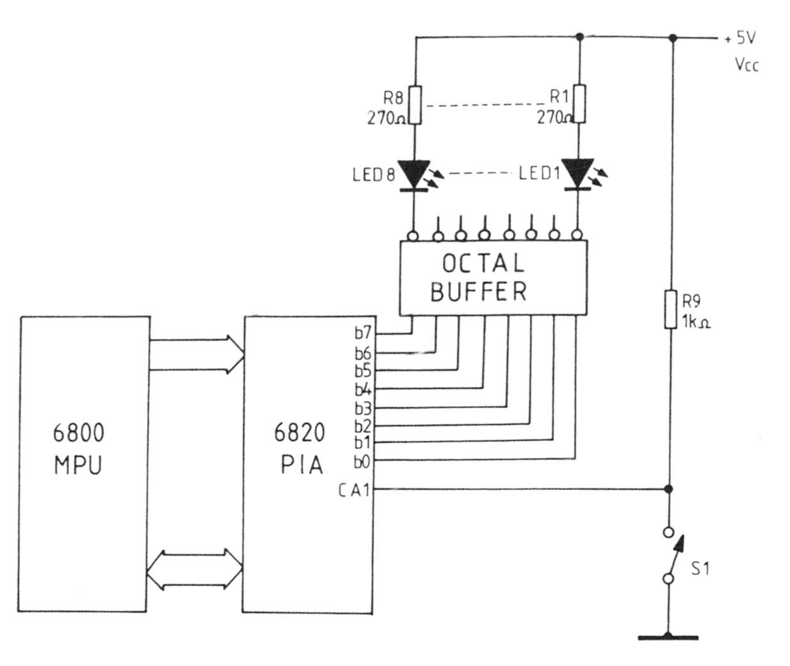

Fig. 34

A suitable **6800** machine code routine to increment the binary counter via CA1 is as follows:

```
                     ;6800 binary counter, using a 6820 PIA with addresses in
                     ;the range $8004 to $8007
                     ;negative edge triggered via CA1 using poll
                     ;
                      DRA   EQU $8004    ;data direction/output register A
                      CRA   EQU $8005    ;control register A
                            ORG $0100
                     ;
0100  4F                    CLR A        ;select data direction
0101  B7 80 05              STA A CRA    ;register A
0104  43                    COM A
0105  B7 80 04              STA A DRA    ;port A all outputs
0108  86 04                 LDA A #$04   ;use 06 for positive edge
010A  B7 80 05              STA A CRA    ;output register/CA1 control word
010D  4F                    CLR A
010E  B7 80 04              STA A DRA    ;update counter
0111  B6 80 05 IRAQ1        LDA A DRA    ;read control register A
0114  2A FB                 BPL IRQA1    ;wait for flag to be set
0116  7C 80 04              INC DRA      ;increment counter
0119  20 F6                 BRA IRQA1    ;wait for next input edge
```

(b) A circuit suitable for implementing the flashing LED is shown in *Fig. 35*. A suitable **6800** machine code routine to cause the LED connected to CA2 to flash on and off at a 1 Hz rate is as follows:

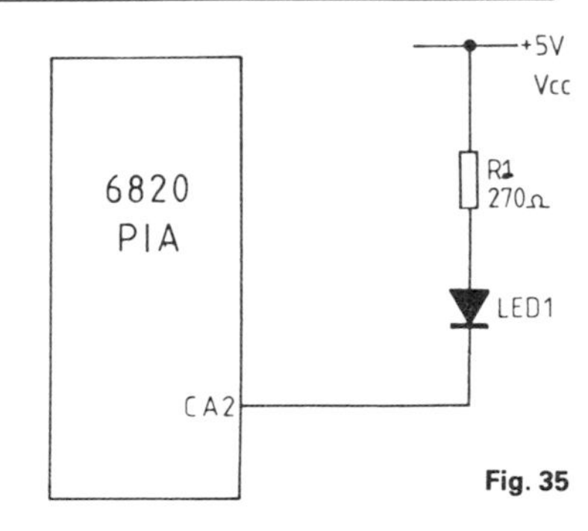

Fig. 35

```
                     ;6820 PIA with addresses in the range
                     ;$8004 to $8007
                     ;LED connected to CA2 (output mode)
                     ;
                      CRA      EQU $8005     ;control register A
                               ORG $0100
                     ;
0100  86 30                    LDA A #$30    ;control word
0102  B7 80 05       FLASH     STA A CRA     ;send to control register A
0105  CE 95 E7                 LDX #$95E7    ;use X as time
0108  09             DELAY     DEX           ;delay loop counter
0109  26 FD                    BNE DELAY
010B  88 08                    EOR A#$08     ;toggle b3 of A
010D  20 F3                    BRA FLASH
```

C FURTHER PROBLEMS ON INTERFACING DEVICES

(a) SHORT ANSWER PROBLEMS

1 A circuit which allows two otherwise incompatible circuits to be interconnected is known as an ....

2 An external device which is connected to a microcomputer is known as a ....

3 A group of parallel conductors in a microcomputer to which an external device may be connected are known collectively as a ....

4 A set of data latches allow information to be passed from a microcomputer .... to ....

5 A set of tri-state buffers allow information to be passed from a .... to a microcomputer ....

6 A special purpose programmable I/O device which enables digital information to be transferred between a microcomputer and its external devices is known as a ....

7 Three methods for transferring data between a microcomputer and its external devices are ...., ... and ....

8 A typical I/O device may contain the following five components: ...., ..., ... and ....

9 Each of the separate components in an I/O device may be treated as an individual entity by ....

10 Serial data transfers between a microcomputer and its external devices involves sending data bits .... along a .... conductor.

11 Parallel to serial data conversion may take place using a ....

12 Additional circuits may be required between a microcomputer I/O devices and external devices to deal with the following three aspects of data transfers: ...., ... and ....

(b) CONVENTIONAL PROBLEMS

1 Explain how the following may be interfaced directly to the data bus of a microcomputer:
(a) a number of two-state input devices
(b) a number of two-state output devices

2 Describe the function of the following components in an interfacing device:
(a) data output register
(b) data direction register
(c) control register
(d) status flags

3 With the aid of a block diagram, show how a PIA (or PIO) may be used to interface a peripheral device to a microcomputer.

4 Describe four aspects of interfacing a peripheral device to a microcomputer which are not handled by a PIA (or PIO).

5 (a) Explain the difference between serial and parallel data transfers.
(b) State one advantage of using serial data transfers.
(c) State one disadvantage of using serial data transfers.

6 Explain how a 6530 RRIOT may be configured such that bits 2, 3 and 7 of port A and bits 1, 5 and 7 of port B act as outputs and all remaining bits act as inputs. The RRIOT may be assumed to be located in the address range 1700_{16} to 1703_{16}.

7 The timer in a 6530 RRIOT is required to generate pulses on bit 7 of port B every 100 ms. Write a 6502 machine code routine to show how the RRIOT is configured and the timer loaded for this application, assuming the RRIOT I/O section is located in the address range 1700_{16} to 1703_{16}, and the timer section is located in the address range 1704_{16} to $170F_{16}$.

8 A 6502 based microcomputer uses a 6522 VIA located in the address range $A000_{16}$ to $A00F_{16}$. Show how the 6522 VIA may be programmed to carry out the following operations:
(a) Generate a 1 kHz output from CA2.
(b) Generate a 1 kHz output from PB7 using timer 1.
(c) Count a hundred input pulses applied to PB6 using timer 2.

9 A 6502 based microcomputer uses a 6522 VIA located in the address range $A000_{16}$ to $A00F_{16}$. Show how 300 bits/s serial data may be loaded into the VIA shift register via CB2 and under the control of timer 2.

10 Explain how the timing of data transfers between a peripheral device and port A of a 6522 VIA may be controlled using CA1 and CA2.

11 Show how a Z80 microcomputer with an 8255 PIA may be programmed to enable the following operations to take place:
(a) Read data in from port A of the 8255, invert the data, and send it back out via port C.
(b) Generate a 1 kHz output from bit 0 of port C using the bit set/reset facility.
(c) Increment the binary number indicated by eight LEDs connected to port A each time that the STB A (PC4) is activated.

12 A Z80 based microcomputer with an 8255 PPI has eight two-state devices connected to port A. The STB input is used to transfer data from the two-state devices into the PPI when it becomes active, and this simultaneously initiates an interrupt.
(a) Draw a simple circuit to show how the Z80, 8255 PPI and the two-state devices are interconnected.
(b) Write a Z80 machine code program to show how the 8255 PPI may be configured for this application.

13 A Z80 based microcomputer has a Z80 PIO located at I/O addresses 04 to 07. Write Z80 machine code routines to show how the following operations may be

carried out:

(a) Read data from eight two-state devices connected to port B and send this data to eight two-state indicators connected to port A.

(b) Read data from four two-state devices connected to bits 0 to 3 of port A, and send this data out to bits 4 to 7 of the same port.

(c) Use bits 5 to 7 of port A as control inputs which cause an interrupt to be initiated when all three are active high. The Z80 MPU must be configured to accept mode 2 interrupts, and the ISR vector may be assumed to be located at address $0D42_{16}$.

14 A Z80-based microcomputer with a Z80 PIO located at I/O addresses 04 to 07 has a single 7-segment LED connected to port A via a suitable interface circuit. A switch is connected to the remaining bit of port A, and is arranged so that the number on the 7-segment LED is incremented each time that the switch is closed.

(a) Draw a diagram to show how the 7-segment display and the switch may be interfaced to the Z80 PIO.

(b) Write a Z80 machine code routine to show how the PIO may be configured to enable the circuit to function as intended.

15 A Z80-based microcomputer has a Z80 CTC located at I/O addresses 00 to 03. Write a Z80 machine code routine to show how the CTC may be configured to count pulses applied to the clock/trigger input of channel 2, and cause an interrupt to occur when 100 pulses have been counted.

16 A Z80 based microcomputer has a Z80 CTC located at I/O addresses 00 to 03. Write a Z80 machine code routine to enable the CTC to generate interrupts at a 1 Hz rate at the output of channel 0 using the internal clock.

17 A 6800 microcomputer has a 6820 PIA located in the address range 8004_{16} to 8007_{17}. Eight two-state input devices are connected to port A of the 6820, and eight two-state indicators are connected to port B. Write a 6800 machine code routine to display the data set up on the input devices on the indicators each time that CA1 becomes active low.

18 A 6800 microcomputer has a 6820 PIA located in the address range 8004_{16} to 8007_{16}. Eight two-state input devices are used to monitor an industrial process, and are connected to port A of the 6820 PIA. The two-state devices normally apply a logical 0 to each of the port I/O lines, but if a logical 1 is detected on any input, an LED connected to CA2 is illuminated.

(a) Draw a diagram to show how the two-state devices and the LED may be connected to the PIA.

(b) Write a 6800 machine code program to enable such a system to be implemented.

6 Assembly language programs

A MAIN POINTS CONCERNED WITH ASSEMBLY LANGUAGE PROGRAMS

1 Programming at machine code level using **hand assembly** is a rather tedious process, since it requires a programmer to spend much time looking up opcodes for each instruction, performing various calculations in binary or hexadecimal, and simultaneously keeping track of many different addresses. It is therefore necessary to use a better method if serious program writing is contemplated, since hand assembly is inevitably both time-consuming and error-prone. Such a method consists of writing programs at **assembly language** level, and using an assembler program to generate all necessary machine code for the microprocessor. The use of assembly language makes programs much easier to follow (hand assembly is normally reserved for very minor alterations to existing programs, i.e., for **patching**).

2 The initial stages in writing an assembly language program involve the preparation of **source code**. Source code consists of a sequence of ASCII characters (or text) that represent instruction mnemonics, symbolic addresses and data for an assembly language program, and is usually stored as a disk file which is subsequently accessed by an assembler to generate the equivalent machine or object code.

3 Initially, source code must be entered via the keyboard in the usual way, but of course, errors may be inadvertently introduced which need subsequent correction. These errors may be obvious, e.g. simple typing errors, and require correcting at the time the source code is being entered, or they may be errors which do not become apparent until after the program has been assembled or debugged. In the case of simple typing errors, these may frequently be corrected by means of keyboard controls such as **backspace** or **delete** provided that this is carried out before the **return** or **enter** key is pressed to transfer the text from the keyboard buffer into the microcomputer. For errors which are only discovered after the source code has been stored in a microcomputer, it is obvious that a facility is required which enables the contents of a text file to be altered or **edited**, since the alternative would be the tedious and time-wasting process of re-entering the whole of the work.

4 This facility, provided by a transient program known as a **text editor**, may also be used to create the original text file. A **word processor**, which may be considered as a more comprehensive text editor, may also be used to create assembly language source code, although when used for this purpose, many of its facilities will never be used. The range of facilities found in different text editors varies considerably, but as a minimum requirement, should be capable of

performing the following operations:

(a) create and save new files;
(b) open and close existing text files;
(c) insert, delete or alter text in new or existing files;
(d) print text files.

5 An assembler is a program which has been written to perform the specific task of converting assembly language source code for a given microprocessor into its equivalent machine or object code. Each assembler generally assembles code for the same type of microprocessor as that on which it runs, but since its main task is essentially one of processing numeric and textual data, it may equally well be written to act as an assembler for a different type of microprocessor. In this case it is then known as a **cross-assembler**.

6 Since each assembler program operates in a manner determined by its creator, it must not be assumed that all assemblers require identical source code formats. Certain assembly language conventions are defined by the manufacturer of each different type of microprocessor and, although for the sake of compatibility it makes sense to adhere to these conventions, software writers often have other ideas. Frequently the differences are only of a minor nature, but in extreme cases, totally non-standard mnemonics may be encountered. It therefore follows that reference to software manuals regarding the precise operating details of a given assembler is a prerequisite for performing successful assembly.

7 An assembler reads assembly language source code sequentially, usually from a disk file, and then performs its assembly processes. Upon completion of this task, output files may be optionally written back on to the disk, but it is usual to postpone writing these output files to disk until an error-free assembly has been achieved. There is little point in storing errors on to disk, particularly where large files are concerned, due to the length of time taken to store them. Two files are usually created by this process.

(a) An object code file containing machine code which corresponds to the original assembly language source program and which may be loaded into the system RAM for execution.
(b) A list file which contains the original source code along with the object code and which may be used for debugging and documentation purposes.

8 Once a program has been correctly assembled, it is necessary to test it for correct functioning, i.e., dynamic testing. The fact that a program assembles without errors is no indication that it is free from logical errors. Such logical errors which prevent a program from operating correctly are known as **bugs**, and the process of detecting and eliminating them is known as **debugging**. Special programs are required which allow dynamic testing of programs under controlled conditions, and these are called **debugging programs** or **debuggers**.

9 A debugging program actually consists of a number of independent test routines, each selectable from within a command loop by a single command letter followed by parameters. Certain routines must be available in a debugging program, but others are desirable rather than being essential. Typically routines may be included to perform the following operations:

(a) inspection/alteration of the contents of specified memory locations;
(b) inspection/alteration of the contents of specified registers;
(c) dumping blocks of memory on the display device in tabular form;

(d) executing a program in real time from a specified address;
(e) single instruction execution or **single stepping**;
(f) setting **break-points** so that a program may be halted once a specified instruction address is encountered;
(g) performing a software trace so that the sequence of instructions leading up to a selected event may be studied;
(h) moving blocks of data from one area of memory to another;
(i) filling blocks of memory with a selected constant;
(j) performing **in-line** assembly;
(k) listing programs as disassembled mnemonics.

10 Program development consists of repeated iterations of the text editor, assembler, debugger cycle as shown in *Fig. 1*.

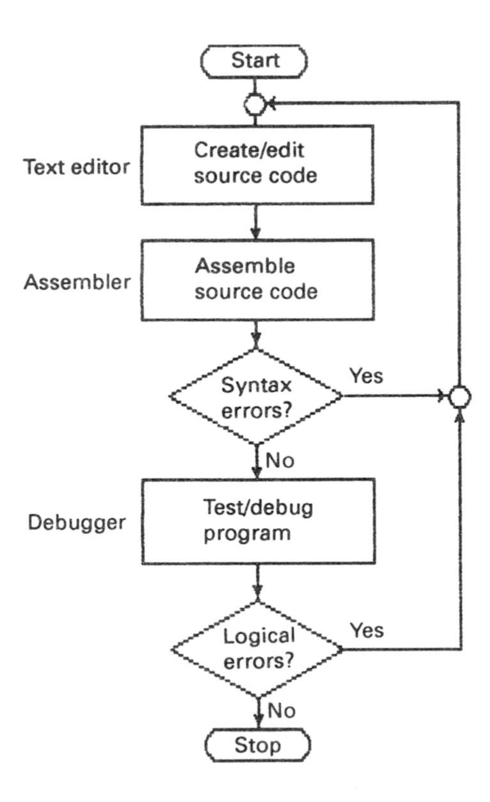

Fig. 1

B WORKED PROBLEMS ON ASSEMBLY LANGUAGE PROGRAMS

Problem 1 List the facilities required in a text editor suitable for the preparation of assembly language source code.

The commands which are likely to be of most use when preparing assembly language source code are as follows:

(a) **Cursor moving commands**
   Right/left character
   Right/left word
   Tab right
   Start/end line
   Start/end file
   Up/down line
   Up/down screen
(b) **Text insert/delete commands**
   Toggle insert (on/off)
   Delete line
   Delete character left/right
   Delete word right
   Delete to end/start of line
(c) **Block commands**
   Mark start/end of block
   Copy/move marked block
   Read file (insert at cursor)
   Write marked block to file
(d) **Search and replace commands**
   Find string
   Find and replace all
   Find/replace next occurrence
(e) **Saving files**
   Save and continue edit
   Save, edit finished
   Save, exit to system
   Save, abandon exit

Problem 2 State the advantages of using assembly language rather than hexadecimal codes for writing microprocessor programs.

The advantage of writing programs in assembly language rather than hexadecimal codes may be summarized as follows:

(a) shorter program writing time;
(b) no need to look up instruction opcodes;

(c) less error-prone;
(d) symbolic addresses may be used to eliminate the need to keep track of absolute addresses;
(e) labels may be assigned to addresses in the program as references for jumps and calls;
(f) relative jump offsets are calculated automatically at assembly time;
(g) references to addresses are automatically readjusted when instructions are added or deleted or when the program is assembled at a new starting address.

Problem 3 Explain why most assemblers perform two passes through the source code.

An assembler usually performs two passes through source code which has previously been read from a disk file. On the first pass, each line of source code is scanned by the assembler, and values are assigned to labels and mnemonics in so far as this is possible. A location counter is initialized by an origin (ORG or *=) directive in the source code, and as each instruction is processed, this location counter is incremented according to the number of bytes required for that particular instruction. Using the location counter, a symbol table is built up which associates each label in a program with the address that it represents. This process is illustrated in *Fig. 2*.

Labels and expressions which are used as operands are also evaluated, although it is not always possible for an assembler to fully evaluate all operands during the first pass, since references may be made to parts of the program which are yet to be processed. Such references are known as **forward references**, and these can only be resolved once the symbol table is complete, thus necessitating a second pass through the source code.

After completion of the second pass, object code is optionally written to disk. Errors detected in the source code during either pass may prevent correct assembly from taking place and will lead to the generation of error messages.

Problem 4 Describe the fields used in a typical line of assembly language source code.

Programs written using assembly language consists of a sequence of statements, and each statement consists of between one and five fields. A field is a group of characters, and fields are separated from each other by at least one space. The fields of a typical assembly language program as as follows:

Line no.	*Label*	*Mnemonic*	*Operand(s)*	*Comment*

and these fields may be identified in the following program segment:

Line no.	*Label*	*Mnemonic*	*Operand(s)*	*Comments*
0000	START:	LD	HL,0200H	;outer loop counter
0001		LD	C,0	;inner loop counter

```
0002    DELAY:  DEC   C           ;count down inner loop
0003            JP    NZ,DELAY    ;loop until timed out
0004            DEC   HL          ;count down outer loop
0005            LD    A,H         ;test for HL=0, (DEC HL
0006            OR    L           ;does not set flags)
0007            JP    NZ,DELAY    ;loop until timed out
```

Note: Line numbering is usually optional and therefore does not appear on all listings.

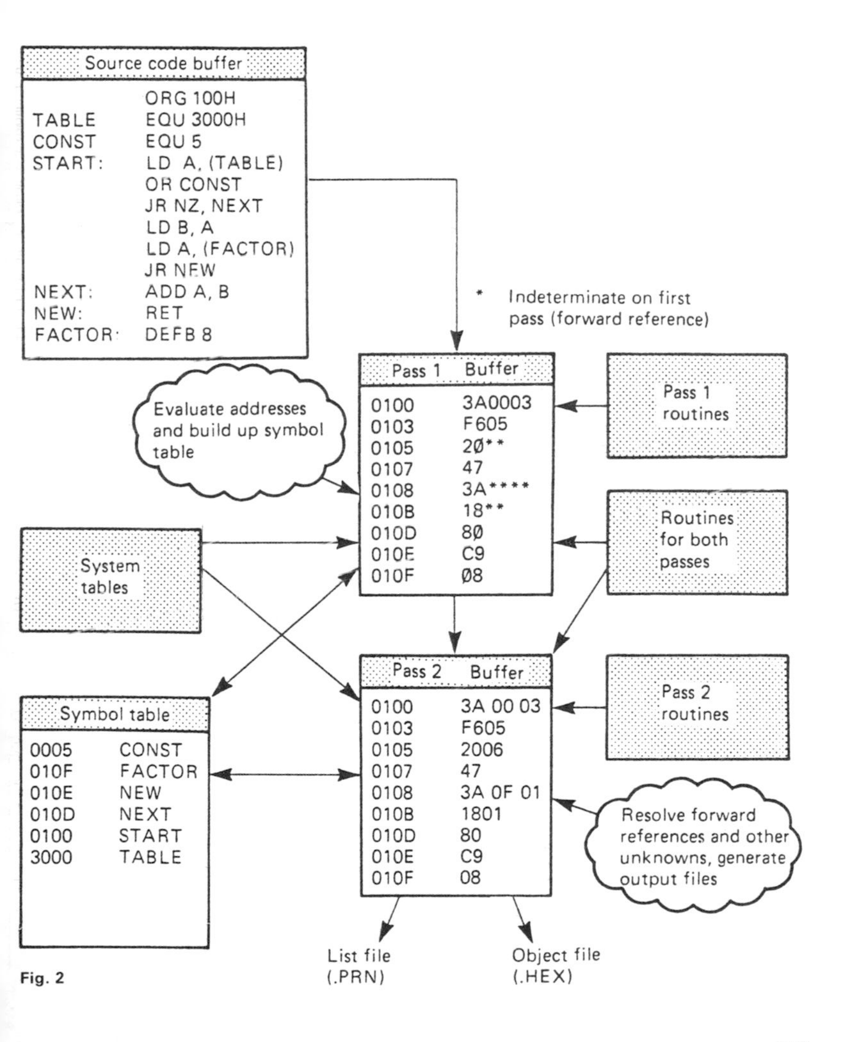

Fig. 2

Problem 5 Explain the function of a **label** in an assembly language program.

A label is a sequence of characters which act as an identifier, and represent an instruction address which may need to be referenced by other instructions within a program. Labels are used instead of the hexadecimal addresses that they represent for two reasons:

(a) at the time of writing the source code, the actual address is probably unknown;
(b) the use of meaningful labels is an aid to following the flow of a program.

As an example of the use of a label, consider the following segment of machine code:

```
0125  0D          DELAY:     DEC C
0126  C2 25 01               JP NZ,DELAY
```

At the time of writing the source code, the address of the DEC C instruction is unknown, but is assigned the label DELAY. During the assembly process, the actual address of the DEC C instruction is evaluated, and the label is recorded as being equivalent to address 0125H. This information is stored by the assembler in a **symbol table**. When a subsequent reference is made to this label, as in the case of the instruction JP NZ,DELAY, the assembler searches the symbol table for an entry DELAY, and once found, determines, the address represented by the label and substitutes this as the operand for the JP instruction. In this case, JP NZ,DELAY is equivalent to JP NZ,0125H.

The number and type of characters permitted in a label varies from one assembler to another. As a general guide, labels may contain up to eight characters, the first of which must be alphabetic. Most assemblers allow the use of a predefined symbol ($ or *) as a label which is assigned the current value of the location counter at the point where it is used. Some assemblers require a colon (:) as a terminator for each label.

Problem 6 Explain the term **pseudo-op** when applied to an assembly language program.

Pseudo-ops or assembler directives are instructions for the assembler rather than for the microprocessor, and they are used to set labels to specified values during assembly, specify starting addresses in the program, define storage areas, perform conditional assembly or control the format of a print file. Different assemblers support various directives, but the following are typical:

6800	6502	Z80	Function
ORG	*=	ORG	set the program or data origin
END	.END	END	end program
EQU	=	EQU	numeric 'equate'
FCB	.BYT	DEFB	define data bytes
FDB	.WOR	DEFW	define data words
FCC	.TXT	DEFM	define text message
RMB	*=*+	DEFS	define space (for data storage)

Problem 7 Explain how the starting address of an assembly language program may be initialized.

An ORG (or *=) directive is used to define the origin or starting address for a program or data area, and it takes the form:

label ORG *expression* (Z80/6800) or

label *= *expression* (6502)

where **label** is an optional program identifier, and **expression** is a 16-bit expression consisting of operands that must be defined before the ORG statement. In effect, an ORG directive presets the assembler location counter, and machine code generation starts at the address specified in the expression. A program may contain any number of ORG statements, but the programmer must ensure that overlapping does not occur, since no checks are normally made by an assembler. If a label is given to an ORG statement, the label assumes the value of the ORG expression.

Problem 8 Explain how an assembly language source program may be terminated.

An END statement is optional in some assemblers and mandatory in others but if present it must be the last statement since all subsequent statements are ignored by an assembler. Various forms of the END directive are:

label END

label .END

label END expression

where **label** is optional. In cases where an expression follows the END directive, the expression is evaluated, and this initializes the program counter.

Problem 9 Explain how values may be assigned to symbols in an assembly language program.

The EQU (equate) or = directive is used to associate labels with particular numeric values. For example, a memory location may be used to store the result of a calculation. In this case the EQU directive may be used to associate the label **RESULT** with the memory address where the result will be stored. In all subsequent references to this address the label **RESULT** may be used in place of its actual hexadecimal value.

The form of this directive is as follows:

label	EQU	*expression*	(Z80/6800) or
label	=	*expression*	(6502)

where the label must be present and must not label any other statement, i.e., labels may not be redefined within a program with the EQU directive (some assemblers support a SET, ASET or DEFL directive for this purpose). The assembler evaluates the expression, and assigns the resulting value to the label. As an aid to following a program, the name used for a label usually relates to its purpose within that program. Examples of the EQU directive are as follows:

TOTAL	**EQU**	**0C80H**	(or TOTAL = $0C80) assign the value 0C80H to the label TOTAL
FINAL	**EQU**	**TOTAL**	(or FINAL = TOTAL) additionally, assign the value already assigned to TOTAL to the label FINAL
DISP	**EQU**	**TOTAL + 10**	(or DISP = TOTAL + 10) assign a value to DISP which is 10 (decimal) more than the value already assigned to TOTAL.

Problem 10 Explain how constant data bytes may be used to initialize selected memory locations in an assembly language program.

The DEFB (define byte), FCB (form constant byte) or . BYTE directive is used when the programmer wishes to store constants directly into defined storage areas as single bytes. For example, a programmer may wish to create a look-up table at the end of a program which performs a code conversion process.

The form of the DEFB directive as as follows:

label:	DEFB	*e1*, *e2*, *e3*, ..., *en*	(Z80) or
label	. BYTE	*e1*, *e2*, *e3*, ..., *en*	(6502) or
label	FCB	*e1*, *e2*, *e3*, ..., *en*	(6800)

where *e1* to *en* are constants or expressions that evaluate to 8-bit numbers or single ASCII characters in quotes. Examples of the use of the DEFB directive are:

DATA:	**DEFB**	**0,1,2,15,20H,0FH,0CAH**	(Z80)
DATA	**.BYTE**	**0,1,2,15,$20,$0F,$CA**	(6502)
DATA	**FCB**	**0,1,2,15,$20,$0F,$CA**	(6800)

Problem 11 Explain how 16-bit addresses may be assigned to selected memory locations in an assembly language program.

The DEFW (define word), FDB (form double byte), or .WORD directive is similar to DEFB directive except that double length (16 bit) words of storage are used. This may be used by a programmer to store addresses directly in a program, for example for the creation of a table of interrupt vectors.

The form of the DEFB directive is as follows:

label	**DEFW**	*e1, e2, e3, ..., en*	(*Z80 – low order first*) or
label	**.WORD**	*e1, e2, e3, ..., en*	(*6502 – low order first*) or
label	**.DBYT**	*e1, e2, e3, ..., en*	(*6502 – high order first*) or
label	**FDB**	*e1, e2, e3, ..., en*	(*6800 – high order first*)

where *e1* to *en* are constants or expressions that evaluate to 16-bit numbers or ASCII strings up to two characters in length. Examples of the use of the DEFW directive are as follows:

VECTR:	**DEFW**	**0FC8H,VECTR+6,255+255**	(Z80)
VECTR	**.WORD**	**$0FC8,VECTR+6,255+255**	(6502)
VECTR	**FDB**	**$0FC8,VECTR+6,255+255**	(6800)

Problem 12 Explain how text in the form of ASCII codes may be stored in selected memory locations in an assembly language program.

The DEFM (define message), FCC (form constant character) or . TEXT directive allows text to be inserted directly into memory as equivalent ASCII codes. This is useful where a program must generate text strings as part of its normal operation, e.g. print error messages on a VDU screen. The use of the DEFM directive avoids having to look up ASCII codes for messages, and also allows the position of text within a program to be readily identified.

The form of the DEFM directive is as follows:

label	DEFM	'*text message*'	(Z80) or
label	. TEXT	'*text message*'	(6502) or
label	FCC	*text message*	(6800)

where the text message may be any sequence of characters, usually with some limit imposed regarding its length.

Problem 13 Explain how memory space may be reserved for a program when writing assembly language source code.

The DEFS (define space), RMB (reserve memory byte) or *=*+ directive is used to reserve space in memory, e.g. for the stack, or for system workspace, and takes the following form:

label DEFS *expression*

Some assemblers allow the DEFS directive to be followed by a constant which then causes the assembler to generate code to fill the reserved space with the selected constant. The assembler continues to create subsequent code at the first address after the area reserved by the DEFS statement. The effect of the DEFS statement is the same as defining a new origin with an ORG directive, e.g.

label	EQU	$
	ORG	$+*expression* or
label	*=*	+*expression*

Problem 14 Describe how operands are formed in an assembly language source code statement.

Many instructions require one or more operands which define the data (or location of data) to be operated on by the instruction. The operand may be a constant, a label or an expression. An example of each of these types of operand is as follows:

0CFH (or $CF) – *numerical constant*
SWITCH – *label*
PORT*5+2 – *expression*

Expressions are formed by combining a number of simple operands (labels and constants) using arithmetic and logic operators. These are evaluated during assembly, and each expression must produce a 16-bit value. After evaluation, the number of significant digits must not exceed the number required for its intended use, i.e., the eight most significant digits must evaluate to zero if the operand is data destined for an 8-bit register.

Z80 instructions usually require source and destination operands, although either may be implied in the instruction mnemonic. Where two operands are specified, they must be separated by a comma (,). The contents of external addresses may be used as operands provided that references to them are enclosed in brackets, e.g. direct reference – (0200H) or indirect reference – (HL).

Problem 15 Describe the different forms used for numeric constants in assembly language source code statements.

A numeric constant is a 16-bit value which may be expressed in one of several different bases. The base or radix is denoted by a preceding or following radix indicator, and those used are:

B (or %)	*binary constant*	(base 2)
D (or none)	*decimal constant*	(base 10)
H (or $)	*hexadecimal constant*	(base 16)
O (or Q,@)	*octal constant*	(base 8)

The omission of a radix indicator causes an assembler to use the default radix which is usually decimal, although it may be possible to change the default radix by including a RADIX pseudo-op at the start of the source code. Hexadecimal constants with a following **H** must begin with a numeric character to avoid confusion with labels, which means that if the constant starts with any of the hexadecimal codes A to F, a leading zero must be added. The following are all examples of the use of numeric constants:

1234 1234D 1011B 1234H 0A9FH 3472O 1265Q (Z80)
1234 %1011 $1234 @3427 (6502 and 6800)

Problem 16 Explain how string constants are formed in assembly language source code statements.

String constants are mainly formed by enclosing ASCII characters within single apostrophe (') characters. The value of a character then becomes its corresponding ASCII code. The string length is restricted to one or two characters, in which case the string becomes an 8- or 16-bit value. The exception to this is where the string constants is used with DEFM, FCC and .TXT pseudo-ops. The following are examples of the use of string constants:

LD A,'A'	which represents	**LD A,41H (Z80)**
LDA #'A		**LDA #$41 (6502)**
LDAA #'A		**LDAA #$41 (6800)**

Problem 17 Explain the term **reserved word** when applied to assembly language.

There are several character sequences which are reserved by an assembler for its own use. In general, all mnemonics, register names and assembler directives are reserved by the assembler and may not be used by the programmer as labels.

Problem 18 Describe the arithmetic operators used in assembly language programs.

The operands described may be combined in normal algebraic notation using any combination of properly formed operands, operators and expressions. The following are examples of arithmetic operators supported by most assemblers:

$a+b$ unsigned arithmetic sum of a and b
$a-b$ unsigned arithmetic difference between a and b
$a * b$ unsigned magnitude multiplication of a and b
a/b unsigned division of a by b

In each case, **a** and **b** represent simple operands (labels, numeric constants, one or two character strings). Further arithmetic and logical operators may be available depending upon the assembler used.

More complex expressions may be formed by using combinations of these operators. All calculations are performed at assembly time as 16-bit unsigned operations, but the expression must evaluate to a number appropriate for its intended use.

Problem 19 Determine the actual value loaded into the accumulator in the following examples.

(a) LD A,3+4

(b) LD A,0CH-4

(c) LD A,3*4

(d) LD A,9/5

Expression	**Represents**
(a) LD A,3+4	LD A,7
(b) LD A,0CH-4	LD A,8
(c) LD A,3*4	LD A,0CH
(d) LD A,9/5	LD A,1

Problem 20 Show how comments may be included in an assembly language program.

This is an optional field which is reserved for the addition of comments which are intended to help follow the logical processes taking place in a program. Comments should not simply repeat the instruction, but should indicate why certain instructions have been included. For example, if the source code contains the instruction LD B,5 then the comment **load register B with 5** is virtually worthless, since this is obvious from the instruction. A more worthwhile comment would be **print 5 characters** or **count 5 pulses** since comments of this type give the reason for loading register B with 5.

e.g.   LD B,5   **;print five characters**

The comment field is always preceded by a semi-colon (;) and all information between the semi-colon and the end of the line is ignored by the assembler. Some assemblers also allow the use of an asterisk (*) to precede the comment field.

Problem 21 Describe the two categories of errors that can occur when writing assembly language programs.

Except for very short and simple programs, it is unlikely that a program will function as intended at the first attempt. In common with all other languages, assembly language may contain two types of error, and these are:

(a) **Syntax errors** which are errors in the grammar or format of assembly language statements which make it impossible for the assembler program to interpret them and hence generate code.
(b) **Logical errors** which do not prevent correct assembly of the program, but which cause the resultant object code to perform in an entirely different manner from that intended.

The method used to report errors in the source code may vary with different assemblers, but frequently an error code is printed against each offending line. The error code is intended to indicate the type of error wherever possible, thus enabling the programmer to locate its exact source.

Problem 22 Explain the importance of proper documentation in an assembly language program.

An assembly language program should be well documented so that it is easy to follow the logical processes used in the program. This enables software to be subsequently maintained, i.e., updated, debugged, installed in different environments or otherwise modified. Without such documentation, it is unlikely that even the original programmer would be able to readily identify the function of each section of program. This requires that all labels and symbolic addresses should, within limits, be set by the assembler, indicate their function in the program. Programs should also be partitioned into individual modules wherever possible, and comments should be liberally added at all stages.

Problem 23 Write a 6502 assembly language subroutine to perform an unsigned multiplication of the contents of the X and A registers and return the 16-bit product in memory locations $0084/$0085.

This routine performs multiplication between two 8-bit numbers using a standard **shift and add** method.

```
0000                      ;
0000                      ;Subroutine to perform 8-bit unsigned
0000                      ;multiplication
0000                      ;using shift and add algorithm
0000                      ;
0000                      ;Entry; X=multiplier, A=multiplicand
0000                      ;Exit: product in memory locations
0000                      ;       $0084/$0085
0000                      ;
0000                      ;
0000          PRODHI      =    $0084
0000          PRODLO      =    PRODHI+1
0000          COUNT       =    8
0000                      ;
0200                     *=    $0200
0200                      ;
0200                      ;initialize registers
0200                      ;
0200   85 84              STA  PRODHI     ;use memory as 16
0202   86 85              STX  PRODLO     ;bit register
0204   A9 00              LDA  #0         ;clear product reg
0206   A2 08              LDX  #COUNT     ;shift/add counter
0208                      ;
0208                      ;multiply routine
0208                      ;
0208   6A     SHIFT       ROR  A          ;shift multiplier
0209   66 85              ROR  PRODLO     ;and product right
020B   90 03              BCC  NOADD      ;check LSB mult
020D   18                 CLC             ;add multiplicand if
020E   65 84              ADC  PRODHI     ;LSB=1 else add 0
0210   CA     NOADD       DEX             :8,7,6,5 . . . . . . . . . .
0211   D0 F5              BNE  SHIFT      ;count down to zero
0213                      ;
0213                      ;correct for loss of first shift
0213                      ;
0213   6A                 ROR  A          ;final shift of
0214   66 85              ROR  PRODLO     ;product
0216   85 84              STA  PRODHI     ;16-bit prod to memory
0218                      ;
0218                      ;return to calling program
0218                      ;
0218   60                 RTS
0219                      ;
0219                      . END
```

Problem 24 Write a 6502 assembly language subroutine to perform an unsigned division of the contents of the A register by the contents of the X. The subroutine should return the quotient in memory location $0086 and the remainder in $0087.

This routine performs division between two 8-bit numbers using a **shift and subtract** or **restoring** method.

```
0000                       ;
0000                       ;Subroutine to perform 8-bit unsigned
0000                       ;division
0000                       ;using shift and subtract algorithm
0000                       ;
0000                       ;Entry; X=divisor, A=dividend
0000                       ;Exit: quotient in $0086
0000                              remainder in $0087
0000                       ,
0000                       ;
0000          QUOTNT       =      $0086
0000          REMNDR       =      QUOTNT+1
0000          COUNT        =      8
0000                       ;
0200                       *=     $0200
0200                       ;
0200                       ;initialize registers
0200                       ;
0200   85 86  DIVIDE       STA    QUOTNT        ;use memory as 16
0202   86 87               STX    REMNDR        ;bit register
0204   A9 00               LDA    #0            ;clear dividend high
0206   A2 08               LDX    #COUNT        ;shift/sub counter
0208                       ;
0208                       ;divide routine
0208                       ;
0208   26 86  SHIFT        ROL    QUOTNT        ;shift dividend
020A   2A                  ROL    A             ;and quotient
020B   38                  SEC                  ;prepare for sub
020C   E5 87               SBC    REMNDR        ;subtract divisor
020E   B0 03               BCS    NOADD         ;skip restore if C=1
0210   65 87               ADC    REMNDR        ;divisor too big
0212   18                  CLC                  ;so restore
0213   CA     NOADD        DEX                  ;8,7,6,5 . . . . . . . . .
0214   D0 F2               BNE    SHIFT         ;count down to zero
0216                       ;
0216                       ;correct for loss of first shift
0216                       ;
0216   26 86               ROL    QUOTNT        ;final shift quotient
0218   85 87               STA    REMNDR        ;store remainder
021A                       ;
```

```
021A                        ;return to calling program
021A                        ;
021A  60                    RTS
021B                        ;
021B                        . END
```

Problem 25 Write a 6502 assembly language subroutine to perform conversion of BCD number in the accumulator into 7-segment format.

When driving 7-segment display devices, it is necessary to convert from the normally available BCD results into equivalent segment codes. Since the logical relationships between BCD and 7-segment codes are quite complex, this routine uses a **look-up table** method of conversion.

```
0000                        ;
0000                        ;BCD to 7-segment subroutine
0000                        ;
0000                        ;Entry: BCD data in A
0000                        ;Exit: segments code in A
0000                        ;
0000                        ;
0200                       *=       $0200
0200                        ;
0200  AA        SEG7        TAX                  ;index table via X
0201  BD 05 02              LDA     SEGTAB,X     ;get segments code
0204  60                    RTS
0205                        ;
0205                        ;segments look-up table
0205                        ;
0205                        ;bit → b7 b6 b5 b4 b3 b2 b1 b0
0205                        ;seg → dp  g  f  e  d  c  b  a
0205                        ;
0205  3F        SEGTAB      . BYTE  %00111111    ;0
0206  06                    . BYTE  %00000110    ;1
0207  5B                    . BYTE  %01011011    ;2
0208  4F                    . BYTE  %01001111    ;3
0209  66                    . BYTE  %01100110    ;4
020A  6D                    . BYTE  %01101101    ;5
020B  7D                    . BYTE  %01111101    ;6
020C  07                    . BYTE  %00000111    ;7
020D  7F                    . BYTE  %01111111    :8
020E  67                    . BYTE  %01100111    ;9
020F                        ;
020F                        . END
```

Problem 26 Write a 6502 assembly language subroutine to convert a pure binary number in A into its decimal equivalent. The subroutine should return the BCD equivalent in memory locations $0089/$008A.

It is often simpler to perform calculations in pure binary when using a microprocessor. For example, the multiplication and division routines included in this chapter could be made to work with BCD numbers but this would undoubtedly introduce complications, since the programmer would then find it necessary to ensure that illegal combinations were not generated. However, BCD results are required for display in most cases, therefore this routine may be used to perform the necessary conversion between 8-bit pure binary and BCD.

A **double and add** method of conversion is used which enables a very simple conversion program to be implemented.

```
0000                         ;
0000                         ;Binary to BCD subroutine
0000                         ;
0000                         ;Entry: binary number in A
0000                         ;Exit: BCD equivalent in memory
0000                         ;      locations $0089/$008A
0000                         ;
0000                         ;
0000             TEMP        =      $0088
0000             BCDLO       =      TEMP+1     ;BCD equivalent
0000             BCDHI       =      TEMP+2     ;(3 digits)
0000             COUNT       =      8
0000                         ;
0200                        *=      $0200
0200                         ;
0200                         ;initialize registers
0200                         ;
0200   F8        BIDEC       SED               ;decimal arithmetic
0201   85 88                 STA    TEMP       ;binary shift reg
0203   8A                    TXA               ;
0204   48                    PHA               ;preserve X reg
0205   A2 08                 LDX    #COUNT     ;count double/adds
0207   A9 00                 LDA    #0         ;clear sub-total
0209   85 89                 STA    BCDLO      ;and its copy
020B                         ;
020B                         ;binary to BCD conversion
020B                         ;
020B   06 88     SHIFT       ASL    TEMP       ;next bit into carry
020D   A5 89                 LDA    BCDLO      ;double subtotal and
020F   65 89                 ADC    BCDLO      ;add carry (decimal!)
0211   85 89                 STA    BCDLO      ;store for next loop
0213   26 8A                 ROL    BCDHI      ;update hundreds
0215   CA                    DEX               ;8,7,6,5 . . . . . . . . .
```

```
0216   D0 F3           BNE    SHIFT        ;down to zero
0218                   ;
0218                   ;restore X register and return
0218                   ;
0218   68              PLA
0219   AA              TAX                 ;restore X reg
021A   D8              CLD                 ;back to binary mode
021B   60              RTS
021C                   ;
021C                   .END
```

Problem 27 Write a Z80 assembly language subroutine to perform an unsigned multiplication of the contents of the H and L registers and return the 16-bit product in HL.

This routine performs multiplication between two 8-bit numbers using a standard **shift and add** method.

```
                       ;
                       ;Subroutine to perform 8-bit unsigned
                       ;multiplication of H and L registers
                       ;using shift and add algorithm
                       ;16-bit product is returned in HL pair
                       ;All other registers are preserved
                       ;
                       ;
0100                   ORG    100H
                       ;
                       ;preserve registers
                       ;
0100  C5      MULT:    PUSH   BC           ;BC and DE registers are
0101  D5               PUSH   DE           ;corrupted by routine
                       ;
                       ;initialize registers
                       ;
0102  5D               LD     E,L          ;transfer multiplicand to E
0103  2E00             LD     L,0          ;product low byte=0
0105  55               LD     D,L          ;multiplicand high byte=0
0106  06 08            LD     B,8          ;8 shifts and adds
                       ;
                       ;main multiplication loop
                       ;
0108  29      SHIFT:   ADD    HL,HL        ;shift HL left into C flag, if
0109  30 01            JR     NC,ADD0      ;C=0, partial product is zero,
010B  19               ADD    HL,DE        ;C=1, add multiplicand
010C  10 FA   ADD0:    DJNZ   SHIFT        ;multiplication complete?
                       ;
```

```
                          ;restore registers and return
                          ;
010E  D1                  POP    DE         ;recover previous contents
010F  C1                  POP    BC         ;of BC and DE registers
0110  C9                  RET               ;go back to main program
                          ;
                          END
```

Problem 28 Write a Z80 assembly language subroutine to perform an unsigned division of the contents of the L register by the contents of the H register. The subroutine should return the quotient in L and the remainder in H.

This routine performs 8-bit by 8-bit division using a **shift and subtract** or **restoring** method.

```
                          ;
                          ;Subroutine to perform 8-bit unsigned
                          ;division of H and L registers
                          ;using shift and subtract algorithm
                          ;
                          ;Entry: divisor in H, dividend in L
                          ;Exit: quotient in L, remainder in H
                          ;
                          ;All other registers are preserved
                          ;
                          ;
0100                      ORG    100H       ;start of TPA
                          ;
                          ;prepare registers
                          ;
0100  F5       DIVIDE:    PUSH   AF         ;preserve AF, BC and DE
0101  C5                  PUSH   BC         ;registers which are
0102  D5                  PUSH   DE         ;corrupted by this routine
                          ;
                          ;initialize registers
                          ;
0103  54                  LD     D,H        ;transfer divisor to E
0104  26 00               LD     H,0        ;clear H and E
0106  5C                  LD     E,H        ;registers
0107  06 08               LD     B,8        ;8 shifts and subtracts
                          ;
                          ;main division loop
                          ;
0109  29       SHIFT:     ADD    HL,HL      ;shift HL left, and
010A  ED 52               SBC    HL,DE      ;try to divide by D
010C  3001                JR     NC,NOADD   ;it goes, so skip restore
010E  19                  ADD    HL,DE      ;D too big, so restore
010F  17       NOADD:     RLA               ;record result in A
```

```
0110   10 F7   ADD0:    DJNZ   SHIFT       ;division completed?
0112   2F               CPL                ;result complemented
0113   6F               LD     L,A         ;and put into L
                        ;
                        ;restore registers and return
                        ;
0114   D1               POP    DE          ;recover previous contents
0115   C1               POP    BC          ;of BC and DE
0116   F1               POP    AF          ;and AF registers
0117   C9               RET                ;go back to main program
                        ;
                        END
```

Problem 29 Write a Z80 assembly language subroutine to perform conversion of BCD number in the accumulator into 7-segment format.

When driving 7-segment display devices, it is necessary to convert from the normally available BCD results into equivalent segment codes. Since the logical relationships between BCD and 7-segment codes are quite complex, this routine uses a **look-up table** method of conversion.

```
                          ;
                          ;BCD to 7-segment subroutine
                          ;
                          ;Entry: BCD data in A
                          ;Exit: segments code in A
                          ;
                          ;
0100                      ORG    100H        ;start of TPA
                          ;
0100   E5         SEG7;   PUSH   HL          ;preserve main HL
0101   21 08 01           LD     HL,TABLE    ;base of code table
0104   85                 ADD    A,L         ;segment index
0105   7E                 LD     A,(HL)      ;get seg code
                          ;
0106   E1                 POP    HL          ;restore main HL
0107   C9                 RET                ;back to main prog
                          ;
                          ;segments look-up table
                          ;
                          ;bit → b7 b6 b5 b4 b3 b2 b1 b0
                          ;seg → dp  g  f  e  d  c  b  a
                          ;
0108   3F         TABLE:  DEFB   00111111B   ;0
0109   06                 DEFB   00000110B   ;1
010A   5B                 DEFB   01011011B   ;2
010B   4F                 DEFB   01001111B   ;3
```

```
010C  66          DEFB   01100110B   ;4
010D  6D          DEFB   01101101B   ;5
010E  7D          DEFB   01111101B   ;6
010F  07          DEFB   00000111B   ;7
0110  7F          DEFB   01111111B   ;8
0111  67          DEFB   01100111B   ;9
                  ;
                  END
```

Problem 30 Write a Z80 assembly language subroutine to convert a pure binary number in A into its decimal equivalent. The subroutine should return the BCD equivalent in HL.

It is often simpler to perform calculations in pure binary when using a microprocessor. For example, the multiplication and division routines included in this chapter could be made to work with BCD numbers but this would undoubtedly introduce complications, since the programmer would then find it necessary to ensure that illegal combinations were not generated. However, BCD results are required for display in most cases, therefore this routine may be used to perform the necessary conversion between 8-bit pure binary and BCD.

A **double and add** method of conversion is used which enables a very simple conversion program to be implemented.

```
                          ;
                          ;Subroutine to perform binary to BCD
                          ;conversion using the double and add
                          ;method
                          ;
                          ;Entry: binary number in A
                          ;Exit: BCD equivalent in HL
                          ;
                          ;All other registers are preserved
                          ;
                          ;
0100                      ORG    100H      ;start of TPA
                          ;
                          ;preserve registers
                          ;
0100  F5       BIDEC:     PUSH   AF        ;preserve AF and BC
0101  C5                  PUSH   BC        ;registers
                          ;
                          ;initialize registers
                          ;
0102  06 08               LD     B,8       ;process 8-bit number
0104  4F                  LD     C,A       ;held in register C
```

```
0105    AF                XOR    A          ;clear working reg
                          ;
                          ;main conversion loop
                          ;
0106    CB11     SHIFT:   RL     C          ;bit by bit into C flag
0108    8F                ADC    A,A        ;double A and add carry
0109    27                DAA               ;convert to valid BCD
010A    CB 14             RL     H          ;update hundreds in H
010C    10 F8             DJNZ   SHIFT      ;repeat for all 8 bits
010E    6F                LD     L,A        ;units and tends into L
                          ;
                          ;restore registers
                          ;
010F    C1                POP    BC         ;recover previous contents
0110    F1                POP    AF         ;of AF and BC registers
0111    C9                RET               ;go back to main program
                          ;
                          END
```

Problem 31 Write a 6800 assembly language subroutine to perform an unsigned multiplication of the contents of the A and B registers and return the 16-bit product in memory locations $0084/$0085.

This routine performs multiplication between two 8-bit numbers using a standard **shift and add** method.

```
0000                      ;
0000                      ;Subroutine to perform 8-bit unsigned
0000                      ;multiplication of A and B registers
0000                      ;using shift and add algorithm
0000                      ;
0000                      ;Entry: B=multiplier, A=multiplicand
0000                      :Exit: product in memory locations
0000                             $0084/$0085
0000                      ;
0000                      ;
0000             PRODHI   EQU    $0084
0000             PRODLO   EQU    PRODHI+1
0000             COUNT    EQU    8
0000                      ;
0200                      ORG    $0200
0200                      ;
0200                      ;initialize registers
0200                      ;
0200    97 84             STAA   PRODHI     ;use memory as 16
0202    D7 85             STAB   PRODLO     ;bit register
0204    4F                CLRA              ;clear product reg
```

```
0205   C6 08                LDAB   #COUNT    ;shift/add counter
0207                        ;
0207                        ;multiply routine
0207                        ;
0207   46         SHIFT     RORA             ;shift multiplier
0208   76 00 85             ROR    PRODLO    ;and product right
020B   24 02                BCC    NOADD     ;check LSB mult
020D   9B 84                ADDA   PRODHI    ;add multiplicand if
020F                                         ;LSB=1 else add 0
020F   5A         NOADD     DECB             ;8,7,6,5 . . . . . . . . . .
0210   26 F5                BNE    SHIFT     ;count down to zero
0212                        ;
0212                        ;correct for loss of first shift
0212                        ;
0212   46                   RORA             ;final shift of
0213   76 00 85             ROR    PRODLO    ;product
0216   97 84                STA    PRODHI    ;16-bit prod to memory
0218                        ;
0218                        ;return to calling program
0218                        ;
0218   39                   RTS
0219                        ;
0219                        END
```

Problem 32 Write a 6800 assembly language subroutine to perform an unsigned division of the contents of the B register by the contents of the A. The subroutine should return the quotient in accumulator A and the remainder in B.

This routine performs multiplication between two 8-bit numbers using a **shift and subtract** or **restoring** method.

```
0000                       ;
0000                       ;Subroutine to perform 8-bit unsigned
0000                       ;multiplication
0000                       ;using shift and subtract algorithm
0000                       ;
0000                       ;Entry: A=divisor, B=dividend
0000                       ;Exit: A=quotient, B=remainder
0000                       ;
0000                       ;
0000          DIVISOR      EQU    $0086
0000          COUNT        EQU    8
0000          COUNTER      EQU    2
0000                       ;
0200                       ORG    $0200
0200                       ;
```

```
0200                       ;initialize registers
0200                       ;
0200  97 86     DIVIDE     STAA    DIVISOR    ;put divisor in memory
0202  86 08                LDAA    #COUNT     ;set up loop counter
0204  97 02                STAA    COUNTER    ;in memory
0206  4F                   CLRA               ;clear remainder reg
0207                       ;
0207                       ;divide routine
0207                       ;
0207  59        SHIFT      ROLB               ;shift dividend/quotient
0208  49                   ROLA               ;one place left, carry
0209                                          ;into A (inverted)
0209  90 86                SUBA    DIVISOR    ;subtract divisor
020B  24 02                BCC     NOADD      ;skip restore if C=1
020D  98 86                ADDA    DIVISOR    ;divisor too big
020F                                          ;so restore
020F  7A 00 02  NOADD      DEC     COUNTER    ;8,7,6,5 . . . . . . . . . .
0212  26 F3                BNE     SHIFT      ;count down to zero
0214                       ;
0214                       ;correct for loss of first shift
0214                       ;
0214  59                   ROLB               ;final shift quotient
0215  53                   COMB               ;correct earlier invert
0216                                          ;of quotient bits
0216                       ;
0216                       ;return to calling program
0216                       ;
0216  39                   RTS
0217                       ;
0217                       END
```

Problem 33 Write a 6800 assembly language subroutine to perform conversion of BCD number in accumulator A into 7-segment format.

When driving 7-segment display devices, it is necessary to convert from the normally available BCD results into equivalent segment codes. Since the logical relationships between BCD and 7-segment codes are quite complex, this routine uses a **look-up table** method of conversion.

```
0000                       ;
0000                       ;BCD to 7-segment subroutine
0000                       ;
0000                       ;Entry: BCD data in acc A
0000                       ;Exit: segments code in acc A
0000                       ;
0000                       ;
0000            XTEMP      EQU     0
```

```
0000                          ;
0200                          ORG    $0200
0200                          ;
0200   DF 00      SEG7        STX    XTEMP          ;preserve X reg
020    CE 02 0D               LDX    #SEGTAB-1      ;X → bottom of table
0205   08         SEGCNT      INX                   ;move X up the table
0206   4A                     DECA                  ;A places
0207   26 FC                  BPL    SEGCNT
0209   A6 00                  LDAA   0,X            ;get segments code
020B   DE 00                  LDX    XTEMP          ;recover X
020D   39                     RTS
020E                          ;
020E                          ;segments look-up table
020E                          ;
020E                          ;bit → b7 b6 b5 b4 b3 b2 b1 b0
020E                          ;seg → dp  g  f  e  d  c  b  a
020E                          ;
020E   3F         SEGTAB      FCB    %00111111      ;0
020F   06                     FCB    %00000110      ;1
0210   5B                     FCB    %01011011      ;2
0211   4F                     FCB    %01001111      ;3
0212   66                     FCB    %01100110      ;4
0213   6D                     FCB    %01101101      ;5
0214   7D                     FCB    %01111101      ;6
0215   07                     FCB    %00000111      ;7
0216   7F                     FCB    %01111111      ;8
0217   67                     FCB    %01100111      ;9
0218                          ;
0216                          END
```

Problem 34 Write a 6800 assembly language subroutine to convert a pure binary number in accumulator A into its decimal equivalent. The subroutine should return the BCD equivalent in memory locations $0089/$008A.

It is often simpler to perform calculations in pure binary when using a microprocessor. For example, the multiplication and division routines included in this chapter could be made to work with BCD numbers but this would undoubtedly introduce complications, since the programmer would then find it necessary to ensure that illegal combinations were not generated. However, BCD results are required for display in most cases, therefore this routine may be used to perform the necessary conversion between 8-bit pure binary and BCD.

A **double and add** method of conversion is used which enables a very simple conversion program to be implemented.

```
0000                          ;
0000                          ;Binary to BCD subroutine
0000                          ;
0000                          ;Entry: binary number in acc A
0000                          ;Exit: BCD equivalent in memory
0000                          ;       locations $0089/$008A
0000                          ;
0000                          ;
0000               TEMP       EQU    $0088
0000               BCDLO      EQU    TEMP+1     ;BCD equivalent
0000               BCDHI      EQU    TEMP+2     ;(3 digits)
0000               COUNT      EQU    8
0000                          ;
0200                          ORG    $0020
0200                          ;
0200                          ;initialize registers
0200                          ;
0200   37          BIDEC      PSHB              ;preserve B reg
0201   C6 08                  LDAB   #COUNT     ;loop counter
0203   97 88                  STAA   TEMP       ;binary shift reg
0205   4F                     CLRA              ;clear subtotal
0206   97 89                  STAA   BCDLO      ;and its copy
0208                          ;
0208                          ;binary to BCD conversion
0208                          ;
0208   78 00 88    SHIFT      ASL    TEMP       ;next bit into carry
020B   96 89                  LDAA   BCDLO      ;double sub-total and
020D   99 89                  ADCA   BCDLO      ;add carry (decimal!)
020F   19                     DAA               ;
0210   97 89                  STAA   BCDLO      ;store for next loop
0212   79 00 8A               ROL    BCDHI      ;update hundreds
0215   5A                     DECB              ;8,7,6,5 ..........
0216   26 F0                  BNE    SHIFT      ;down to zero
0218                          ;
0218                          ;restore B register and return
0218                          ;
0218   33                     PULB              ;restore B reg
0219   39                     RTS
021A                          ;
021A                          END
```

C FURTHER PROBLEMS ON ASSEMBLY LANGUAGE PROGRAMS

(a) SHORT ANSWER PROBLEMS

1 The initial stages of writing an assembly language program involve preparing the .... code.

2 Initially the program code is prepared with the aid of a .... program.

3 One advantage of writing machine code programs using assembly language is ....

4 A sequence of characters which identify the address of an instruction in a program is known as a ....

5 In an assembly language program, an instruction for the assembler rather than for the microprocessor is known as ....

6 The starting address of an assembly language program is set by the use of an .... statement.

7 In an assembly language program, 8-bit constants may be stored directly into memory by the use of a .... statement.

8 In an assembly language program, 16-bit constants may be stored directly into memory by the use of a .... statement.

9 In an assembly language program, text may be stored directly into memory by the use of a .... statement.

10 In an assembly language program, memory space may be reserved by the use of a .... statement.

11 A value may be assigned to a symbol in an assembly language program by the use of an .... statement.

12 The operands of an assembly language instruction may be formed by the use of ...., ... or ....

13 The default radix of an assembler is usually ....

14 The digits 101 may be expressed in an assembly language program in: (a) binary as ...., (b) octal as ...., and (c) hexadecimal as ....

15 A character sequence which is used by an assembler for its own internal use, and which is therefore not available to the programmer, is known as a ....

16 In an assembly language program, a semicolon (;) is often used to denote the start of a ....

17 Two types of error that may occur when writing assembly language programs are .... and ....

18 Operands in an assembly language program may be joined by the use of .... or .... operators.

(b) CONVENTIONAL PROBLEMS

1 Explain why a text editor is required in order to create assembly language source code.

2 Describe **five** essential features of a text editor suitable for producing assembly language source code.

3 State the main advantages of using an assembler rather than programming directly in machine code.

4 (a) Explain the term **pseudo-op**,
(b) For the 6502, Z80 or 6800 microprocessors, list **four** pseudo-ops and describe a use for each of them.

5 State the advantages of using **labels** and **symbolic addresses** to replace hexadecimal notation in assembly language programs.

6 If an assembly language program contains the following lines of source code:

BUFFER = $2000 (or **BUFFER EQU 2000H**)
COUNTER = $200A (or **COUNTER EQU 200AH**)

When loading the accumulator from memory location $200A, show how the instruction operand may be written as:

(a) a constant (use 6502, Z80 or 6800 assembly language);
(b) a label;
(c) an expression.

7 Signal degradation

A MAIN POINTS CONCERNED WITH SIGNAL DEGRADATION

1 The term **signal degradation** is used to describe the fact that changes in the amplitude or shape of a signal have taken place as a result of that signal being subjected to some unwanted process while being transferred through the system. Sometimes such degrading is unavoidable. However, good design practices should minimize much signal degrading.

2 A logic circuit diagram is used to show the theoretical organization and interconnections between the various logic devices used in the construction of a microelectronic system. It does not normally indicate the actual physical layout of the system, but is arranged so that the operation of the system as a whole may be more clearly followed. In fact, parts of the *same* IC, e.g. gates or buffers, are often widely separated on a circuit diagram.

3 For a given logic circuit it is possible to have very many different physical layouts, i.e., position and orientation of devices, size of circuit board, etc. The ideal physical layouts are those that keep the interconnecting wiring to a minimum with conductors of the shortest possible length, but factors such as the need for testing and fault finding may call for slight deviations from the ideal. Given these constraints, however, there usually remains a large number of different physical layouts which provide equal electrical performance.

4 Many different construction techniques are available, e.g. stripboard, wire wrap, printed circuit. Some of these techniques are more appropriate for design and development work. Virtually all production circuits are constructed using printed circuit board (PCB) techniques.

5 The circuit diagram for a microcomputer memory expansion board which consists of three 4 K × 8 bit EPROMS is shown in *Fig. 1*.

The circuit is constructed using a double-sided PCB for which the physical layouts are shown in *Fig. 2(a) and (b)*.

Fig. 2(a) shows the component side of the board and *Fig. 2(b)* shows the solder side of the board. This is just one of the many possible layouts for this simple circuit, and the relationship between logic circuit diagram and physical layout may be clearly seen.

6 Physical circuits such as those shown in *Fig. 2* contain many **hidden** components which are not obvious when considering only the logic circuit diagram. Each interconnecting conductor on the PCB has associated resistance and inductance; stray capacitance exists between adjacent conductors. Circuit components also have hidden L, C and R, as shown in *Fig. 3*.

7 The amount of L, C and R associated with each conductor is very small, but microelectronic circuits operate at very high switching rates (in excess of

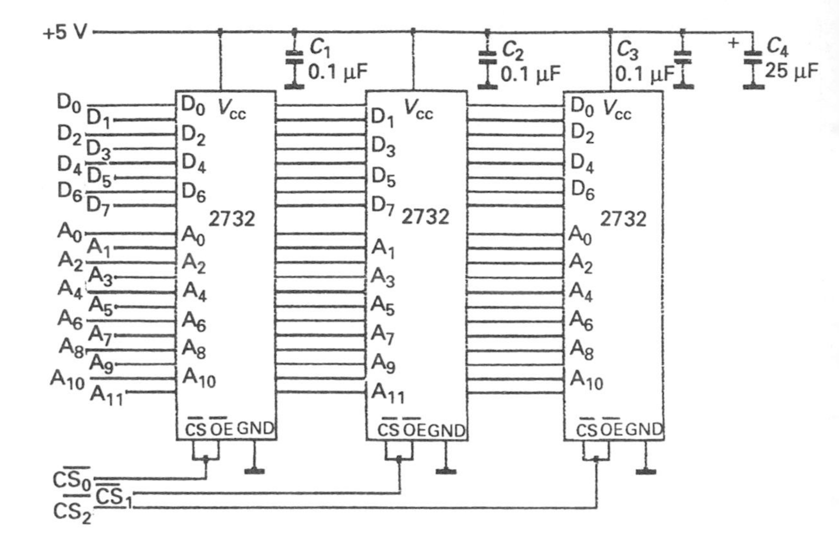

Fig. 1

1 MHz). Switching pulses which operate at these rates have very fast rise times and therefore contain very high frequency harmonic components. To such signals, even minute values of *L*, *C* and *R* become significant and may result in severe signal degradation.

8 The types of signal degradation commonly seen are shown in *Fig. 4*.

Combinations of *R* and *C* cause degradation of the signal in the form of **rounding** of the signals. The degree of rounding depends upon the ratio of *R* to *C* (see *Fig. 4*(*b*) *and* (*c*)). Small amounts of rounding can be tolerated, but if this form of degradation becomes severe it can lead to timing errors.

Combinations of *L* and *C* form **tuned circuits** which can oscillate at their resonant frequency if provided with energy by fast changing logic signals (see *Fig. 4*(*d*)). This effect is known as **ringing**, which if severe, may cause a single transition of the logic signal to generate multiple switching transitions (see *Fig. 5*).

Cross talk is another form of degradation which is the result of one signal causing disturbance of another signal. This may be caused by inductive or capacitive coupling between adjacent conductors, or may be caused by unwanted coupling via the power supply tracks.

9 Ringing in the logic circuits may be eliminated by **damping** the tuned circuits formed by *L* and *C*. This may be achieved by driving each PCB conductor from a low resistance source. Buffer ICs are available which have low resistance outputs, and this output resistance is used to damp the *L-C* circuit, as shown in *Fig. 6*.

10 Much cross talk occurs due to coupling via the power supply tracks. This problem may be kept to a minimum by:

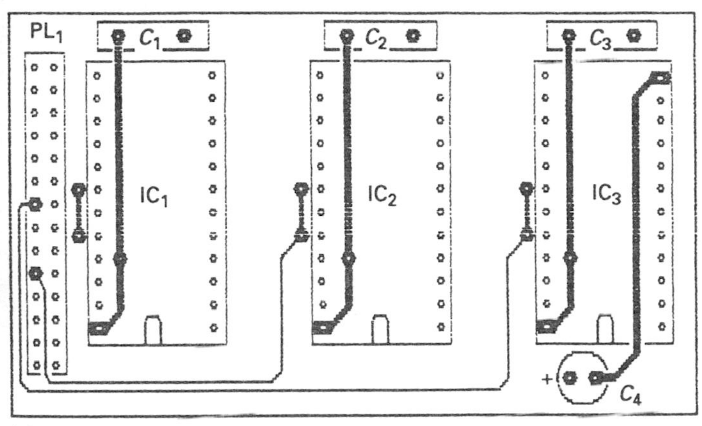

(a)

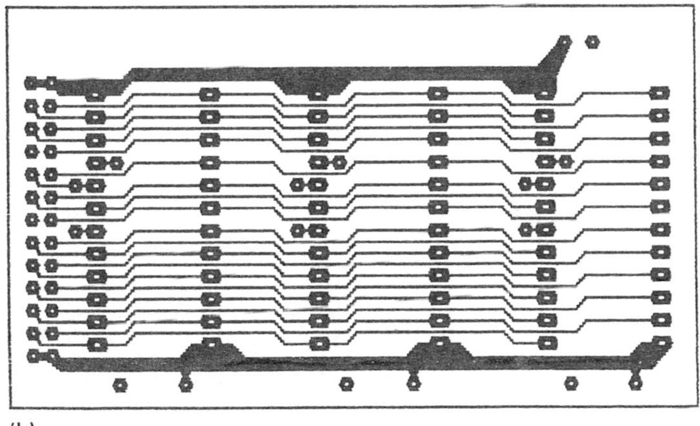
(b)

Fig. 2

(a) using wider PCB tracks to lower their resistance, and
(b) liberal use of decoupling capacitors, possibly as many as one for each IC.

11 Most PCs are designed using CAD equipment which simplifies the design of complex track patterns and allows more compact circuits to be developed. As a general rule, all ICs on a board share the same orientation, and tracks on the component side of the board run at 90° to those on the solder side.

12 The outputs of a typical microprocessor are only capable of driving a single **TTL load**. Each memory or I/O device that is connected to the buses of a

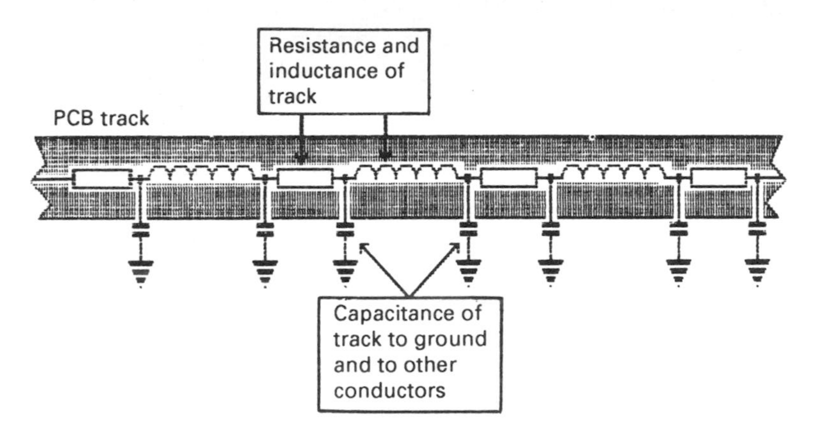

Fig. 3

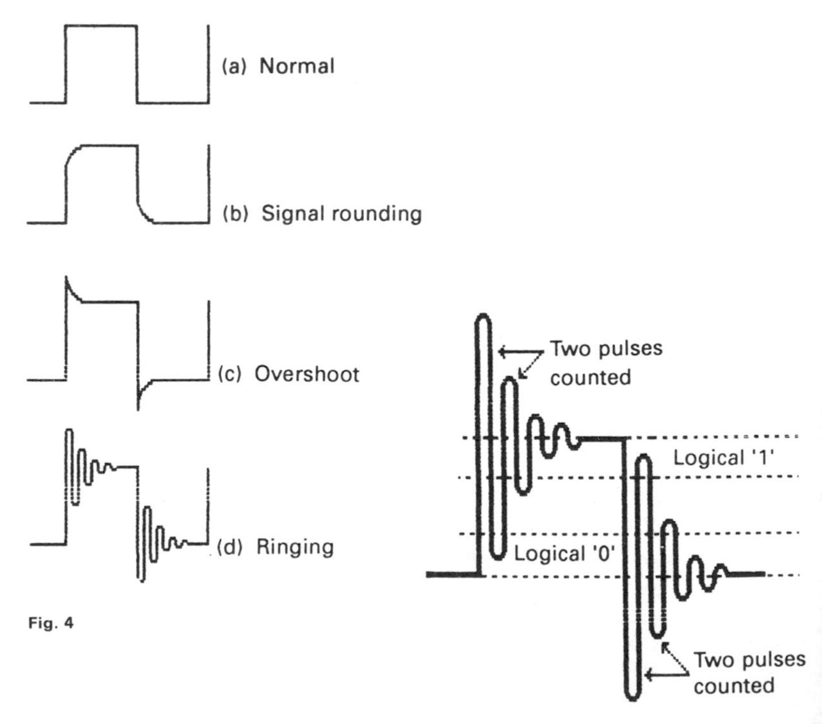

Fig. 4

Fig. 5

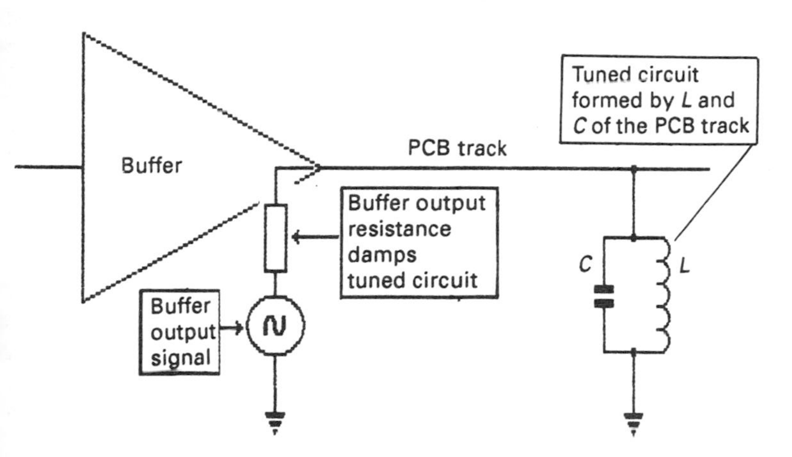

Fig. 6

microcomputer represents a **capacitive load** on each line of the bus. This is particularly so with MOS devices, since these have high input and output capacitances (typically 5 pF input capacitance and 10 pF output capacitance). The effect of excessive capacitance on the buses of a microcomputer is to slow down the rate at which data can change and thus prevent correct timing of operations.

13 Where many memory chips are connected to the buses of a microcomputer, it becomes necessary to connect a **buffer** circuit between the microprocessor and the buses, whose function is to supply large drive currents to the buses, yet impose very little loading on the microprocessor to which it is connected. Such a circuit is known as a **bus driver**. A unidirectional bus driver may be used for the address bus, but a bidirectional bus driver is required for the data bus to permit data transfers between the MPU and its memory devices to take place in either direction. The use of bus drivers is shown in *Fig. 7*.

B WORKED PROBLEMS ON SIGNAL DEGRADATION

Problem 1 Explain why **cross-talk** occurs in a microcomputer via the power supply lines and show how this problem may be eliminated.

When the output of logic ICs change state, for a very short period of time, both output devices conduct simultaneously (see Chapter 3, PROBLEM 14). This momentary **overlap** causes a large pulse of current to flow through the power supply tracks of the PCB. Due to the resistance of these PCB tracks, a volt-drop

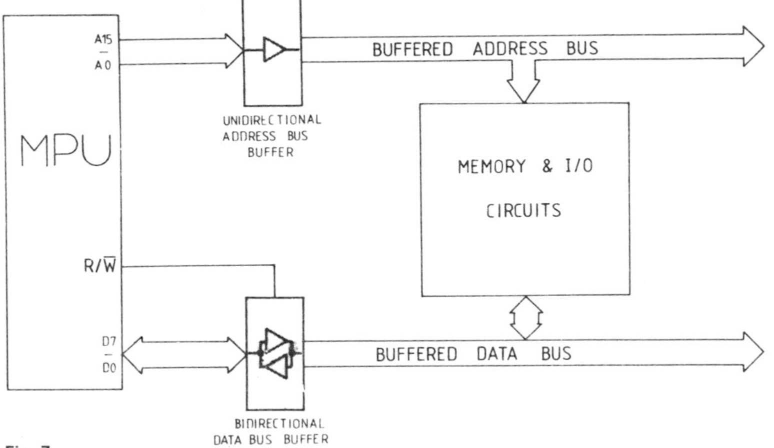

Fig. 7

occurs and all other ICs on the board are pulsed via the power supply. Therefore each IC interacts with all of the other ICs, i.e., **cross-talk** occurs which causes spurious switching of the logic circuits. This effect is shown in *Fig. 8*.

Problems associated with cross-talk of this nature may be eliminated by stabilizing the supply potentials to each device. There are two practical methods of achieving this in the PCB design, and these are:

(a) use of wide PCB tracks for the power supplies, and
(b) liberal use of decoupling capacitors.

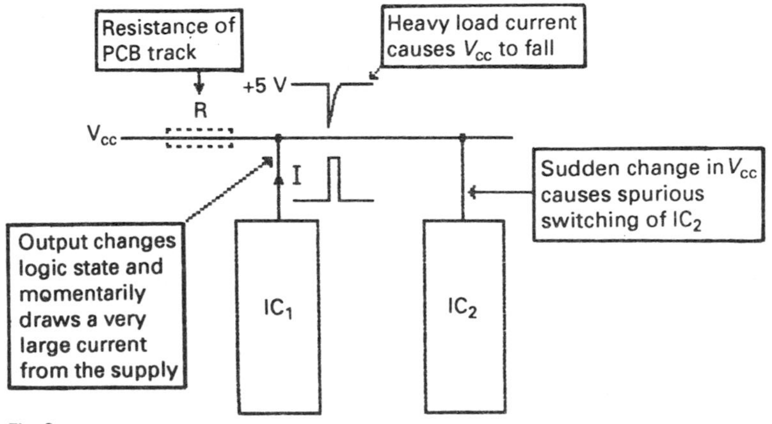

Fig. 8

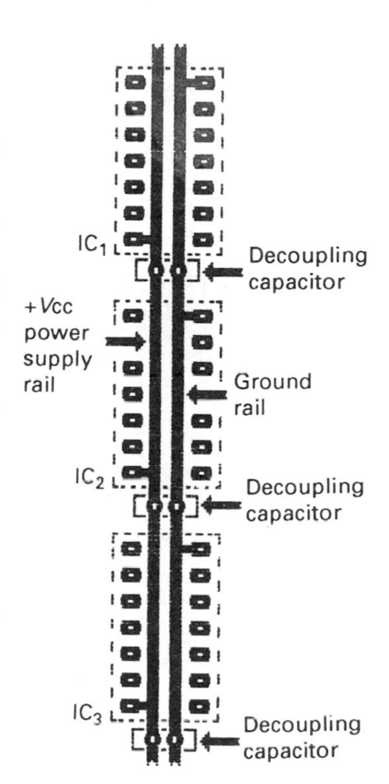

Fig. 9

All PCB tracks have resistance. Power supply tracks must normally pass more current than signal tracks, therefore steps must be taken to minimize their resistance by making them wider. This also helps to minimize cross-talk effects.

A small decoupling capacitor is mounted across the power supply tracks close to each IC. This capacitor stores energy, which it gives up when the IC switches so that the power supply potential is maintained at its correct level (see *Fig. 9*).

It is advisable to use a liberal number of decoupling capacitors on a circuit board, preferably one per IC. Power supply tracks may usually be identified on a PCB by their width, and they often pass between the IC pins as shown in *Fig. 9*. This layout facilitates the connection of decoupling capacitors and enables lead lengths to be kept short to avoid the effects of lead inductance.

Problem 2 Explain the difference between **static** and **dynamic** loading of the buses of a microcomputer.

Static loading is the **resistive** load placed upon the buses of a microcomputer by memory, I/O and decoding devices. This form of loading causes current to flow along the bus conductors and changes the logic levels which are obtained. This form of loading is a particular problem where many TTL devices are connected to the buses of a microcomputer.

Dynamic loading is the **capacitive** load placed upon the buses of a microcomputer by memory, I/O and decoding devices. Whenever the logic levels on the buses are changed, it is necessary to charge or discharge this capacitance. This has the effect of slowing down the voltage transitions, and may result in timing errors if excessive capacitance is present. This form of loading may be a problem where MOS devices are used, due to their relatively high input capacitances.

Problem 3 Describe the characteristics of devices which are suitable for **address bus buffering.**

The main characteristics of a bus buffer or bus driver device which is suitable for use on the address bus of a microcomputer are as follows:

(a) **unidirectional operation**;
(b) **minimal loading of the bus** (typically 200 μA per input);
(c) **high output current capacility** (typically 50 mA current sink);
(d) **capable of driving a large capacitive load** (typically up to 300 pF); and
(e) **high switching speed** (typically 5 to 20 ns).

Examples of typical bus buffer devices are shown in *Fig. 10.*

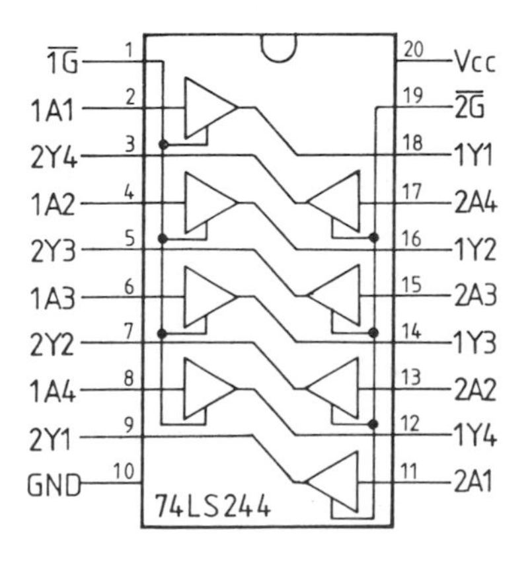

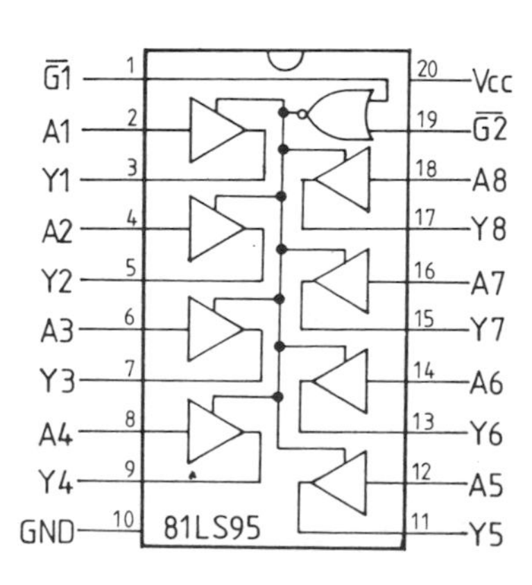

Fig. 10

Problem 4 Describe the characteristics of devices which are suitable for **data bus buffering**.

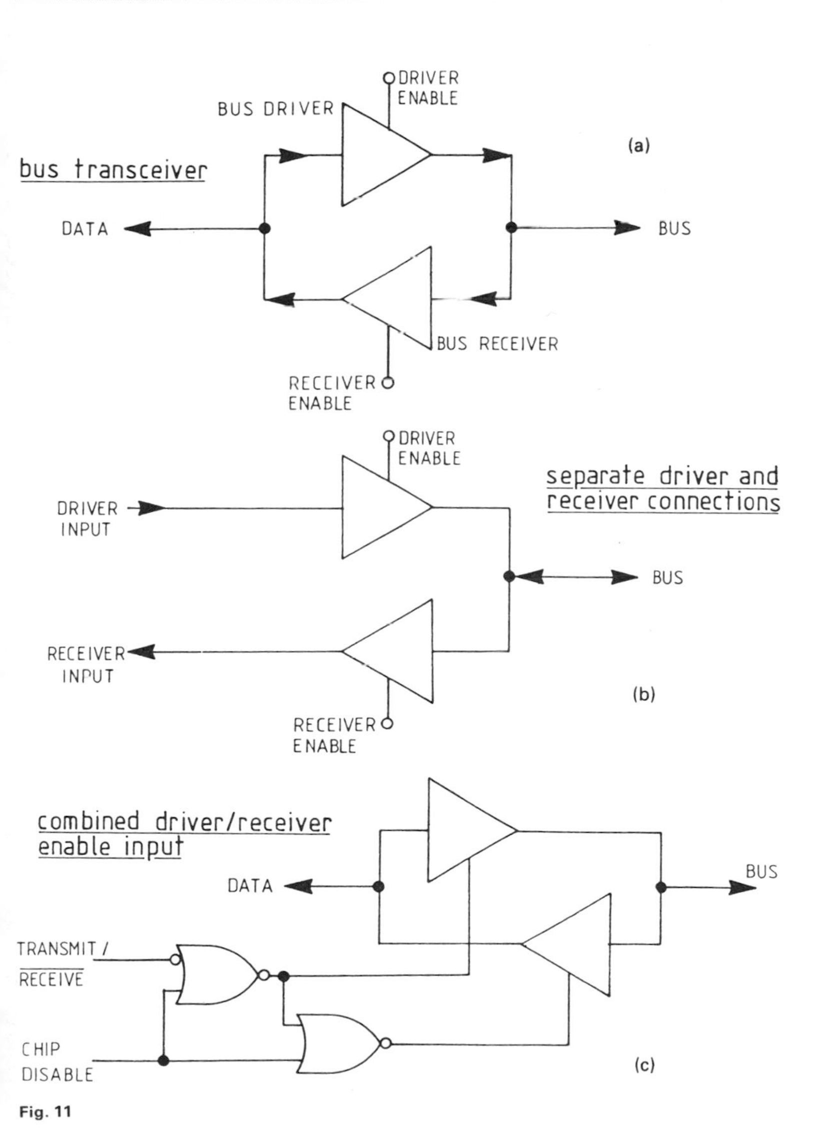

Fig. 11

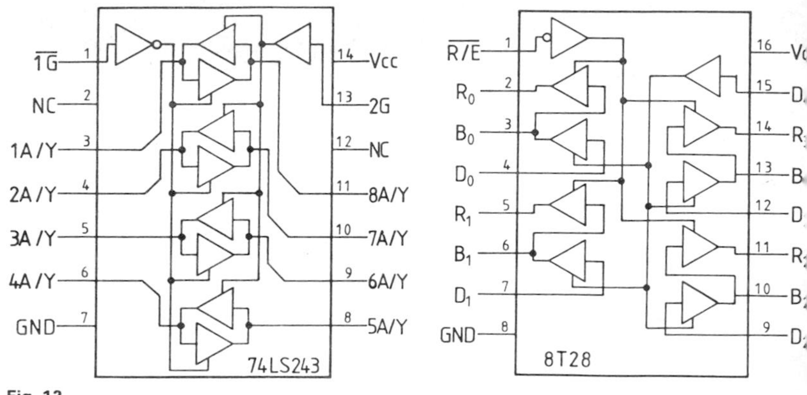

$\overline{1G}$ 1
NC 2
1A/Y 3
2A/Y 4
3A/Y 5
4A/Y 6
GND 7
14 Vcc
13 2G
12 NC
11 8A/Y
10 7A/Y
9 6A/Y
8 5A/Y
74LS243
$\overline{R/E}$ 1
R_0 2
B_0 3
D_0 4
R_1 5
B_1 6
D_1 7
GND 8
16
15
14
13
12
11
10
9
8T28

Fig. 12

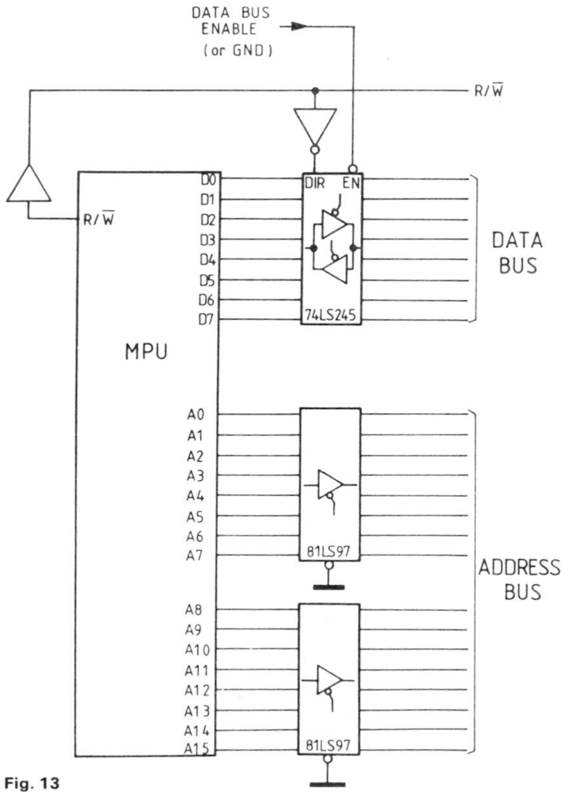

DATA BUS ENABLE (or GND)
R/$\overline{W}$
R/$\overline{W}$
MPU
D0
D1
D2
D3
D4
D5
D6
D7
DIR
EN
74LS245
DATA BUS
A0
A1
A2
A3
A4
A5
A6
A7
81LS97
A8
A9
A10
A11
A12
A13
A14
A15
81LS97
ADDRESS BUS

Fig. 13

The data bus of a microcomputer is **bidirectional**, therefore a buffering device suitable for this bus must also be bidirectional. Two sections are required in a data bus buffer device, and these are:

(a) a **driver** or **transmitter** section;
(b) a **receiver** section.

Together these two sections form a **bus transceiver**, and the organization of typical devices of this type is shown in *Fig. 11(a) to (c)*.

The characteristics of each section of a bus transceiver are similar to those given in *Problem 3* for an address bus buffer, except that the receiver section is not normally required to drive such a heavy load (16 mA typical). Examples of typical bus transceiver are shown in *Fig. 12*.

Problem 5 With the aid of a diagram, show how the data and address buses of a microcomputer may be buffered.

A suitable buffering circuit for the data and address buses of a microcomputer is shown in *Fig. 13*.

C FURTHER PROBLEMS ON SIGNAL DEGRADATION

(a) SHORT ANSWER PROBLEMS

1 Signal degradation results in changes in .... and .... of a digital signal.

2 Signal degradation may cause .... errors in a microcomputer circuit.

3 Signal degradation occurs due to ....

4 'Cross talk' may be caused in a microcomputer by unwanted coupling via the ....

5 'Ringing' on a digital signal is caused by ....

6 'Ringing' may be eliminated by the use of ....

7 A resistive load on the buses of a microcomputer is known as a ....

8 A capacitive load on the buses of a microcomputer is known as a ....

9 In order to reduce the loading on the buses of a microcomputer, a .... circuit must be used.

10 A bus transceiver is a device which may be used to ....

(b) CONVENTIONAL PROBLEMS

1 With the aid of diagrams, explain how the physical layout of a PCB may influence its operation.

2 List the main causes of signal degradation and show how the effects of each of these may be eliminated.

Appendix: Instruction sets for the 6502, Z80 and 6800 microprocessors

The following microprocessor instruction sets are included in this Appendix:

MS 6502 (pages 261 and 262) from MOS Technology Inc.

Z80 (pages 263 to 273) from Mostek UK Ltd.

MC 6800 (pages 274 and 275) from Motorola Semiconductor Products Inc.

The author and publisher would like to thank the manufacturers concerned for their permission to publish this information.

THE MCS 6502 INSTRUCTION SET

INSTRUCTIONS		IMMEDIATE			ABSOLUTE			ZERO PAGE			ACCUM			IMPLIED		
MNEMONIC	OPERATION	OP	N	#	OP	N	#	OP	N	#	OP	N	#	OP	N	#
A D C	A+M+C → A (4) (1)	69	2	2	6D	4	3	65	3	2						
A N D	A ∧ M → A (1)	29	2	2	2D	4	3	25	3	2						
A S L	C ← [7 … 0] ← 0				0E	6	3	06	5	2	0A	2	1			
B C C	BRANCH ON C=0 (2)															
B C S	BRANCH ON C=1 (2)															
B E Q	BRANCH ON Z=1 (2)															
B I T	A ∧ M				2C	4	3	24	3	2						
B M I	BRANCH ON N=1 (2)															
B N E	BRANCH ON Z=0 (2)															
B P L	BRANCH ON N=0 (2)															
B R K	(See Fig. 1)													00	7	1
B V C	BRANCH ON V=0 (2)															
B V S	BRANCH ON V=1 (2)															
C L C	0 → C													18	2	1
C L D	0 → D													D8	2	1
C L I	0 → I													58	2	1
C L V	0 → V													B8	2	1
C M P	A − M (1)	C9	2	2	CD	4	3	C5	3	2						
C P X	X − M	E0	2	2	EC	4	3	E4	3	2						
C P Y	Y − M	C0	2	2	CC	4	3	C4	3	2						
D E C	M − 1 → M				CE	6	3	C6	5	2						
D E X	X − 1 → X													CA		
D E Y	Y − 1 → Y													88		
E O R	A ⊻ M → A (1)	49	2	2	4D	4	3	45	3	2						
I N C	M + 1 → M				EE	6	3	E6	5	2						
I N X	X + 1 → X													E8	2	1
I N Y	Y + 1 → Y													C8	2	1
J M P	JUMP TO NEW LOC				4C	3	3									
J S R	(See Fig. 2) JUMP SUB				20	6	3									
L D A	M → A (1)	A9	2	2	AD	4	3	A5	3	2						

INSTRUCTIONS	(IND, X)			(IND), Y			Z, PAGE, X			ABS, X			ABS, Y			RELATIVE			INDIRECT			Z, PAGE, Y		
MNEMONIC	OP	N	#	OP	N	#	OP	N	#	OP	N	#	OP	N	#	OP	N	#	OP	N	#	OP	N	#
A D C	61	6	2	71	5	2	75	4	2	7D	4	3	79	4	3									
A N D	21	6	2	31	5	2	35	4	2	3D	4	3	39	4	3									
A S L							16	6	2	1E	7	3												
B C C																90	2	2						
B C S																B0	2	2						
B E Q																F0	2	2						
B I T																								
B M I																30	2	2						
B N E																D0	2	2						
B P L																10	2	2						
B R K																								
B V C																50	2	2						
B V S																70	2	2						
C L C																								
C L D																								
C L I																								
C L V																								
C M P	C1	6	2	D1	5	2	D5	4	2	DD	4	3	D9	4	3									
C P X																								
C P Y																								
D E C							D6	6	2	DE	7	3												
D E X																								
D E Y																								
E O R	41	6	2	51	5	2	55	4	2	5D	4	3	59	4	3									
I N C							F6	6	2	FE	7	3												
I N X																								
I N Y																								
J M P																			6C	5	3			
J S R																								
L D A	A1	6	2	B1	5	2	B5	4	2	BD	4	3	B9	4	3									

INSTRUCTIONS	CONDITION CODES					
MNEMONIC	N	Z	C	I	D	V
A D C	✓	✓	✓	–	–	✓
A N D	✓	✓	–	–	–	–
A S L	✓	✓	✓	–	–	–
B C C	–	–	–	–	–	–
B C S	–	–	–	–	–	–
B E Q	–	–	–	–	–	–
B I T	M_7	✓	–	–	–	M_6
B M I	–	–	–	–	–	–
B N E	–	–	–	–	–	–
B P L	–	–	–	–	–	–
B R K	–	–	–	1	–	–
B V C	–	–	–	–	–	–
B V S	–	–	–	–	–	–
C L C	–	–	0	–	–	–
C L D	–	–	–	–	0	–
C L I	–	–	–	0	–	–
C L V	–	–	–	–	–	0
C M P	✓	✓	✓	–	–	–
C P X	✓	✓	✓	–	–	–
C P Y	✓	✓	✓	–	–	–
D E C	✓	✓	–	–	–	–
D E X	✓	✓	–	–	–	–
D E Y	✓	✓	–	–	–	–
E O R	✓	✓	–	–	–	–
I N C	✓	✓	–	–	–	–
I N X	✓	✓	–	–	–	–
I N Y	✓	✓	–	–	–	–
J M P	–	–	–	–	–	–
J S R	–	–	–	–	–	–
L D A	✓	✓	–	–	–	–

THE MCS 6502 INSTRUCTION SET

MNEMONIC	OPERATION	IMMEDIATE			ABSOLUTE			ZERO PAGE			ACCUM			IMPLIED		
		OP	N	#	OP	N	#	OP	N	#	OP	N	#	OP	N	#
L D X	M → X (1)	A2	2	2	AE	4	3	A6	3	2						
L D Y	M → Y (1)	A0	2	2	AC	4	3	A4	3	2						
L S R	0 → [7 … 0] → C				4E	6	3	46	5	2	4A	2	1			
N O P	NO OPERATION													EA	2	1
O R A	A ∨ M → A	09	2	2	0D	4	3	05	3	2						
P H A	A → Ms S − 1 → S													48	3	1
P H P	P → Ms S − 1 → S													08	3	1
P L A	S + 1 → S Ms → A													68	4	1
P L P	S + 1 → S Ms → P													28	4	1
R O L	← [7 … 0] ← [C] ←				2E	6	3	26	5	2	2A	2	1			
R O R	→ [C] → [7 … 0] →				6E	6	3	66	5	2	6A	2	1			
R T I	(See Fig 1) RTRN INT													40	6	1
R T S	(See Fig 2) RTRN SUB													60	6	1
S B C	A − M − C̄ → A (1)	E9	2	2	ED	4	3	E5	3	2						
S E C	1 → C													38	2	1
S E D	1 → D													F8	2	1
S E I	1 → I													78	2	1
S T A	A → M				8D	4	3	85	3	2						
S T X	X → M				8E	4	3	86	3	2						
S T Y	Y → M				8C	4	3	84	3	2						
T A X	A → X													AA	2	1
T A Y	A → Y													A8	2	1
T S X	S → X													BA	2	1
T X A	X → A													8A	2	1
T X S	X → S													9A	2	1
T Y A	Y → A													98	2	1

MNEMONIC	(IND, X)			(IND), Y			Z, PAGE, X			ABS, X			ABS, Y			RELATIVE			INDIRECT			Z, PAGE, Y			CONDITION CODES					
	OP	N	#	OP	N	#	OP	N	#	OP	N	#	OP	N	#	OP	N	#	OP	N	#	OP	N	#	N	Z	C	I	D	V
L D X													BE	4	3							B6	4	2	✓	✓	-	-	-	-
L D Y							B4	4	2	BC	4	3													✓	✓	-	-	-	-
L S R							56	6	2	5E	7	3													0	✓	✓	-	-	-
N O P																									-	-	-	-	-	-
O R A	01	6	2	11	5	2	15	4	2	1D	4	3	19	4	3										✓	✓	-	-	-	-
P H A																									-	-	-	-	-	-
P H P																									-	-	-	-	-	-
P L A																									✓	✓	-	-	-	-
P L P																									(RESTORED)					
R O L							36	6	2	3E	7	3													✓	✓	✓	-	-	-
R O R							76	6	2	7E	7	3													✓	✓	✓	-	-	-
R T I																									(RESTORED)					
R T S																									-	-	-	-	-	-
S B C	E1	6	2	F1	5	2	F5	4	2	FD	4	3	F9	4	3										✓	✓	(3)	-	-	✓
S E C																									-	-	1	-	-	-
S E D																									-	-	-	-	1	-
S E I																									-	-	-	1	-	-
S T A	81	6	2	91	6	2	95	4	2	9D	5	3	99	5	3										-	-	-	-	-	-
S T X																						96	4	2	-	-	-	-	-	-
S T Y							94	4	2																-	-	-	-	-	-
T A X																									✓	✓	-	-	-	-
T A Y																									✓	✓	-	-	-	-
T S X																									✓	✓	-	-	-	-
T X A																									✓	✓	-	-	-	-
T X S																									-	-	-	-	-	-
T Y A																									✓	✓	-	-	-	-

(1) ADD 1 TO "N" IF PAGE BOUNDRY IS CROSSED

(2) ADD 1 TO "N" IF BRANCH OCCURS TO SAME PAGE
ADD 2 TO "N" IF BRANCH OCCURS TO DIFFERENT PAGE.

(3) CARRY NOT = BORROW.

(4) IF IN DECIMAL MODE Z FLAG IS INVALID
ACCUMULATOR MUST BE CHECKED FOR ZERO RESULT.

X INDEX X
Y INDEX Y
A ACCUMULATOR
M MEMORY PER EFFECTIVE ADDRESS
M_S MEMORY PER STACK POINTER

\+ ADD
— SUBTRACT
∧ AND
∨ OR

⊻ EXCLUSIVE OR
✓ MODIFIED
\- NOT MODIFIED
M_7 MEMORY BIT 7
M_6 MEMORY BIT 6

N NO CYCLES
\# NO BYTES

Z80 INSTRUCTION SET

8-BIT LOAD GROUP

Mnemonic	Symbolic Operation	S	Z		H		P/V	N	C	Op-Code 76 543 210	Hex	No. of Bytes	No. of M Cycles	No. of T States	Comments	
LD r, s	r ← s	•	•	X	•	X	•	•	•	01 r s		1	1	4	r, s	Reg.
LD r, n	r ← n	•	•	X	•	X	•	•	•	00 r 110		2	2	7	000	B
										← n →					001	C
LD r, (HL)	r ← (HL)	•	•	X	•	X	•	•	•	01 r 110		1	2	7	010	D
LD r, (IX+d)	r ← (IX+d)	•	•	X	•	X	•	•	•	11 011 101	DD	3	5	19	011	E
										01 r 110					100	H
										← d →					101	L
LD r, (IY+d)	r ← (IY+d)	•	•	X	•	X	•	•	•	11 111 101	FD	3	5	19	111	A
										01 r 110						
										← d →						
LD (HL), r	(HL) ← r	•	•	X	•	X	•	•	•	01 110 r		1	2	7		
LD (IX+d), r	(IX+d) ← r	•	•	X	•	X	•	•	•	11 011 101	DD	3	5	19		
										01 110 r						
										← d →						
LD (IY+d), r	(IY+d) ← r	•	•	X	•	X	•	•	•	11 111 101	FD	3	5	19		
										01 110 r						
										← d →						
LD (HL), n	(HL) ← n	•	•	X	•	X	•	•	•	00 110 110	36	2	3	10		
										← n →						
LD (IX+d), n	(IX+d) ← n	•	•	X	•	X	•	•	•	11 011 101	DD	4	5	19		
										00 110 110	36					
										← d →						
										← n →						
LD (IY+d), n	(IY+d) ← n	•	•	X	•	X	•	•	•	11 111 101	FD	4	5	19		
										00 110 110	36					
										← d →						
										← n →						
LD A, (BC)	A ← (BC)	•	•	X	•	X	•	•	•	00 001 010	0A	1	2	7		
LD A, (DE)	A ← (DE)	•	•	X	•	X	•	•	•	00 011 010	1A	1	2	7		
LD A, (nn)	A ← (nn)	•	•	X	•	X	•	•	•	00 111 010	3A	3	4	13		
										← n →						
										← n →						
LD (BC), A	(BC) ← A	•	•	X	•	X	•	•	•	00 000 010	02	1	2	7		
LD (DE), A	(DE) ← A	•	•	X	•	X	•	•	•	00 010 010	12	1	2	7		
LD (nn), A	(nn) ← A	•	•	X	•	X	•	•	•	00 110 010	32	3	4	13		
										← n →						
										← n →						
LD A, I	A ← I	↕	↕	X	0	X	IFF	0	•	11 101 101	ED	2	2	9		
										01 010 111	57					
LD A, R	A ← R	↕	↕	X	0	X	IFF	0	•	11 101 101	ED	2	2	9		
										01 011 111	5F					
LD I, A	I ← A	•	•	X	•	X	•	•	•	11 101 101	ED	2	2	9		
										01 000 111	47					
LD R, A	R ← A	•	•	X	•	X	•	•	•	11 101 101	ED	2	2	9		
										01 001 111	4F					

Notes: r, s means any of the registers A, B, C, D, E, H, L
IFF the content of the interrupt enable flip-flop (IFF) is copied into the P/V flag

Flag Notation: • = flag not affected, 0 = flag reset, 1 = flag set, X = flag is unknown,
↕ = flag is affected according to the result of the operation.

Z80 INSTRUCTION SET

16-BIT LOAD GROUP

Mnemonic	Symbolic Operation	Flags								Op-Code		No. of Bytes	No. of M Cycles	No. of T States	Comments
		S	Z		H		P/V	N	C	76 543 210	Hex				
LD dd, nn	dd ← nn	•	•	X	•	X	•	•	•	00 dd0 001		3	3	10	dd Pair
										– n –					00 BC
										– n –					01 DE
LD IX, nn	IX ← nn	•	•	X	•	X	•	•	•	11 011 101	DD	4	4	14	10 HL
										00 100 001	21				11 SP
										– n –					
										– n –					
LD IY, nn	IY ← nn	•	•	X	•	X	•	•	•	11 111 101	FD	4	4	14	
										00 100 001	21				
										– n –					
										– n –					
LD HL, (nn)	H ← (nn+1)	•	•	X	•	X	•	•	•	00 101 010	2A	3	5	16	
	L ← (nn)									– n –					
										– n –					
LD dd, (nn)	dd_H ← (nn+1)	•	•	X	•	X	•	•	•	11 101 101	ED	4	6	20	
	dd_L ← (nn)									01 dd1 011					
										– n –					
										– n –					
LD IX, (nn)	IX_H ← (nn+1)	•	•	X	•	X	•	•	•	11 011 101	DD	4	6	20	
	IX_L ← (nn)									00 101 010	2A				
										– n –					
										– n –					
LD IY, (nn)	IY_H ← (nn+1)	•	•	X	•	X	•	•	•	11 111 101	FD	4	6	20	
	IY_L ← (nn)									00 101 010	2A				
										– n –					
										– n –					
LD (nn), HL	(nn+1) ← H	•	•	X	•	X	•	•	•	00 100 010	22	3	5	16	
	(nn) ← L									– n –					
										– n –					
LD (nn), dd	(nn+1) ← dd_H	•	•	X	•	X	•	•	•	11 101 101	ED	4	6	20	
	(nn) ← dd_L									01 dd0 011					
										– n –					
										– n –					
LD (nn), IX	(nn+1) ← IX_H	•	•	X	•	X	•	•	•	11 011 101	DD	4	6	20	
	(nn) ← IX_L									00 100 010	22				
										– n –					
										– n –					
LD (nn), IY	(nn+1) ← IY_H	•	•	X	•	X	•	•	•	11 111 101	FD	4	6	20	
	(nn) ← IY_L									00 100 010	22				
										– n –					
										– n –					
LD SP, HL	SP ← HL	•	•	X	•	X	•	•	•	11 111 001	F9	1	1	6	
LD SP, IX	SP ← IX	•	•	X	•	X	•	•	•	11 011 101	DD	2	2	10	
										11 111 001	F9				
LD SP, IY	SP ← IY	•	•	X	•	X	•	•	•	11 111 101	FD	2	2	10	
										11 111 001	F9				qq Pair
PUSH qq	(SP-2) ← qq_L	•	•	X	•	X	•	•	•	11 qq0 101		1	3	11	00 BC
	(SP-1) ← qq_H														01 DE
PUSH IX	(SP-2) ← IX_L	•	•	X	•	X	•	•	•	11 011 101	DD	2	4	15	10 HL
	(SP-1) ← IX_H									11 100 101	E5				11 AF
PUSH IY	(SP-2) ← IY_L	•	•	X	•	X	•	•	•	11 111 101	FD	2	4	15	
	(SP-1) ← IY_H									11 100 101	E5				
POP qq	qq_H ← (SP+1)	•	•	X	•	X	•	•	•	11 qq0 001		1	3	10	
	qq_L ← (SP)														
POP IX	IX_H ← (SP+1)	•	•	X	•	X	•	•	•	11 011 101	DD	2	4	14	
	IX_L ← (SP)									11 100 001	E1				
POP IY	IY_H ← (SP+1)	•	•	X	•	X	•	•	•	11 111 101	FD	2	4	14	
	IY_L ← (SP)									11 100 001	E1				

Notes: dd is any of the register pairs BC, DE, HL, SP
qq is any of the register pairs AF, BC, DE, HL
$(PAIR)_H$, $(PAIR)_L$ refer to high order and low order eight bits of the register pair respectively.
e.g. BC_L = C, AF_H = A

Flag Notation: • = flag not affected, 0 = flag reset, 1 = flag set, X = flag is unknown,
↕ flag is affected according to the result of the operation.

Z80 INSTRUCTION SET

EXCHANGE GROUP AND BLOCK TRANSFER AND SEARCH GROUP

Mnemonic	Symbolic Operation	S	Z		H		P/V	N	C	Op-Code 76 543 210	Hex	No. of Bytes	No. of M Cycles	No. of T States	Comments
EX DE, HL	DE↔HL	•	•	X	•	X	•	•	•	11 101 011	EB	1	1	4	
EX AF, AF'	AF↔AF'	•	•	X	•	X	•	•	•	00 001 000	08	1	1	4	
EXX	BC↔BC' DE↔DE' HL↔HL'	•	•	X	•	X	•	•	•	11 011 001	D9	1	1	4	Register bank and auxiliary register bank exchange
EX (SP), HL	H↔(SP+1) L↔(SP)	•	•	X	•	X	•	•	•	11 100 011	E3	1	5	19	
EX (SP), IX	IX_H↔(SP+1) IX_L↔(SP)	•	•	X	•	X	•	•	•	11 011 101 11 100 011	DD E3	2	6	23	
EX (SP), IY	IY_H↔(SP+1) IY_L↔(SP)	•	•	X	•	X	•	•	•	11 111 101 11 100 011	FD E3	2	6	23	
LDI	(DE)←(HL) DE ← DE+1 HL ← HL+1 BC ← BC-1	•	•	X	0	X	(1) ↕	0	•	11 101 101 10 100 000	ED A0	2	4	16	Load (HL) into (DE), increment the pointers and decrement the byte counter (BC)
LDIR	(DE)←(HL) DE ← DE+1 HL ← HL+1 BC ← BC-1 Repeat until BC = 0	•	•	X	0	X	0	0	•	11 101 101 10 110 000	ED B0	2 2	5 4	21 16	If BC ≠ 0 If BC = 0
LDD	(DE)←(HL) DE ← DE-1 HL ← HL-1 BC ← BC-1	•	•	X	0	X	(1) ↕	0	•	11 101 101 10 101 000	ED A8	2	4	16	
LDDR	(DE)←(HL) DE ← DE-1 HL ← HL-1 BC ← BC-1 Repeat until BC = 0	•	•	X	0	X	0	0	•	11 101 101 10 111 000	ED B8	2 2	5 4	21 16	If BC ≠ 0 If BC = 0
CPI	A – (HL) HL ← HL+1 BC ← BC-1	↕	(2) ↕	X	↕	X	(1) ↕	1	•	11 101 101 10 100 001	ED A1	2	4	16	
CPIR	A – (HL) HL ← HL+1 BC ← BC-1 Repeat until A = (HL) or BC = 0	↕	(2) ↕	X	↕	X	(1) ↕	1	•	11 101 101 10 110 001	ED B1	2 2	5 4	21 16	If BC ≠ 0 and A ≠ (HL) If BC = 0 or A = (HL)
CPD	A – (HL) HL ← HL-1 BC ← BC-1	↕	(2) ↕	X	↕	X	(1) ↕	1	•	11 101 101 10 101 001	ED A9	2	4	16	
CPDR	A – (HL) HL ← HL-1 BC ← BC-1 Repeat until A = (HL) or BC = 0	↕	(2) ↕	X	↕	X	(1) ↕	1	•	11 101 101 10 111 001	ED B9	2 2	5 4	21 16	If BC ≠ 0 and A ≠ (HL) If BC = 0 or A = (HL)

Notes: (1) P/V flag is 0 if the result of BC-1 = 0, otherwise P/V = 1
(2) Z flag is 1 if A = (HL), otherwise Z = 0.

Flag Notation: • = flag not affected, 0 = flag reset, 1 = flag set, X = flag is unknown,
↕ = flag is affected according to the result of the operation.

Z80 INSTRUCTION SET

8-BIT ARITHMETIC AND LOGICAL GROUP

Mnemonic	Symbolic Operation	Flags S	Z		H		P/V	N	C	Op-Code 76 543 210	Hex	No. of Bytes	No. of M Cycles	No. of T States	Comments
ADD A, r	A ← A + r	↕	↕	X	↕	X	V	0	↕	10 [000] r		1	1	4	r Reg.
ADD A, n	A ← A + n	↕	↕	X	↕	X	V	0	↕	11 [000] 110		2	2	7	000 B
										– n –					001 C
															010 D
ADD A, (HL)	A ← A+(HL)	↕	↕	X	↕	X	V	0	↕	10 [000] 110		1	2	7	011 E
ADD A, (IX+d)	A ← A+(IX+d)	↕	↕	X	↕	X	V	0	↕	11 011 101	DD	3	5	19	100 H
										10 [000] 110					101 L
										– d –					111 A
ADD A, (IY+d)	A ← A+(IY+d)	↕	↕	X	↕	X	V	0	↕	11 111 101	FD	3	5	19	
										10 [000] 110					
										– d –					
ADC A, s	A ← A+s+CY	↕	↕	X	↕	X	V	0	↕	[001]					s is any of r, n,
SUB s	A ← A − s	↕	↕	X	↕	X	V	1	↕	[010]					(HL), (IX+d),
SBC A, s	A ← A − s − CY	↕	↕	X	↕	X	V	1	↕	[011]					(IY+d) as shown for
AND s	A ← A ∧ s	↕	↕	X	1	X	P	0	0	[100]					ADD instruction.
OR s	A ← A ∨ s	↕	↕	X	0	X	P	0	0	[110]					The indicated bits
XOR s	A ← A ⊕ s	↕	↕	X	0	X	P	0	0	[101]					replace the [000] in
CP s	A − s	↕	↕	X	↕	X	V	1	↕	[111]					the ADD set above.
INC r	r ← r + 1	↕	↕	X	↕	X	V	0	•	00 r [100]		1	1	4	
INC (HL)	(HL) ← (HL)+1	↕	↕	X	↕	X	V	0	•	00 110 [100]		1	3	11	
INC (IX+d)	(IX+d) ← (IX+d)+1	↕	↕	X	↕	X	V	0	•	11 011 101	DD	3	6	23	
										00 110 [100]					
										– d –					
INC (IY+d)	(IY+d) ← (IY+d)+1	↕	↕	X	↕	X	V	0	•	11 111 101	FD	3	6	23	
										00 110 [100]					
										– d –					
DEC s	s ← s − 1	↕	↕	X	↕	X	V	1	•	[101]					s is any of r, (HL), (IX+d), (IY+d) as shown for INC. DEC same format and states as INC. Replace [100] with [101] in OP Code.

Notes: The V symbol in the P/V flag column indicates that the P/V flag contains the overflow of the result of the operation. Similarly the P symbol indicates parity. V = 1 means overflow, V = 0 means not overflow, P = 1 means parity of the result is even, P = 0 means parity of the result is odd.

Flag Notation: • = flag not affected, 0 = flag reset, 1 = flag set, X = flag is unknown.
↕ = flag is affected according to the result of the operation.

Z80 INSTRUCTION SET

GENERAL PURPOSE ARITHMETIC AND CPU CONTROL GROUPS

Mnemonic	Symbolic Operation	S	Z		H		P/V	N	C	Op-Code 76 543 210	Hex	No. of Bytes	No. of M Cycles	No. of T States	Comments
DAA	Converts acc. content into packed BCD following add or subtract with packed BCD operands	↕	↕	X	↕	X	P	•	↕	00 100 111	27	1	1	4	Decimal adjust accumulator
CPL	$A \leftarrow \overline{A}$	•	•	X	1	X	•	1	•	00 101 111	2F	1	1	4	Complement accumulator (One's complement)
NEG	$A \leftarrow \overline{A}+1$	↕	↕	X	↕	X	V	1	↕	11 101 101 01 000 100	ED 44	2	2	8	Negate acc. (two's complement)
CCF	$CY \leftarrow \overline{CY}$	•	•	X	X	X	•	0	↕	00 111 111	3F	1	1	4	Complement carry flag
SCF	$CY \leftarrow 1$	•	•	X	0	X	•	0	1	00 110 111	37	1	1	4	Set carry flag
NOP	No operation	•	•	X	•	X	•	•	•	00 000 000	00	1	1	4	
HALT	CPU halted	•	•	X	•	X	•	•	•	01 110 110	76	1	1	4	
DI*	$IFF \leftarrow 0$	•	•	X	•	X	•	•	•	11 110 011	F3	1	1	4	
EI*	$IFF \leftarrow 1$	•	•	X	•	X	•	•	•	11 111 011	FB	1	1	4	
IM 0	Set interrupt mode 0	•	•	X	•	X	•	•	•	11 101 101 01 000 110	ED 46	2	2	8	
IM 1	Set interrupt mode 1	•	•	X	•	X	•	•	•	11 101 101 01 010 110	ED 56	2	2	8	
IM 2	Set interrupt mode 2	•	•	X	•	X	•	•	•	11 101 101 01 011 110	ED 5E	2	2	8	

Notes: IFF indicates the interrupt enable flip-flop
CY indicates the carry flip-flop

Flag Notation: • = flag not affected, 0 = flag reset, 1 = flag set, X = flag is unknown,
↕ = flag is affected according to the result of the operation.

*Interrupts are not sampled at the end of EI or DI

Z80 INSTRUCTION SET

16-BIT ARITHMETIC GROUP

Mnemonic	Symbolic Operation	Flags								Op-Code		No. of Bytes	No. of M Cycles	No. of T States	Comments	
		S	Z		H		P/V	N	C	76 543 210	Hex					
ADD HL, ss	HL ← HL+ss	•	•	X	X	X	•	0	↕	00 ss1 001		1	3	11	ss	Reg.
															00	BC
ADC HL, ss	HL ← HL+ss+CY	↕	↕	X	X	X	V	0	↕	11 101 101	ED	2	4	15	01	DE
										01 ss1 010					10	HL
															11	SP
SBC HL, ss	HL ← HL−ss−CY	↕	↕	X	X	X	V	1	↕	11 101 101	ED	2	4	15		
										01 ss0 010						
ADD IX, pp	IX ← IX + pp	•	•	X	X	X	•	0	↕	11 011 101	DD	2	4	15	pp	Reg.
										00 pp1 001					00	BC
															01	DE
															10	IX
															11	SP
ADD IY, rr	IY ← IY + rr	•	•	X	X	X	•	0	↕	11 111 101	FD	2	4	15	rr	Reg.
										00 rr1 001					00	BC
															01	DE
															10	IY
															11	SP
INC ss	ss ← ss + 1	•	•	X	•	X	•	•	•	00 ss0 011		1	1	6		
INC IX	IX ← IX + 1	•	•	X	•	X	•	•	•	11 011 101	DD	2	2	10		
										00 100 011	23					
INC IY	IY ← IY + 1	•	•	X	•	X	•	•	•	11 111 101	FD	2	2	10		
										00 100 011	23					
DEC ss	ss ← ss − 1	•	•	X	•	X	•	•	•	00 ss1 011		1	1	6		
DEC IX	IX ← IX − 1	•	•	X	•	X	•	•	•	11 011 101	DD	2	2	10		
										00 101 011	2B					
DEC IY	IY ← IY − 1	•	•	X	•	X	•	•	•	11 111 101	FD	2	2	10		
										00 101 011	2B					

Notes: ss is any of the register pairs BC, DE, HL, SP
pp is any of the register pairs BC, DE, IX, SP
rr is any of the register pairs BC, DE, IY, SP.

Flag Notation: • = flag not affected, 0 = flag reset, 1 = flag set, X = flag is unknown.
↕ = flag is affected according to the result of the operation.

Z80 INSTRUCTION SET

ROTATE AND SHIFT GROUP

Mnemonic	Symbolic Operation	S	Z		H		P/V	N	C	Op-Code 76 543 210	Hex	No. of Bytes	No. of M Cycles	No. of T States	Comments
RLCA	CY ← 7←0 ← (circular) A	•	•	X	0	X	•	0	↕	00 000 111	07	1	1	4	Rotate left circular accumulator
RLA	CY ← 7←0 ← (through CY) A	•	•	X	0	X	•	0	↕	00 010 111	17	1	1	4	Rotate left accumulator
RRCA	7→0 → CY (circular) A	•	•	X	0	X	•	0	↕	00 001 111	0F	1	1	4	Rotate right circular accumulator
RRA	7→0 → CY (through CY) A	•	•	X	0	X	•	0	↕	00 011 111	1F	1	1	4	Rotate right accumulator
RLC r		↕	↕	X	0	X	P	0	↕	11 001 011 00 000 r	CB	2	2	8	Rotate left circular register r
RLC (HL)		↕	↕	X	0	X	P	0	↕	11 001 011 00 000 110	CB	2	4	15	r Reg 000 B
RLC (IX+d)	CY ← 7←0 ← (circular) r,(HL),(IX+d),(IY+d)	↕	↕	X	0	X	P	0	↕	11 011 101 11 001 011 – d – 00 000 110	DD CB	4	6	23	001 C 010 D 011 E 100 H 101 L 111 A
RLC (IY+d)		↕	↕	X	0	X	P	0	↕	11 111 101 11 001 011 – d – 00 000 110	FD CB	4	6	23	
RL s	CY ← 7←0 ← (through CY) s ≡ r,(HL),(IX+d),(IY+d)	↕	↕	X	0	X	P	0	↕	010					Instruction format and states are as shown for RLC's. To form new Op-Code replace 000 of RLC's with shown code
RRC s	7→0 → CY (circular) s ≡ r,(HL),(IX+d),(IY+d)	↕	↕	X	0	X	P	0	↕	001					
RR s	7→0 → CY (through CY) s ≡ r,(HL),(IX+d),(IY+d)	↕	↕	X	0	X	P	0	↕	011					
SLA s	CY ← 7←0 ← 0 s ≡ r,(HL),(IX+d),(IY+d)	↕	↕	X	0	X	P	0	↕	100					
SRA s	7→0 → CY (bit 7 retained) s ≡ r,(HL),(IX+d),(IY+d)	↕	↕	X	0	X	P	0	↕	101					
SRL s	0 → 7→0 → CY s ≡ r,(HL),(IX+d),(IY+d)	↕	↕	X	0	X	P	0	↕	111					
RLD	A 7-4 3-0 7-4 3-0 (HL)	↕	↕	X	0	X	P	0	•	11 101 101 01 101 111	ED 6F	2	5	18	Rotate digit left and right between the accumulator and location (HL)
RRD	A 7-4 3-0 7-4 3-0 (HL)	↕	↕	X	0	X	P	0	•	11 101 101 01 100 111	ED 67	2	5	18	The content of the upper half of the accumulator is unaffected

Flag Notation: • = flag not affected, 0 = flag reset, 1 = flag set, X = flag is unknown,
↕ = flag is affected according to the result of the operation.

Z80 INSTRUCTION SET

BIT SET, RESET AND TEST GROUP

Mnemonic	Symbolic Operation	Flags S	Z		H		P/V	N	C	Op-Code 76 543 210	Hex	No. of Bytes	No. of M Cycles	No. of T States	Comments	
BIT b, r	$Z \leftarrow \overline{r}_b$	X	↕	X	1	X	X	0	•	11 001 011	CB	2	2	8	r	Reg.
										01 b r					000	B
BIT b, (HL)	$Z \leftarrow \overline{(HL)}_b$	X	↕	X	1	X	X	0	•	11 001 011	CB	2	3	12	001	C
										01 b 110					010	D
BIT b, (IX+d)$_b$	$Z \leftarrow \overline{(IX+d)}_b$	X	↕	X	1	X	X	0	•	11 011 101	DD	4	5	20	011	E
										11 001 011	CB				100	H
										- d -					101	L
										01 b 110					111	A
															b	Bit Tested
BIT b, (IY+d)$_b$	$Z \leftarrow \overline{(IY+d)}_b$	X	↕	X	1	X	X	0	•	11 111 101	FD	4	5	20	000	0
										11 001 011	CB				001	1
										- d -					010	2
										01 b 110					011	3
															100	4
															101	5
															110	6
															111	7
SET b, r	$r_b \leftarrow 1$	•	•	X	•	X	•	•	•	11 001 011	CB	2	2	8		
										[11] b r						
SET b, (HL)	$(HL)_b \leftarrow 1$	•	•	X	•	X	•	•	•	11 001 011	CB	2	4	15		
										[11] b 110						
SET b, (IX+d)	$(IX+d)_b \leftarrow 1$	•	•	X	•	X	•	•	•	11 011 101	DD	4	6	23		
										11 001 011	CB					
										- d -						
										[11] b 110						
SET b, (IY+d)	$(IY+d)_b \leftarrow 1$	•	•	X	•	X	•	•	•	11 111 101	FD	4	6	23		
										11 001 011	CB					
										- d -						
										[11] b 110						
RES b, s	$s_b \leftarrow 0$ s ← r, (HL), (IX+d), (IY+d)	•	•	X	•	X	•	•	•	[10]					To form new Op Code replace [11] of SET b, s with [10]. Flags and time states for SET instruction	

Notes: The notation s_b indicates bit b (0 to 7) or location s.

Flag Notation: • = flag not affected, 0 = flag reset, 1 = flag set, X = flag is unknown,
↕ = flag is affected according to the result of the operation.

Z80 INSTRUCTION SET

JUMP GROUP

Mnemonic	Symbolic Operation	S	Z		H		P/V	N	C	Op-Code 76 543 210	Hex	No. of Bytes	No. of M Cycles	No. of T States	Comments
JP nn	PC ← nn	•	•	X	•	X	•	•	•	11 000 011	C3	3	3	10	
										← n →					
										← n →					
JP cc, nn	If condition cc is true PC ← nn, otherwise continue	•	•	X	•	X	•	•	•	11 cc 010		3	3	10	
										← n →					
										← n →					
JR e	PC ← PC + e	•	•	X	•	X	•	•	•	00 011 000	18	2	3	12	
										← e-2 →					
JR C, e	If C = 0, continue	•	•	X	•	X	•	•	•	00 111 000	38	2	2	7	If condition not met
										← e-2 →					
	If C = 1, PC ← PC+e											2	3	12	If condition is met
JR NC, e	If C = 1, continue	•	•	X	•	X	•	•	•	00 110 000	30	2	2	7	If condition not met
										← e-2 →					
	If C = 0, PC ← PC+e											2	3	12	If condition is met
JR Z, e	If Z = 0 continue	•	•	X	•	X	•	•	•	00 101 000	28	2	2	7	If condition not met
										← e-2 →					
	If Z = 1, PC ← PC+e											2	3	12	If condition is met
JR NZ, e	If Z = 1, continue	•	•	X	•	X	•	•	•	00 100 000	20	2	2	7	If condition not met
										← e-2 →					
	If Z = 0, PC ← PC+e											2	3	12	If condition is met
JP (HL)	PC ← HL	•	•	X	•	X	•	•	•	11 101 001	E9	1	1	4	
JP (IX)	PC ← IX	•	•	X	•	X	•	•	•	11 011 101	DD	2	2	8	
										11 101 001	E9				
JP (IY)	PC ← IY	•	•	X	•	X	•	•	•	11 111 101	FD	2	2	8	
										11 101 001	E9				
DJNZ, e	B ← B-1 If B = 0, continue	•	•	X	•	X	•	•	•	00 010 000	10	2	2	8	If B = 0
										← e-2 →					
	If B ≠ 0, PC ← PC+e											2	3	13	If B ≠ 0

cc	Condition
000	NZ non zero
001	Z zero
010	NC non carry
011	C carry
100	PO parity odd
101	PE parity even
110	P sign positive
111	M sign negative

Notes: e represents the extension in the relative addressing mode.

e is a signed two's complement number in the range ⟨126, 129⟩

e-2 in the op-code provides an effective address of pc+e as PC is incremented by 2 prior to the addition of e.

Flag Notation: • = flag not affected, 0 = flag reset, 1 = flag set, X = flag is unknown,
↕ = flag is affected according to the result of the operation.

Z80 INSTRUCTION SET

CALL AND RETURN GROUP

Mnemonic	Symbolic Operation	S	Z		H		P/V	N	C	76 543 210	Hex	No. of Bytes	No. of M Cycles	No. of T States	Comments
CALL nn	$(SP-1) \leftarrow PC_H$	•	•	X	•	X	•	•	•	11 001 101	CD	3	5	17	
	$(SP-2) \leftarrow PC_L$									← n →					
	$PC \leftarrow nn$									← n →					
CALL cc, nn	If condition	•	•	X	•	X	•	•	•	11 cc 100		3	3	10	If cc is false
	cc is false									← n →					
	continue,									← n →		3	5	17	If cc is true
	otherwise														
	same as														
	CALL nn														
RET	$PC_L \leftarrow (SP)$	•	•	X	•	X	•	•	•	11 001 001	C9	1	3	10	
	$PC_H \leftarrow (SP+1)$														
RET cc	If condition	•	•	X	•	X	•	•	•	11 cc 000		1	1	5	If cc is false
	cc is false														
	continue,											1	3	11	If cc is true
	otherwise														
	same as														
	RET														
RETI	Return from	•	•	X	•	X	•	•	•	11 101 101	ED	2	4	14	
	interrupt									01 001 101	4D				
RETN[1]	Return from	•	•	X	•	X	•	•	•	11 101 101	ED	2	4	14	
	non maskable									01 000 101	45				
	interrupt														
RST p	$(SP-1) \leftarrow PC_H$	•	•	X	•	X	•	•	•	11 t 111		1	3	11	
	$(SP-2) \leftarrow PC_L$														
	$PC_H \leftarrow 0$														
	$PC_L \leftarrow p$														

cc	Condition	
000	NZ	non zero
001	Z	zero
010	NC	non carry
011	C	carry
100	PO	parity odd
101	PE	parity even
110	P	sign positive
111	M	sign negative

t	p
000	00H
001	08H
010	10H
011	18H
100	20H
101	28H
110	30H
111	38H

[1] RETN loads $IFF_2 \leftarrow IFF_1$

Flag Notation: • = flag not affected, 0 = flag reset, 1 = flag set, X = flag is unknown,
↕ = flag is affected according to the result of the operation.

Z80 INSTRUCTION SET

INPUT AND OUTPUT GROUP

Mnemonic	Symbolic Operation	S	Z		H		P/V	N	C	Op-Code 76 543 210	Hex	No. of Bytes	No. of M Cycles	No. of T States	Comments
IN A, (n)	A ← (n)	•	•	X	•	X	•	•	•	11 011 011 ← n →	DB	2	3	11	n to $A_0 \sim A_7$ Acc to $A_8 \sim A_{15}$
IN r, (C)	r ← (C) if r = 110 only the flags will be affected	↕	↕	X	↕	X	P	0	•	11 101 101 01 r 000	ED	2	3	12	C to $A_0 \sim A_7$ B to $A_8 \sim A_{15}$
INI	(HL) ← (C) B ← B − 1 HL ← HL + 1	X	↕[1]	X	X	X	X	1	X	11 101 101 10 100 010	ED A2	2	4	16	C to $A_0 \sim A_7$ B to $A_8 \sim A_{15}$
INIR	(HL) ← (C) B ← B − 1 HL ← HL + 1 Repeat until B = 0	X	1	X	X	X	X	1	X	11 101 101 10 110 010	ED B2	2 2	5 (If B ≠ 0) 4 (If B = 0)	21 16	C to $A_0 \sim A_7$ B to $A_8 \sim A_{15}$
IND	(HL) ← (C) B ← B − 1 HL ← HL − 1	X	↕[1]	X	X	X	X	1	X	11 101 101 10 101 010	ED AA	2	4	16	C to $A_0 \sim A_7$ B to $A_8 \sim A_{15}$
INDR	(HL) ← (C) B ← B − 1 HL ← HL − 1 Repeat until B = 0	X	1	X	X	X	X	1	X	11 101 101 10 111 010	ED BA	2 2	5 (If B ≠ 0) 4 (If B = 0)	21 16	C to $A_0 \sim A_7$ B to $A_8 \sim A_{15}$
OUT (n), A	(n) ← A	•	•	X	•	X	•	•	•	11 010 011	D3	2	3	11	n to $A_0 \sim A_7$ Acc to $A_8 \sim A_{15}$
OUT (C), r	(C) ← r	•	•	X	•	X	•	•	•	11 101 101 01 r 001	ED	2	3	12	C to $A_0 \sim A_7$ B to $A_8 \sim A_{15}$
OUTI	B ← B − 1 (C) ← (HL) HL ← HL + 1	X	↕[1]	X	X	X	X	1	X	11 101 101 10 100 011	ED A3	2	4	16	C to $A_0 \sim A_7$ B to $A_8 \sim A_{15}$
OTIR	B ← B − 1 (C) ← (HL) HL ← HL + 1 Repeat until B = 0	X	1	X	X	X	X	1	X	11 101 101 10 110 011	ED B3	2 2	5 (If B ≠ 0) 4 (If B = 0)	21 16	C to $A_0 \sim A_7$ B to $A_8 \sim A_{15}$
OUTD	(C) ← (HL) B ← B − 1 HL ← HL − 1	X	↕[1]	X	X	X	X	1	X	11 101 101 10 101 011	ED AB	2	4	16	C to $A_0 \sim A_7$ B to $A_8 \sim A_{15}$
OTDR	(C) ← (HL) B ← B − 1 HL ← HL − 1 Repeat until B = 0	X	1	X	X	X	X	1	X	11 101 101 10 111 011	ED BB	2 2	5 (If B ≠ 0) 4 (If B = 0)	21 16	C to $A_0 \sim A_7$ B to $A_8 \sim A_{15}$

Notes: 1 If the result of B − 1 is zero the Z flag is set, otherwise it is reset.

Flag Notation: • = flag not affected, 0 = flag reset, 1 = flag set, X = flag is unknown,
↕ = flag is affected according to the result of the operation.

MC 6800 INSTRUCTION SET

ACCUMULATOR AND MEMORY INSTRUCTIONS

OPERATIONS	MNEMONIC	IMMED OP	~	#	DIRECT OP	~	#	INDEX OP	~	#	EXTND OP	~	#	IMPLIED OP	~	#	BOOLEAN/ARITHMETIC OPERATION (All register labels refer to contents)	H (5)	I (4)	N (3)	Z (2)	V (1)	C (0)
Add	ADDA	8B	2	2	9B	3	2	AB	5	2	BB	4	3				A + M → A	↕	•	↕	↕	↕	↕
	ADDB	CB	2	2	DB	3	2	EB	5	2	FB	4	3				B + M → B	↕	•	↕	↕	↕	↕
Add Acmltrs	ABA													1B	2	1	A + B → A	↕	•	↕	↕	↕	↕
Add with Carry	ADCA	89	2	2	99	3	2	A9	5	2	B9	4	3				A + M + C → A	↕	•	↕	↕	↕	↕
	ADCB	C9	2	2	D9	3	2	E9	5	2	F9	4	3				B + M + C → B	↕	•	↕	↕	↕	↕
And	ANDA	84	2	2	94	3	2	A4	5	2	B4	4	3				A · M → A	•	•	↕	↕	R	•
	ANDB	C4	2	2	D4	3	2	E4	5	2	F4	4	3				B · M → B	•	•	↕	↕	R	•
Bit Test	BITA	85	2	2	95	3	2	A5	5	2	B5	4	3				A · M	•	•	↕	↕	R	•
	BITB	C5	2	2	D5	3	2	E5	5	2	F5	4	3				B · M	•	•	↕	↕	R	•
Clear	CLR							6F	7	2	7F	6	3				00 → M	•	•	R	S	R	R
	CLRA													4F	2	1	00 → A	•	•	R	S	R	R
	CLRB													5F	2	1	00 → B	•	•	R	S	R	R
Compare	CMPA	81	2	2	91	3	2	A1	5	2	B1	4	3				A − M	•	•	↕	↕	↕	↕
	CMPB	C1	2	2	D1	3	2	E1	5	2	F1	4	3				B − M	•	•	↕	↕	↕	↕
Compare Acmltrs	CBA													11	2	1	A − B	•	•	↕	↕	↕	↕
Complement, 1's	COM							63	7	2	73	6	3				$\bar{M}$ → M	•	•	↕	↕	R	S
	COMA													43	2	1	$\bar{A}$ → A	•	•	↕	↕	R	S
	COMB													53	2	1	$\bar{B}$ → B	•	•	↕	↕	R	S
Complement, 2's (Negate)	NEG							60	7	2	70	6	3				00 − M → M	•	•	↕	↕	①	②
	NEGA													40	2	1	00 − A → A	•	•	↕	↕	①	②
	NEGB													50	2	1	00 − B → B	•	•	↕	↕	①	②
Decimal Adjust, A	DAA													19	2	1	Converts Binary Add. of BCD Characters into BCD Format	•	•	↕	↕	↕	③
Decrement	DEC							6A	7	2	7A	6	3				M − 1 → M	•	•	↕	↕	4	•
	DECA													4A	2	1	A − 1 → A	•	•	↕	↕	4	•
	DECB													5A	2	1	B − 1 → B	•	•	↕	↕	4	•
Exclusive OR	EORA	88	2	2	98	3	2	A8	5	2	B8	4	3				A ⊕ M → A	•	•	↕	↕	R	•
	EORB	C8	2	2	D8	3	2	E8	5	2	F8	4	3				B ⊕ M → B	•	•	↕	↕	R	•
Increment	INC							6C	7	2	7C	6	3				M + 1 → M	•	•	↕	↕	⑤	•
	INCA													4C	2	1	A + 1 → A	•	•	↕	↕	⑤	•
	INCB													5C	2	1	B + 1 → B	•	•	↕	↕	⑤	•
Load Acmltr	LDAA	86	2	2	96	3	2	A6	5	2	B6	4	3				M → A	•	•	↕	↕	R	•
	LDAB	C6	2	2	D6	3	2	E6	5	2	F6	4	3				M → B	•	•	↕	↕	R	•
Or, Inclusive	ORAA	8A	2	2	9A	3	2	AA	5	2	BA	4	3				A + M → A	•	•	↕	↕	R	•
	ORAB	CA	2	2	DA	3	2	EA	5	2	FA	4	3				B + M → B	•	•	↕	↕	R	•
Push Data	PSHA													36	4	1	A → M_{SP}, SP − 1 → SP	•	•	•	•	•	•
	PSHB													37	4	1	B → M_{SP}, SP − 1 → SP	•	•	•	•	•	•
Pull Data	PULA													32	4	1	SP + 1 → SP, M_{SP} → A	•	•	•	•	•	•
	PULB													33	4	1	SP + 1 → SP, M_{SP} → B	•	•	•	•	•	•
Rotate Left	ROL							69	7	2	79	6	3				M	•	•	↕	↕	⑥	↕
	ROLA													49	2	1	A } C ← b7 ← b0 (rotate through carry)	•	•	↕	↕	⑥	↕
	ROLB													59	2	1	B	•	•	↕	↕	⑥	↕
Rotate Right	ROR							66	7	2	76	6	3				M	•	•	↕	↕	⑥	↕
	RORA													46	2	1	A } C → b7 → b0 (rotate through carry)	•	•	↕	↕	⑥	↕
	RORB													56	2	1	B	•	•	↕	↕	⑥	↕
Shift Left, Arithmetic	ASL							68	7	2	78	6	3				M	•	•	↕	↕	⑥	↕
	ASLA													48	2	1	A } C ← b7 ← b0 ← 0	•	•	↕	↕	⑥	↕
	ASLB													58	2	1	B	•	•	↕	↕	⑥	↕
Shift Right, Arithmetic	ASR							67	7	2	77	6	3				M	•	•	↕	↕	⑥	↕
	ASRA													47	2	1	A } b7 → b0 → C (b7 retained)	•	•	↕	↕	⑥	↕
	ASRB													57	2	1	B	•	•	↕	↕	⑥	↕
Shift Right, Logic	LSR							64	7	2	74	6	3				M	•	•	R	↕	⑥	↕
	LSRA													44	2	1	A } 0 → b7 → b0 → C	•	•	R	↕	⑥	↕
	LSRB													54	2	1	B	•	•	R	↕	⑥	↕
Store Acmltr	STAA				97	4	2	A7	6	2	B7	5	3				A → M	•	•	↕	↕	R	•
	STAB				D7	4	2	E7	6	2	F7	5	3				B → M	•	•	↕	↕	R	•
Subtract	SUBA	80	2	2	90	3	2	A0	5	2	B0	4	3				A − M → A	•	•	↕	↕	↕	↕
	SUBB	C0	2	2	D0	3	2	E0	5	2	F0	4	3				B − M → B	•	•	↕	↕	↕	↕
Subtract Acmltrs	SBA													10	2	1	A − B → A	•	•	↕	↕	↕	↕
Subtr. with Carry	SBCA	82	2	2	92	3	2	A2	5	2	B2	4	3				A − M − C → A	•	•	↕	↕	↕	↕
	SBCB	C2	2	2	D2	3	2	E2	5	2	F2	4	3				B − M − C → B	•	•	↕	↕	↕	↕
Transfer Acmltrs	TAB													16	2	1	A → B	•	•	↕	↕	R	•
	TBA													17	2	1	B → A	•	•	↕	↕	R	•
Test, Zero or Minus	TST							6D	7	2	7D	6	3				M − 00	•	•	↕	↕	R	R
	TSTA													4D	2	1	A − 00	•	•	↕	↕	R	R
	TSTB													5D	2	1	B − 00	•	•	↕	↕	R	R

LEGEND

- OP Operation Code (Hexadecimal)
- ~ Number of MPU Cycles
- \# Number of Program Bytes
- \+ Arithmetic Plus
- − Arithmetic Minus
- • Boolean AND
- M_{SP} Contents of memory location pointed to be Stack Pointer
- \+ Boolean Inclusive OR
- ⊙ Boolean Exclusive OR
- $\bar{M}$ Complement of M
- → Transfer Into
- 0 Bit = Zero
- 00 Byte = Zero

Note – Accumulator addressing mode instructions are included in the column for IMPLIED addressing

CONDITION CODE SYMBOLS

- H Half carry from bit 3
- I Interrupt mask
- N Negative (sign bit)
- Z Zero (byte)
- V Overflow, 2's complement
- C Carry from bit 7
- R Reset Always
- S Set Always
- ↕ Test and set if true, cleared otherwise
- • Not Affected

MC 6800 INSTRUCTION SET

INDEX REGISTER AND STACK MANIPULATION INSTRUCTIONS

POINTER OPERATIONS	MNEMONIC	IMMED OP	IMMED ~	IMMED =	DIRECT OP	DIRECT ~	DIRECT #	INDEX OP	INDEX ~	INDEX #	EXTND OP	EXTND ~	EXTND #	IMPLIED OP	IMPLIED ~	IMPLIED =	BOOLEAN/ARITHMETIC OPERATION	5 H	4 I	3 N	2 Z	1 V	0 C
Compare Index Reg	CPX	8C	3	3	9C	4	2	AC	6	2	BC	5	3				$X_H - M, X_L - (M+1)$	•	•	⑦	↕	⑧	•
Decrement Index Reg	DEX													09	4	1	$X - 1 \rightarrow X$	•	•	•	↕	•	•
Decrement Stack Pntr	DES													34	4	1	$SP - 1 \rightarrow SP$	•	•	•	•	•	•
Increment Index Reg	INX													08	4	1	$X + 1 \rightarrow X$	•	•	•	↕	•	•
Increment Stack Pntr	INS													31	4	1	$SP + 1 \rightarrow SP$	•	•	•	•	•	•
Load Index Reg	LDX	CE	3	3	DE	4	2	EE	6	2	FE	5	3				$M \rightarrow X_H, (M+1) \rightarrow X_L$	•	•	⑨	↕	R	•
Load Stack Pntr	LDS	8E	3	3	9E	4	2	AE	6	2	BE	5	3				$M \rightarrow SP_H, (M+1) \rightarrow SP_L$	•	•	⑨	↕	R	•
Store Index Reg	STX				DF	5	2	EF	7	2	FF	6	3				$X_H \rightarrow M, X_L \rightarrow (M+1)$	•	•	⑨	↕	R	•
Store Stack Pntr	STS				9F	5	2	AF	7	2	BF	6	3				$SP_H \rightarrow M, SP_L \rightarrow (M+1)$	•	•	⑨	↕	R	•
Indx Reg → Stack Pntr	TXS													35	4	1	$X - 1 \rightarrow SP$	•	•	•	•	•	•
Stack Pntr → Indx Reg	TSX													30	4	1	$SP + 1 \rightarrow X$	•	•	•	•	•	•

JUMP AND BRANCH INSTRUCTIONS

OPERATIONS	MNEMONIC	RELATIVE OP	RELATIVE ~	RELATIVE #	INDEX OP	INDEX ~	INDEX #	EXTND OP	EXTND ~	EXTND #	IMPLIED OP	IMPLIED ~	IMPLIED #	BRANCH TEST	5 H	4 I	3 N	2 Z	1 V	0 C
Branch Always	BRA	20	4	2										None	•	•	•	•	•	•
Branch If Carry Clear	BCC	24	4	2										C = 0	•	•	•	•	•	•
Branch If Carry Set	BCS	25	4	2										C = 1	•	•	•	•	•	•
Branch If = Zero	BEQ	27	4	2										Z = 1	•	•	•	•	•	•
Branch If ≥ Zero	BGE	2C	4	2										N ⊕ V = 0	•	•	•	•	•	•
Branch If > Zero	BGT	2E	4	2										Z + (N ⊕ V) = 0	•	•	•	•	•	•
Branch If Higher	BHI	22	4	2										C + Z = 0	•	•	•	•	•	•
Branch If ≤ Zero	BLE	2F	4	2										Z + (N ⊕ V) = 1	•	•	•	•	•	•
Branch If Lower Or Same	BLS	23	4	2										C + Z = 1	•	•	•	•	•	•
Branch If < Zero	BLT	2D	4	2										N ⊕ V = 1	•	•	•	•	•	•
Branch If Minus	BMI	2B	4	2										N = 1	•	•	•	•	•	•
Branch If Not Equal Zero	BNE	26	4	2										Z = 0	•	•	•	•	•	•
Branch If Overflow Clear	BVC	28	4	2										V = 0	•	•	•	•	•	•
Branch If Overflow Set	BVS	29	4	2										V = 1	•	•	•	•	•	•
Branch If Plus	BPL	2A	4	2										N = 0	•	•	•	•	•	•
Branch To Subroutine	BSR	8D	8	2										See Special Operations	•	•	•	•	•	•
Jump	JMP				6E	4	2	7E	3	3				See Special Operations	•	•	•	•	•	•
Jump To Subroutine	JSR				AD	8	2	BD	9	3				See Special Operations	•	•	•	•	•	•
No Operation	NOP										01	2	1	Advances Prog. Cntr. Only	•	•	•	•	•	•
Return From Interrupt	RTI										3B	10	1		⑩					
Return From Subroutine	RTS										39	5	1	See Special Operations	•	•	•	•	•	•
Software Interrupt	SWI										3F	12	1	See Special Operations	•	•	•	•	•	•
Wait for Interrupt*	WAI										3E	9	1	See Special Operations	•	⑪	•	•	•	•

*WAI puts Address Bus, R/W, and Data Bus in the three-state mode while VMA is held low.

CONDITION CODE REGISTER MANIPULATION INSTRUCTIONS

OPERATIONS	MNEMONIC	IMPLIED OP	IMPLIED ~	IMPLIED =	BOOLEAN OPERATION	5 H	4 I	3 N	2 Z	1 V	0 C
Clear Carry	CLC	0C	2	1	0 → C	•	•	•	•	•	R
Clear Interrupt Mask	CLI	0E	2	1	0 → I	•	R	•	•	•	•
Clear Overflow	CLV	0A	2	1	0 → V	•	•	•	•	R	•
Set Carry	SEC	0D	2	1	1 → C	•	•	•	•	•	S
Set Interrupt Mask	SEI	0F	2	1	1 → I	•	S	•	•	•	•
Set Overflow	SEV	0B	2	1	1 → V	•	•	•	•	S	•
Acmltr A → CCR	TAP	06	2	1	A → CCR	⑫					
CCR → Acmltr A	TPA	07	2	1	CCR → A	•	•	•	•	•	•

CONDITION CODE REGISTER NOTES (Bit set if test is true and cleared otherwise)

1 (Bit V) Test: Result = 10000000?
2 (Bit C) Test: Result = 00000000?
3 (Bit C) Test: Decimal value of most significant BCD Character greater than nine? (Not cleared if previously set.)
4 (Bit V) Test: Operand = 10000000 prior to execution?
5 (Bit V) Test: Operand = 01111111 prior to execution?
6 (Bit V) Test: Set equal to result of N⊙C after shift has occurred.
7 (Bit N) Test: Sign bit of most significant (MS) byte = 1?
8 (Bit V) Test: 2's complement overflow from subtraction of MS bytes?
9 (Bit N) Test: Result less than zero? (Bit 15 = 1)
10 (All) Load Condition Code Register from Stack. (See Special Operations)
11 (Bit I) Set when interrupt occurs. If previously set, a Non Maskable Interrupt is required to exit the wait state.
12 (All) Set according to the contents of Accumulator A.

Index